Bacteriophage Biochemistry

Bacteriophage Biochemistry

ACS MONOGRAPH

CHRISTOPHER K. MATHEWS
Department of Biochemistry
University of Arizona College of Medicine
Tucson, Arizona

VNR

VAN NOSTRAND REINHOLD COMPANY
New York Cincinnati Toronto London Melbourne

Van Nostrand Reinhold Company Regional Offices:
New York Cincinnati Chicago Millbrae Dallas

Van Nostrand Reinhold Company International Offices:
London Toronto Melbourne

Library of Congress Catalog Card Number: 70-143542

Manufactured in the United States of America

Published by Van Nostrand Reinhold Company
450 West 33rd Street, New York, N.Y. 10001

Published simultaneously in Canada by
Van Nostrand Reinhold Ltd.

15 14 13 12 11 10 9 8 7 6 5 4 3 2 1

GENERAL INTRODUCTION

American Chemical Society's Series of Chemical Monographs

By arrangement with the interallied Conference of Pure and Applied Chemistry, which met in London and Brussels in July, 1919, the American Chemical Society undertook the production and publication of Scientific and Technologic Monographs on chemical subjects. At the same time it was agreed that the National Research Council, in cooperation with the American Chemical Society and the American Physical Society, should undertake the production and publication of Critical Tables of Chemical and Physical Constants. The American Chemical Society and the National Research Council mutually agreed to care for these two fields of chemical progress.

The Council of the American Chemical Society, acting through its Committee on National Policy, appointed editors and associates (the present list of whom appears at the close of this sketch) to select authors of competent authority in their respective fields and to consider critically the manuscripts submitted. Since 1944 the scientific and Technologic Monographs have been combined in the Series. The first Monograph appeared in 1921, and, up to 1967, 162 treatises have enriched the Series.

These Monographs are intended to serve two principal purposes: first to make available to chemists a thorough treatment of a selected area in form usable by persons working in more or less unrelated fields to the end that they may correlate their own work with a larger area of physical science;

secondly, to stimulate further research in the specific field treated. To implement this purpose the authors of Monographs give extended references to the literature.

American Chemical Society

F. Marshall Beringer

Editor of Monographs

Preface

This book is intended as a review of the biochemistry of bacterial viruses. Emphasis is upon the metabolism of the infected cell, with supporting information being provided in the overlapping areas of phage structure, viral genetics, and microbial physiology. I have limited this review to phages about which significant biochemical information is available, realizing that many interesting systems have been neglected.

Reflecting my own research background and interests, I have reviewed the biochemistry of the T-even phages in the greatest detail. The other major classes of phages—T-odd phages, temperate phages, single-stranded DNA phages, *Bacillus subtilis* phages, and RNA phages—have been discussed in somewhat less depth. However, I have tried to highlight the major research problems encountered in each area and to present, insofar as possible, the current status of research on each group of phages.

A number of people have provided material assistance in the preparation of this monograph. Primarily, I must acknowledge the forebearance of my wife, Kate, and children, Lawrence and Anne, during the unexpectedly long period when I was not available to fully participate in family life. But for their loving and patience, this book would not have come into existence.

I am grateful to several scientific colleagues, whose names are cited in figure legends, for permission to reproduce photographic material and for providing me with copies of this material for the present volume. Conversations with colleagues and students have been indispensable in shaping my present ideas, and in particular I must acknowledge the contributions of Drs. Robert Murray, Byron Ballou, Harris Bernstein, and David Mount. John Hostetter rendered capable assistance in preparation of some of the illustrations, and Mrs. Judy Yoshimoto rendered many hours of willing and cheerful labor in the preparation of the manuscript.

The research program of my own laboratory during the past seven years has received financial support through research grants awarded by the National Institute of Allergy and Infectious Diseases (Grant No. AI-

08230) and by the American Heart Association (Grant No. 68811). The work of my colleagues and myself, much of which is described in this monograph, could not have been done without this support.

Finally, I must acknowledge a special debt of gratitude to Dr. Seymour S. Cohen, the first "bacteriophage biochemist." It was my good fortune to spend two years as a postdoctoral fellow in Dr. Cohen's laboratory. Dr. Cohen set standards for eclecticism of thought and rigor in interpretation of scientific data which would be difficult for most scientists to meet, but which have provided an ideal toward which I have tried to strive. I hope that the present volume has succeeded, in some small measure, in meeting the test of that ideal.

C. K. Mathews

Contents

1

The Goals of Biochemical Virology

Most scientific monographs, such as the present volume, attempt to define and describe a "field" of study. This will not be the goal of the present volume, for I do not really believe in the existence of "bacteriophage biochemistry" as a distinct field of endeavor. The many types of scientists who have chosen bacterial viruses, or bacteriophages, as their experimental material—including geneticists, microbiologists, cell physiologists, physical scientists, and biochemists—share a common objective: to comprehend the events of bacteriophage replication in molecular terms. The goal of this monograph is to summarize the present state of progress toward this objective, as seen through the eyes of a biochemist.

The above-stated goal of phage research would appear to equate it with a branch of "molecular biology." I do not wish to belabor the distinction between biochemistry and molecular biology, but I must acknowledge that it exists, at least in the minds of many biological scientists, and has formed an important component of the intellectual history of phage research (cf. Cairns, Stent, and Watson, eds., 1966; Cohen, 1968). Several major American universities have separate departments of biochemistry and molecular biology, and in the past there has been considerable antagonism between proponents of the two fields, even though one cannot clearly distinguish between them. Many who consider themselves to be molecular biologists seem to regard biochemists as rather unimaginative creatures, and the study of biochemistry an activity to be avoided whenever possible. Not a few biochemists, on the other hand, view molecular biology as a haven for those who would, like themselves, study life processes at the molecular level, but who are too lazy or impatient to first become familiar with the immense body of knowledge already cultivated by themselves.

Another reason for acknowledging the distinction between biochemistry and molecular biology is to point out that there is today very little basis for hostility between adherents to the two fields—if, indeed, a distinction between the two fields still exists in 1970. This point is clearly illustrated by the spectacular development over the past 25 years of our knowledge of bacteriophage replication and of the larger field of biochemical virology. As I hope will be evident throughout this volume, our present detailed knowledge of the metabolism of virus-infected bacteria has arisen through the performance of critical experiments by many workers with many different orientations. This has led in turn to the illumination of many fundamental biological problems not directly related to bacteriophages. It has also led to a scientific climate wherein today's most productive phage workers use both biochemical and nonbiochemical approaches to their research, and show no hesitation in teaching themselves unfamiliar experimental techniques wherever necessary to solve the problem at hand.

Among the several levels of biological organization at which one can study life processes, the historical approach of biochemistry has been to observe biological processes below the level of organization of the living cell. Students in elementary biochemistry courses are told of the triumph of the Buchner brothers, who first demonstrated alcoholic fermentation of sugar in a cell-free extract of yeast, and of Sumner, who crystallized the first enzyme. These were landmark achievements in the development of enzymology, a field which for many years had as a primary goal the understanding of how purified enzymes act as chemical reagents. However, enzymologists also have the responsibility of proving that their observations with purified systems are biologically significant, i.e., that what is observed *in vitro* also occurs *in vivo*. This latter, and perhaps more difficult, aspect of enzymology has received ever-increasing attention in recent years.

The approach of many biologists, on the other hand, has been to design experiments with more-or-less intact organisms and to try to make inferences about intracellular events at the molecular level. The performance of a genetic cross is an example of this approach. One can construct a genetic map of an organism, based upon recombination frequencies determined by observations on the offspring resulting from a series of carefully controlled matings. The geneticist would like to believe that the map has some relation to the structure of the chromosome, but this necessitates isolation of the chromosome and verification of genetic inferences by some physical criteria. Thus, the geneticist has to turn to biochemistry, just as the enzymologist has had to consider the biology of the system he is studying. As we shall see, these developments have been especially fruitful in the study of the biochemistry of the virus-infected bacterium.

In order to completely describe an organism in molecular biological terms, one must be able to completely correlate genes with gene functions for that organism. In other words, one must identify the product of each gene and its metabolic role. For even the simplest of cellular organisms, this proves to be a horrendous task, which is, even now, only being distantly approached. Consider the intestinal bacterium *Escherichia coli,* surely the best-understood cellular organism, from both the biochemical and genetic standpoints. Virtually the entire complement of genetic information, or genome, of this single-celled organism* consists of a single molecule of double-stranded DNA, having a molecular weight of some 2.8 billion. Assuming that a sequence of three successive nucleotide pairs can code for one amino acid, and assuming that the average protein molecule contains about 500 amino acid residues, then the *E. coli* genome contains sufficient information to code for two to three thousand different protein molecules. In order to reach the above-described goal for *E. coli,* one must map the gene specifying the structure of each protein† and describe the biological function of every gene product. Since Watson (1965) estimated that only one fifth to one third of all metabolic reactions in *E. coli* are known, it is clear that a complete molecular biological description of this organism is still quite a remote attainment.

The problem of correlating genes with gene functions becomes even less approachable in the case of multicellular organisms. Cells of higher plants and animals are much larger than those of bacteria, and they contain far more DNA. One can imagine the huge amount of genetic information which must be stored in a single cell of a highly differentiated organism, merely by considering the numerous specialized properties of cells from different organs or tissues of such an organism. To add to the complexity of the problem, many genes in higher organisms occur more than once in the haploid content of genetic information. On the basis of DNA-DNA hybridization studies, Britten and Kohne (1968) have found that "in general, more than one-third of the DNA of higher organisms is made up of sequences which recur anywhere from a thousand to a million times per cell." Whereas this is probably not all true gene duplication, the existence of duplicated genes enormously complicates the biochemical-genetic analysis of these organisms. One can obtain a mutant defective in a particular gene locus. However, such a mutant could never be characterized if dupli-

* With the exception of a small amount of information in extrachromosomal determinants, the plasmids and episomes (cf. Hayes, 1968).

† For simplicity this discussion ignores the fact that the products of some genes are not proteins but nucleic acids and that some regulatory genes do not appear to have diffusible gene products (Jacob and Monod, 1961).

cate genes in the cell can provide the defective gene product. Indeed, it is reasonable to suppose that gene duplication arose during evolution at least in part to protect organisms against loss of essential functions through chance mutations.

In light of the above rather pessimistic remarks, the viruses assume special attractiveness as experimental systems for combined biochemical and genetic analysis. Viruses are obligate intracellular parasites, that is, they can only replicate by using the metabolic machinery of an appropriate host cell. As such, a virus needs relatively little genetic information in order to exist. Indeed, the smallest viruses contain only enough nucleic acid to code for between three and ten average-sized proteins. Even the largest known viruses probably contain no more than about 200 genes. Thus, the problem of correlating genes with gene functions assumes manageable proportions.

Although considerable diversity exists among the viruses, they all share certain structural features. A central core of nucleic acid (either DNA or RNA) is surrounded by a protein coat. Although the coat may be quite complex, consisting of several types of proteins, its sole function, with very few exceptions, appears to be to protect the viral genome and to facilitate its entry into a host cell. Thus, the virus particle, or virion, contains no metabolic apparatus of its own, but only the set of instructions for viral replication (the genome) and a vehicle (the coat) for transport of these instructions to cellular sites where they can be read. In order to understand how a virus replicates, therefore, one must have knowledge both of viral genes and gene products and of how the viral genome interacts with the metabolic machinery of the host. Specifically, how is the biosynthetic apparatus of the host diverted to the manufacture of viral components? A complete answer to this question for any virus-host system demands considerable knowledge of the gene products of both the virus and its host cell. In general, the large viruses, containing a relatively large amount of genetic information, direct the appearance of new virus-specific metabolic pathways, while the smaller viruses must depend to a greater extent upon preexisting cellular pathways. Thus, although some of the smallest viruses are almost completely characterized as regards correlation of genes with gene functions, a full understanding of the replication of these viruses will require much more information than we presently have on the contribution of cellular gene products.

Biochemical virology can properly be said to have begun shortly after World War II, when Seymour S. Cohen decided to study the chemical events associated with infection of *E. coli* with T-even bacteriophages. Certainly, viruses had been studied from a biochemical standpoint earlier than that time. The crystallization of tobacco mosaic virus and its chemical characterization, for example, was an outstanding contribution reported a decade earlier. Moreover, bacteriophages were being recognized as attractive systems for genetic investigation by a small but rapidly-enlarging group led by Max Delbrück. However, Cohen was the first to apply the methods of biochemistry to virus replication. His initial efforts were so successful that within seven years many other phage workers, most of whom were not biochemically oriented, began to recognize the necessity of doing biochemistry in order to understand virus replication. These exciting times have been detailed in two absorbing books. Cohen (1968) has described his own contributions in relation to the intellectual development of biochemical virology. As one might expect, Cohen espouses an uncompromisingly pro-biochemical viewpoint. The opposite orientation is evident in most of the contributions to *Phage and the Origins of Molecular Biology,* (Cairns, Stent, and Watson, eds., 1966) a collection of retrospective essays by former students of Delbrück. Most of these workers had been attracted to phage because of the rich opportunities these systems presented for understanding the gene. Although several contributors to the volume acknowledge some aversion to biochemistry, a reading of both volumes reveals the importance to phage research of both genetics, and biochemistry.

Because the early history of phage biochemistry has been so well described by the workers who shaped it, this volume will not emphasize that aspect of the field, although some attempt will be made to convey the major developments chronologically. In addition, this volume will have relatively little to say about the early period of phage biology, dating from the discoveries of Twort and d'Herelle, for this has been dealt with superbly in the first chapter of Stent's textbook (Stent, 1963).

2

General Biology of Bacteriophages

Why bacteriophages? If one grants the attractiveness of viruses for biochemical-genetic study, as discussed in Chapter 1, what are the special advantages of bacterial viruses, as compared to animal or plant viruses? To answer this question, consider the state of animal virology some two to three decades ago. At that time a considerable number of viruses had been isolated and recognized as infectious agents, capable of causing certain well-characterized diseases in infected animals. However, as we shall see repeatedly throughout this book, any quantitative study of virus infection requires that one be able to *assay* the virus at will, i.e., to determine the concentration of biologically active agents. The only way this could be done was to prepare various dilutions of the virus preparation, apply these to a series of experimental animals in some standardized fashion, and wait to see how many of the animals died or contracted symptoms of the disease. Obviously, this type of assay is cumbersome, expensive, and time-consuming.

Because virus infection could only be studied at the whole-animal level, it was difficult, if not impossible, to study biochemical changes in infected cells. Upon infection of an animal by a virus preparation, only a tiny fraction of all the cells in the organism become infected. One could explore the indirect effects of virus infection on the animal, but an examination of virus-infected cells was impossible. In addition, infection at the whole-animal level is asynchronous, i.e., the initiation of infection in different cells occurs at different times. One simply could not explore the metabolic changes associated with infection of a particular cell population against the biochemical background of uninfected cells and cells which became infected at different times.

Similar problems plagued the field of plant virology.

The special appeal of bacteriophages as experimental objects becomes apparent in light of the preceding remarks. The host cells for these viruses are bacteria. It has long been possible to isolate bacteria in pure culture and to grow them in chemically defined media. Thus, one can easily prepare large populations of genetically homogeneous, susceptible host cells, under well-defined nutritional conditions. Furthermore, most phages adsorb rather rapidly to host cells under proper experimental conditions, such that one can synchronize the infection of a large culture. Thus, by sampling an infected culture at various times, one can isolate relatively homogeneous cell populations at definite stages of infection, and use these to study timing of virus-associated metabolic changes. In addition, the time required for a cycle of phage replication is short compared to the growth time for animal viruses—typically, of the order of twenty minutes to a few hours, as compared with many hours or days for various animal viruses. Thus, the bacterial virologist can design and perform experiments at a faster rate than his competitor in animal virology. This natural advantage of the phage worker is further increased by the availability of a superb method for quantitation of bacteriophages—the plaque assay. This assay, developed in what is essentially its present form by Gratia (1936), depends upon the ability of an active phage particle to clear a small portion of a "lawn" of infected bacteria, which has been seeded on a nutrient agar plate. In practice one mixes, in a tube of melted nutrient agar, a dilution of phage containing between 100 and 1000 infectious particles and an amount of bacteria sufficient to grow so as to smoothly cover the surface of the plate, after it has been poured on. After the agar has hardened, the plate is incubated at an appropriate temperature. As we shall see, the phage infective cycle normally terminates with *lysis,* or rupture, of the infected cell, with release of the newly formed virus progeny. Since diffusion of newly formed virus is prevented in the solid agar medium, the progeny infect hitherto uninfected cells in their immediate vicinity. Elsewhere the bacterial lawn grows undisturbed, giving an opaque appearance to most of the plate. Each plaque is a clear area, resulting from lysis of all of the cells in the immediate vicinity of an infectious particle.* The appearance of such a plate is shown in Figure 2–1.

* However, not all phage plaques are clear, and, as we shall see, much of the development of phage genetics has depended upon an investigator's ability to recognize differences in plaque morphology.

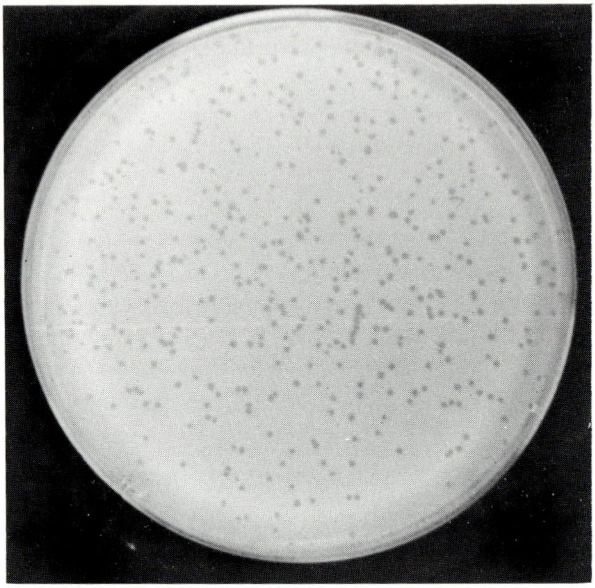

FIGURE 2-1 A Petri plate showing plaques of phage T4 on a lawn of *E. coli B*.

Usually the "development" of phage plaques on a plate is complete within a few hours, or overnight. Thus, the phage worker cannot only perform experiments rapidly, but he can be promptly informed of the results —a type of "reinforcement therapy" whose effect was not wasted upon early workers in the field.

A further advantage of phage systems relates to the ease of genetic manipulation of both the viruses and their hosts. Many types of bacterial and viral mutants can be isolated with little effort. Mutations can be mapped, either genetically or physically, and the consequences of particular mutations can be explored biochemically. Indeed, as stated in Chapter 1, it was the potential of phages for genetic investigation which originally attracted Delbrück and his followers. Later, the development of conditional lethal phage mutants provided tremendous impetus to the field of phage biochemistry (Epstein *et al,* 1963). The interlocking of phage genetics and biochemistry will be illustrated repeatedly throughout this book.

Finally, because one can grow large quantities of host cells, and because the infective cycle is rapid, one can easily prepare and purify large quantities of phage. Provided one has a proper centrifuge, one can easily grow and purify 10^{15} T4 bacteriophage particles, or about half a gram of material, in a single day. This has made it possible to study the structural components of phage at the same time as metabolic changes in infected cells were being elucidated. These studies have complemented one another, allowing conclusions to be drawn about the functional roles of structural components.

"All right," the reader may well ask, "so bacteriophages are attractive as experimental biological systems. Does this fact alone justify the enormous amount of phage work which has been done? After all, phages may be simply biological freaks—laboratory curiosities with little relation to the rest of the biological world." This question has occupied phage workers throughout the relatively short history of the field. It should be recalled that initial interest in bacteriophages, shortly after their discovery some fifty years ago, centered about an intensely practical aspect of these viruses—namely, their use in the treatment of bacterial infections. The excitement generated by this possibility was brought back to life by Sinclair Lewis in his novel, "Arrowsmith." As it turned out, this possibility could not be realized, partly because bacteria could mutate to become resistant to the phages which were being used as therapeutic agents. As a result, there was not a great deal of activity in the field until Delbrück and his colleagues realized that phages, being part of the biological world, probably obey the same genetic rules as do other organisms. In particular, the discovery of genetic recombination in bacteriophages (Delbrück and Bailey, 1946) spurred the study of these viruses in an effort to reveal universal genetic truths. During this era, Cohen and others were becoming aware that phage might be suitable models for studying host-virus interactions at the biochemical level.

To what extent were the early workers correct in their judgments about the generality of the principles of phage biology? Certainly, many aspects of phage are unique. I know of no other viruses, for example, which inject their nucleic acid into infected cells through a contractile tail, as do many bacteriophages. On the other hand, the great extent to which phage do fit into the rest of the biological world must have provided a great deal of pleasure to even the most optimistic of the early workers. Certainly the types of experiments one can do with phage are a source of great intellectual satisfaction *per se*. That the results of such experiments have some general significance can only serve to intensify the pleasure of those who design them and carry them out. Throughout this book I shall attempt to emphasize those areas where phage biology has had the greatest impact on the rest of the world. However, it is perhaps appropriate to list a few of these areas here.

The discovery of Avery, Macleod, and McCarty (1944) that DNA is the repository of hereditary information was confirmed by Hershey and Chase (1952), using the bacteriophage T2. Infected cells could be treated in a Waring blendor, to give release of viral protein, which did not enter the cell. The cells, containing viral DNA, could be separated from the stripped-off phage protein coat and, upon further incubation, could release a crop of infective progeny phage. Thus, the phage DNA carried into the infected cell sufficient information to direct the formation of complete virus particles. The extension of this proof of the genetic role of DNA in phage to the rest of the biological spectrum is too well known to discuss here.

Many features of phage infection are common to the infective process with plant and animal viruses. Indeed, the development of biochemical animal virology has proceeded along lines very similar to those of bacterial virology. Development of this field, for reasons mentioned above, required methods for the large-scale preparation of genetically homogeneous populations of host cells. The necessary impetus was provided by tissue culture techniques, which allowed one to clone single cells (i.e., to grow homogeneous cell lines from single cells), and to grow them in reasonably well-defined media. As a result, we now realize the existence of several features common to all virus-host systems. Some of these were discussed in Chapter 1. An additional feature is the existence, for any virus-host system, of a well-timed replication cycle. In general, this involves as an early event the induction of virus-specific enzymes, creating new metabolic pathways for the biosynthesis of viral components. At a later time, the structural components of the virus begin to be synthesized, and still later, these structural components are assembled into mature virions, which are ultimately released. Clearly, the existence of a precisely timed replication cycle must involve some type of sequential gene activation and inactivation. Thus, developmental biologists are interested in phage as a possible key to understanding the mechanism of cellular differentiation. It must be stated, however, that the general significance of mechanisms of viral "differentiation" are still not known.

A further point of contact between phage and the rest of the world relates to the temperate bacteriophages. Unlike virulent phages, in which infection is invariably fatal to the host, these viruses do not always initiate a cycle of viral growth in the host cell. Rather, the viral chromosome is integrated into the bacterial genome, whence it is maintained as a stable genetic entity, able to direct a viral growth cycle if appropriate environmental stimuli are applied. The nature of the stably integrated, or lysogenic condition, is under intense study and is becoming reasonably well understood. At the same time, tumor virologists are asking how oncogenic viruses can transform certain cell lines to a neoplastic state. Such a condition persists through many generations, even though no infectious virus can be recovered from a transformed cell. However, sensitive DNA-RNA hybridization techniques have demonstrated the presence, in the transformed cell, of virus-specific nucleic acids. The masking of viral genes in a transformed cell has many similarities to the masking of phage genes in a lysogenized bacterium, and it is hoped that studies on lysogeny will help to reveal the relationship between viruses and cancer.

MEASURING BACTERIOPHAGE GROWTH

Many of the fundamental questions about bacteriophage replication can be posed in the context of the results of a *one-step growth experiment.* This procedure, devised by Ellis and Delbrück (1939), represents the earliest known method for studying virus growth quantitatively. A bacterial culture growing in liquid medium is infected with sufficient phage to infect a small fraction of the cells. Within a few minutes, the culture is diluted by a large factor to prevent initiation of further infections by unadsorbed phage or newly released progeny virus. Samples are removed at various times, diluted further, and aliquots plated for assay of plaque-forming units (PFU) per ml of culture. Under the conditions of Ellis and Delbrück, the plaque-forming titer was seen to remain relatively constant for about 25 minutes at 37°, whereupon it abruptly increased some one hundredfold and did not subsequently change (see Figure 2–2). These results were seen as verifying the earlier prediction of d'Herelle (1926) that there are three phases in the phage life cycle: (1) attachment to the host bacterium; (2) multiplication inside the cell; and (3) rupture of the cell, with concomitant release of progeny phage.

What is the physical significance of a plaque-forming unit? Since plaque-forming titer is directly proportional to amount of a phage preparation plated, one can conclude that a single phage particle is capable of giving rise to a plaque. Furthermore, one can show that a single infected cell can give rise to a single plaque, when plated sufficiently early in infection. This is because the progeny phage released from a single cell on an agar plate remain trapped in the solid medium, and can only infect cells in the immediate vicinity of the original infected cell. Thus, when development of the plate is complete, only a single cleared area, or plaque, is seen for a single infected cell. But what fraction of the phage particles and infected cells initiate plaque formation? In other words, what is the *efficiency of plating?* One can enumerate the phage particles in a purified preparation most easily by electron microscopy. By comparing the total particle count with the plaque-forming titer, it is seen that, under optimal conditions, nearly every particle can give rise to a plaque, i.e., the efficiency of plating is close to 1.0. However, even when the plating efficiency is less than 1.0, it is about the same for all of the samples plated during an experiment, so that the results have internal consistency.

With the above in mind, let us consider how Ellis and Delbrück analyzed the results of their first one-step growth curves. The first 25 minutes, the period in which no change in phage titer is seen, was called

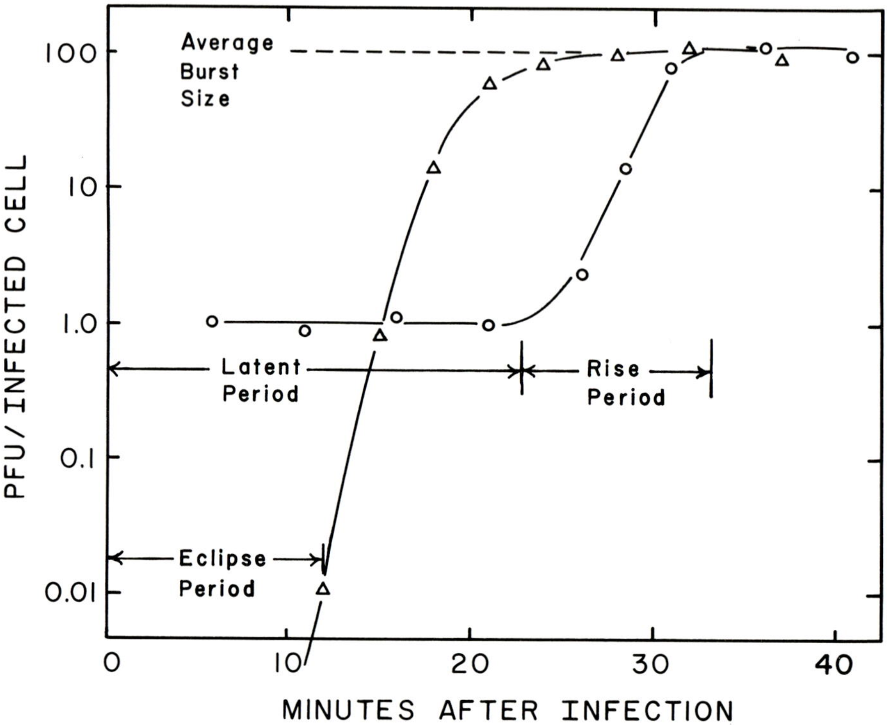

FIGURE 2-2 A one-step growth curve for bacteriophage T4 on *E. coli B* (data obtained in the author's laboratory). The curve indicated by the circles was obtained by plating aliquots of the infected cell culture directly upon plates seeded with *E. coli B,* while the curve indicated by the triangles was obtained by premature lysis of each aliquot of culture with chloroform before plating.

the *latent period;* this is the time of adsorption and intracellular growth. The following period, during which the titer rose abruptly, was called the *rise period* and was believed to correspond to the time of lysis of the infected cell. At the end of this time no further increase in titer is seen, because the culture has been diluted to such an extent that essentially no adsorption of progeny phage to uninfected bacteria occurs. The ratio of final titer to original titer was called the *burst size* and represents the average yield of progeny phage per infected cell. This assumes that adsorption was complete, i.e., that each plaque-forming unit seen during the latent period was an infected cell.

The experiment described here does not directly demonstrate the initial process, that of adsorption. However, as we shall see, there are many ways to demonstrate that adsorption occurs, and to measure its rate.

One can make many variations in the design of a one-step growth experiment, depending upon those facets of phage growth which are under study. For example, if one wants to remove unadsorbed phage, one can centrifuge down the bacteria early in the latent period and resuspend them in fresh medium; more directly, one can treat the culture with phage antiserum before dilution. This will inactivate free phage but will not affect intracellular phage growth. After dilution the antiserum is at too low a concentration to inactivate progeny phage. Possibly one would wish to study phage growth under conditions where essentially all of the cells are infected. One then adds phage in severalfold excess over bacteria (multiple infection, as opposed to single infection). The proportion of uninfected bacteria in a culture can be calculated from the Poisson distribution: $P_0 = e^{-m}$, where P_0 is the probability that a given cell will adsorb no phage, e is the base of natural logarithms, and m is the multiplicity of infection (average number of phage added per bacterium). In performing such an experiment, it is usually wise to verify the expected fraction of uninfected bacteria by spreading an aliquot on an agar plate and determining its ability to grow and form colonies on the plate. Since infection kills the host cell,* only the uninfected bacteria will survive to form colonies.

One of the most important variations on the one-step growth technique, first devised by Doermann (1948a), allows one to follow the intracellular development of infectious phage. Doermann devised techniques for artificial lysis of infected cells, before the end of the latent period. Doermann's original procedure involved the addition of cyanide to a portion of an infected culture of *E. coli,* along with a massive amount of bacteriophage T6. This led to almost immediate rupture of the infected cells, by a process called "lysis from without" (Delbrück, 1940). The culture was then diluted and plated on an indicator strain of bacteria which is resistant to T6, so that only the phage whose development is being studied form plaques. Thus, one can determine the average number of infectious units per infected cell at any desired time.

Because the T6-cyanide procedure does not give immediate lysis, other methods have been sought. The most common procedure in current use involves brief shaking of an aliquot of culture with chloroform (Séchaud and Kellenberger, 1956). This kills all uninfected cells and causes immediate lysis of infected cells, but has no effect on the plaque-forming ability of phage, once the phage have been diluted away from the small amount of chloroform remaining in the aqueous phase. Figure 2–2 shows data from a one-step growth experiment for phage T4 carried out with and without chloroform lysis. During the first 12 minutes after infection, no infectious phage are detectable inside the host cell. At this time the formation of infectious units begins until, at about 15

* An exception to this statement is the establishment of lysogeny, a situation which is discussed in detail in Chapter 10.

minutes, there is an average of 1 plaque-forming unit per infected cell. The first 12 minutes, when no infective particles can be seen inside the infected cells, is called the *eclipse period.* A major problem of bacterial virology is an understanding of the eclipse period. In other words, how does the viral genome interact with the metabolic machinery of the host cell in such a way that infectivity is lost and then regained?

The chloroform lysis technique presents a simple method for following phage adsorption. If samples are removed and chloroform lysed at rapid intervals early in infection, only the unadsorbed phage survive to form plaques. Many phages attach quite rapidly to their hosts, being essentially completely adsorbed within a minute or two. Thus, an infection of the type depicted in Figure 2–2 is essentially synchronous, since virtually all of the cycles of infection began simultaneously. For slower-adsorbing phages, or when even more precise synchrony is desired, there are many techniques for simultaneously initiating the processes of infection. Preadsorption in the cold is a commonly-used method. At ice-water temperature, adsorption is not impaired, but essentially no metabolic reactions occur. Once adsorption is judged to be complete, phage growth is started synchronously by shifting the culture to a higher temperature. Similarly, adsorption can be carried out in the absence of nutrients, and infected cell metabolism begun by the addition of a carbon source. A drawback to this method is the difficulty of completely "starving" the host cells; some metabolism may occur through breakdown of cellular substances. Another common method is to preadsorb in the presence of a metabolic inhibitor such as cyanide or azide, and to initiate infection by diluting away from the inhibitor. This is satisfactory if one wants merely to follow the formation of plaque-forming units, or if one is doing a genetic experiment, such as a cross. However, for most biochemical experiments this procedure does not work, for one must keep the culture sufficiently concentrated that samples withdrawn for analysis contain sufficient material for the measurements being performed. This usually means designing experiments to be carried out at cell concentrations of 10^8 to 10^9 ml^{-1}. Thus, the biochemist cannot usually afford the dilution factor of 10^3 or 10^4 necessary to obviate the effects of cyanide or azide.

A final point to note from Figure 2–2 is that, at the end of the latent period, the phage titers seen with and without chloroform lysis are equal. This is consistent with the conclusion of Ellis and Delbrück that the end of the rise period corresponds to the time at which every infected cell has lysed. At this time, every phage particle in the culture is an extracellular phage.

The genetic cross can be considered a special type of growth experiment. One infects cells mixedly with phage containing two or more different genotypes. By use of selective plating techniques, discussed further in Chapter 5, one enumerates the progeny containing both and/or neither parental genotype, and, by comparing these figures with the total phage yield, calculates the frequency of recombination.

It is beyond the scope of this book to provide detailed instructions for performance of growth and biochemical experiments with phage. An excellent compendium of standard laboratory techniques for isolation of phage, preparation and assay of phage stocks, measurement of growth, etc., can be found in the textbook of Adams (1959). This book, which was completed in 1959 by several other authors after the untimely death of Adams, is a classic work in phage biology. Although the growth of this field has been spectacular in the past few years, it is perhaps comforting to know that there is at least one work in the field which is of such lasting value that it has been recently reprinted without revision, some ten years after the original date of publication.

SOME COMMENTS ON BACTERIOPHAGE SOCIOLOGY

There exist different types of bacteriophages capable of infecting almost all known bacterial species. In general, one finds a bacteriophage in the same habitat occupied by its natural host. For example, the most widely studied bacteriophages are those infecting the intestinal bacterium *E. coli.* Because of the environment of the host, most coliphages have been searched for, and found, in raw sewage or in feces. The isolation simply involves filtration of a suspension of the material, followed by dilution and plating with an excess of susceptible bacteria. A phage stock is then prepared by picking a single plaque with a sterile needle, using this to inoculate a small culture growing in liquid medium and waiting until repeated cycles of phage growth result in lysis of the entire culture.

Because of the ease of isolation of new phages, and because of the large number of different phages in existence, the field of bacterial virology was in some disarray during the early years of its existence. Each investigator in the field isolated and studied his own phage-host system, thus making comparison of results obtained in different laboratories difficult or impossible. Fortunately, an agreement was reached in the mid-1940s by American workers in the then-fledgling field to concentrate their efforts on the T-series of bacteriophages whose common host is *E. coli.* There are 7 virulent phages in this series, namely T1, T2, T3, T4, T5, T6, and T7. Within this series, the closely related T-even phages, T2, T4, and T6, have received the greatest attention. The choice of these phages was based largely on technical grounds: the host is extremely easy to cultivate, the phages adsorb rapidly and plate with high efficiency, and so forth. In some ways, the choice of the T-even phages was a lucky accident. It was thought that these would provide relatively simple "model" viruses. As it happens, the T-even phages are among the largest and most complex of all known viruses. There still remain many major areas of ignorance about the mode of replication of these phages. On the other hand, the complexity of the T-even phages, has made the biology of most other viruses seem simple by comparison. In addition, some unique aspects of T-even phages have provided "handles" for making discoveries of general virological interest. For example, the discovery of Wyatt and Cohen (1953), that T-even phages contain the unique pyrimidine 5-hydroxymethylcytosine in their DNA in place of cytosine made it possible to study the metabolism of phage DNA independently from that of the host cell (Hershey, 1953). Furthermore, this fact led, a few years later, to the discovery, by Flaks and Cohen (1957), that bacteriophages can induce the formation of new enzymes in infected cells, enzymes necessary for the synthesis of viral components. This discovery has been proven to be quite general in virology. Thus, viruses, which were earlier considered to be inert particles wholly dependent upon the metabolic machinery of infected cells for their replication, were seen to be capable of profoundly altering the metabolism of the host cell.

Because of the early concentration of American phage workers on the T-even phages, many important aspects of phage biology have been neglected until relatively recently. For example, the nature of lysogeny, wherein a temperate bacteriophage can stably integrate its genome into that of an infected cell, was not conclusively demonstrated until 1950 by Lwoff, a French scientist (Lwoff and Gutmann, 1950), although the existence of this phenomenon had been known since the 1920s. Phages with single-stranded DNA were not recognized as such until the late 1950s (Sinsheimer, 1959), and RNA-containing phages were not discovered until two years later (Loeb and Zinder, 1961). Little was known about the mode of replication of any of these viruses until the mid-1960s. At any rate, the historical framework provided by work on the T-even phages makes it convenient to concentrate on these viruses in the early parts of this book, and then to present material concerned with other phages of biochemical interest.

3

General Chemistry of Bacteriophages

In the previous chapter, we explored some characteristics of a bacteriophage life cycle. Although we have so far had little to say about bacteriophage structure, it is obvious that we cannot explain the events of infection in molecular terms without first describing the phage as a physical and chemical entity. The earliest studies in this area, carried out in the 1930s, were concerned primarily with the size of phage particles and with overall chemical composition. It had long been known that phages are very small compared to bacteria, for they pass through filters which retain bacteria. This fact was explored quantitatively by Elford and Andrewes (1932), who prepared a series of collodion filters with known pore sizes and measured the ability of various phages to pass through the filters. The minimum pore size which allowed a particular virus to pass through could be used to estimate the particle diameter of the virus. The range of values determined for different phages, from 21 $m\mu$ for phage S13 to 90 $m\mu$ for D12 (a close relative of the T-even phages), is in reasonable agreement with values determined more recently by electron microscopy. Similar values were obtained by Schlesinger (1932), who studied the rate of sedimentation of phages in a centrifugal field. In these early experiments, phage suspensions were centrifuged for known times and at known gravitational forces, and the infectious titers of the supernatants determined after careful decantation.

With the development of the modern analytical ultracentrifuge it became possible to obtain more accurate physical data on bacteriophage particles. Cummings and Kozloff (1959) determined the sedimentation constant and the diffusion coefficient for T2L (the "Luria strain" of T2). From these data they calculated a particle weight of 215 million daltons, or about 3.3×10^{-16} grams per particle. However, by far the most powerful tool in elucidating the structure of phages and other viruses has been the electron microscope, as we shall see presently.

PREPARATION AND PURIFICATION OF PHAGE STOCKS

Examination of phage by either analytical ultracentrifugation or electron microscopy, as well as by the classical methods of chemistry, requires the availability of highly purified virus preparations. Let us briefly examine the most common methods used in reaching this goal.

First, one must be able to prepare a high-titer phage stock, i.e., a relatively concentrated suspension of infectious particles. This can be done by the plate method, in which a Petri plate is seeded with bacteria and sufficient phage to give confluent lysis, or clearing of the entire surface of the plate by viral growth (except for the occasional colonies of phage-resistant bacterial mutants which arise during development of the plate). About 10^5 or more T-even phage particles suffice to give confluent lysis on a standard-sized Petri plate of 15 cm in diameter. One then scrapes off the layer of soft top agar, bearing the phage and resistant bacteria. The agar is extracted with an aqueous buffer and centrifuged at low speed to sediment the agar, leaving most of the phage particles in the supernatant.

Less messy and more amenable to large-scale preparative procedures, is the liquid culture method, in which cells are infected and allowed to lyse in liquid medium. Typically, a bacterial culture in the early exponential phase of growth is infected at an average multiplicity of infection (MOI) of about 0.01 phage per bacterium. About 1% of the cells are infected, while the rest continue growing. At the end of the latent period, the infected cells rupture, and the progeny phage released are sufficiently numerous to infect essentially all of the remaining cells, which are still in the exponential phase of growth (a necessity for successful infection). The culture is then incubated until it lyses. Chloroform is added to kill any uninfected cells and to rupture any infected cells which have failed to lyse. By this technique one can add phage at about 10^6 plaque-forming units (PFU) per ml and obtain a lysate containing more than 10^{11} PFU per ml: in other words, a hundred thousandfold return on one's investment within just a few hours. This method can also be used for preparation of genetically homogeneous stocks from single plaques. One stabs a plaque with a sterile needle, shakes the needle tip in a young bacterial culture, and continues incubation. Inasmuch as the number of phage particles clinging to the needle tip is quite variable, this technique requires some practice and knowledge of the phage-host system before an investigator can routinely obtain high-titer lysates starting from single plaques.

Once a high-titer phage lysate has been prepared, the simplest and most reliable procedure for purifying the phage is differential centrifugation. One centrifuges the lysate at low speed (e.g., 5000 $\times$ *g* for 10 minutes) to remove bacterial debris, and the supernatant is spun at high speed (typically, 60,000 $\times$ *g* for 30 minutes for a large phage), to sediment the virus. After resuspension of the pellet, an additional low-speed spin is used to remove more particulate matter. This can be followed by treatment with deoxyribonuclease to digest any DNA not packaged into intact virions, and another cycle of differential centrifugation. The purity of the preparation is monitored by measuring the infective titer and the absorbance at 260 mμ, the absorption maximum for most phages. The lower the ratio A_{260}/PFU, the purer the phage. Typically, for highly purified T-even phages, A_{260}/PFU is between 0.7 and 0.8. Values for several other phages are presented in Thomas (1966).

Other methods are available for phage purification. Preparative ultracentrifugation, either through sucrose gradients or preformed cesium chloride gradients (Matthews, 1960), is effective for the preparation of nearly pure phage. Column chromatographic methods have also been used with some success. An example is the purification of λ phage by chromatography on hydroxylapatite (Miyazawa and Thomas, 1965).

CHEMICAL COMPOSITION OF PHAGES

Once one has obtained a phage preparation which is relatively free of incomplete phage particles and bacterial debris, one can examine the structure of the virus, using either the methods of quantitative chemical analysis, or electron microscopy, or a combination of the two. The earliest chemical analyses were carried out in the mid-1930s by Schlesinger (1934, 1936), using phage WLL which had been purified by differential filtration through graded collodion membranes. From the results of quantitative elementary analysis, and positive qualitative tests for DNA and protein, Schlesinger concluded that the phage is a deoxyribonucleoprotein. Subsequent studies with other phages tended to confirm this conclusion; for many years all phages which were analyzed chemically were found to consist almost entirely of DNA and protein. It was not until 1961 that RNA-containing phages were discovered (Loeb and Zinder, 1961), and several such phages are now known. However, except for a small amount of carbohydrate present in the DNA of some phages and other small molecules present in minute amounts (see below), all known phages contain only nucleic acid and protein. Nearly all the protein is in the *capsid,* a general term for that part of a virus particle which surrounds the nucleic acid and facilitates its entry into host cells. Data on the chemical composition of some phages are presented in Table 3–1. Aspects of the structure of phage proteins and nucleic acids are discussed later in this chapter.

TABLE 3–1
SIZE AND NUCLEIC ACID CONTENT OF SOME PHAGES

Phage	Host	Temperate or Virulent	Type of Nucleic Acid	Particle weight of Virion, Daltons $\times 10^{-6}$	Nucleic Acid per Virion, Daltons $\times 10^{-6}$
T-even	*E. coli*	Virulent	DNA	215	130
T5	*E. coli*	Virulent	DNA	—	76
T7	*E. coli*	Virulent	DNA	50	24
ϕx174	*E. coli*	Virulent	DNA	6.2	1.7
fd	*E. coli* (Male only)	Virulent	DNA	11.3	1.3
λ	*E. coli*	Temperate	DNA	57	32
Plkc	*E. coli*	Temperate	DNA	—	58
R17	*E. coli*	Virulent	RNA	3.6	1.1
MS2	*E. coli*	Virulent	RNA	4.0	1.1
P22	*Salmonella typhimurium*	Temperate	DNA	—	26
SP8	*Bacillus subtilis*	Virulent	DNA	—	69

Data whose sources are not listed in the text are from Fraenkel-Conrat (1968) or Thomas and MacHattie (1967).

PHAGE STRUCTURE AS DETERMINED BY ELECTRON MICROSCOPY

Although the first electron microscopic examination of a bacteriophage was conducted as early as 1940 (Ruska, 1940), details of phage structure at the molecular level have been forthcoming only within the past decade or so, pursuant to the development of improved techniques for shadow-casting and staining samples for electron microscopy. Early studies by Luria and Anderson (1942) had revealed that T-even phages are possessed of both a head and a tail, but only recently has the electron microscope begun to reveal the large number of different components, particularly in the tail, and their intricate interrelationships. Concurrent with the development of improved electron microscopic techniques, methods have become available for the dissection of phage particles, using both chemical (cf. Brenner et al., 1959) and genetic (cf. Epstein et al., 1963) methods. Thus, one can study the architecture of isolated phage components at both the anatomical and chemical levels. A logical outgrowth of this work, which is being pursued in several laboratories, is the question of the physical and chemical forces binding these components together in the intact virion. To date this question has received relatively little attention, but it is now being vigorously pursued at the chemical level primarily by Kozloff and Cummings and their colleagues at Colorado (cf. Kozloff, 1968) and at the genetic level principally in the laboratories of Edgar and Wood (cf. Wood et al., 1968). Whatever the nature of these forces, it now appears probable that no covalent bonds are involved in binding T-even phage substructures together, for intact virions can be completely dissociated by agents such as urea, guanidine hydrochloride, detergents, and extremes of pH (To, Kellenberger, and Eisenstark, 1969). In addition, the failure of reducing agents to disrupt phage structures would appear to rule out the existence of disulfide bonds.

The earliest electron micrographs of T-even phages showed them to contain a head nearly 1000 Å long, which appeared hexagonal in two-dimensional pictures, and a tail of approximately equal length, attached to one of the vertices (see Figure 3–1). The phage DNA is packed inside the head, as shown by experiments involving osmotic shock (Anderson, 1950, 1953). When phage are rapidly diluted from an environment of high ionic strength into distilled water, they lose their infectivity. Electron microscopy shows that phage heads have ruptured, and the DNA has been released from the head into the medium (Figure 3–2). Very little protein is released from the DNA-less particles, which are called ghosts. Ghosts retain some biological properties of intact phage, including the ability to adsorb to susceptible bacteria and to kill them, although, as we shall see in Chapter 4, the lethal effect of ghosts is quite different from that of whole phage.

Before we discuss the detailed substructure of the components of a T-even phage, let us briefly examine the overall structure, referring both to Figure 3–1 and to the schematic diagram shown in Figure 3–3. In addition to the DNA-containing head, we see that the tail is of remarkably complex design. It consists of two coaxial cylindrical structures, an outer sheath and an inner core, or tube. At the proximal end, near the point of attachment to the head, is a collar structure, and at the distal end is a hexagonal structure, the baseplate, to which are attached six tail fibers and several shorter structures, the tail pins. As we shall see below, a good deal is known about the roles of some of the tail

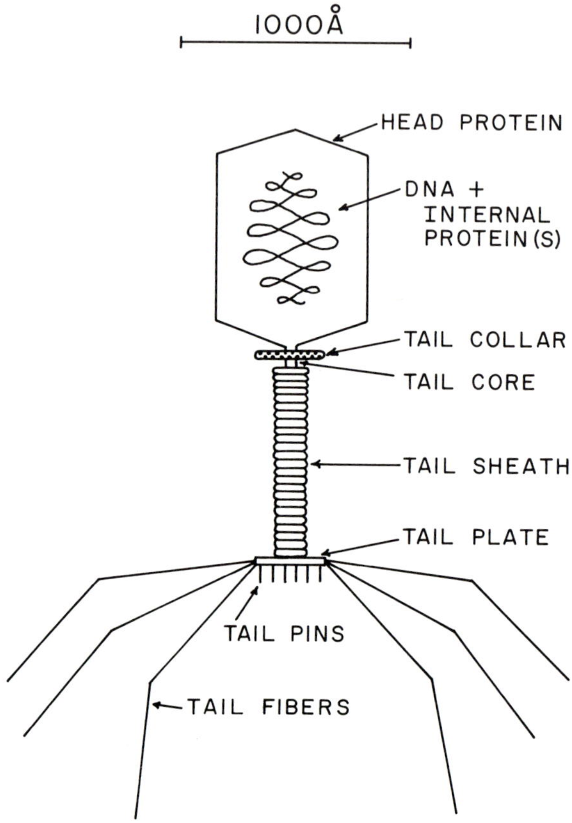

FIGURE 3-3 Schematic drawing of a T-even phage particle.

STRUCTURE OF THE HEAD

Until recently, only one type of protein molecule had been found in T-even phage head membranes. Originally reported to have a molecular weight of 80,000 (Van Vunakis, Baker, and Brown, 1958; Brenner et al., 1959), the head protein was subsequently reported to exist as a molecule of 42,000 daltons which could readily undergo dimerization (Cummings, 1963a, b). Since there are about 80 million daltons of protein in a single phage head, it is clear that this protein must participate as a monomer in some type of repeating subunit structure. Oddly enough, however, no clear indication of the existence of a subunit structure has yet been obtained from electron microscopy. In fact, even the overall details of shape and symmetry of the head have remained elusive. In part, this is due to the fragility of the head, which tends to become distorted under the conditions applied for negative staining (cf. discussion in Kozloff, 1968). [Note added in proof. Bayer and Remsen (1970) have recently observed the structure of T2 by means of the freeze-etching technique. Their pictures show a definite repeating structure in the head, with subunits, or capsomeres, of about 10 mμ in diameter.]

Recently, Forrest and Cummings (1970) have demonstrated the existence of more than one protein species in T-even phage heads, as shown following disruption of purified heads in 6M guanidine hydrochloride. Heads from T2H (the "Hershey" strain), and T6 all contain at least three proteins, including entities of molecular weights 40,000 18,000, and 11,000, while T2L contains but two detectable proteins, the 40,000- and 18,000-dalton moieties. In all cases the 40,000-dalton protein constitutes some 80% of the entire protein of the head.

A further feature of head structure in phage T2L which must be accounted for in any model proposed for the head is its existence in two interchangeable conformations. Ultracentrifugal analysis by Cummings and Kozloff (1959) showed that this phage sediments in two separate forms, with sedimentation coefficients of 700S and 1000S, respectively. Subsequent analysis in the electron microscope showed that the head of the slower-sedimenting form is some 15% longer than that of the rapidly-sedimenting form, and that it is also more permeable to small molecules. The phage must be in the short-head form to inject its DNA. It is unlikely that a transition of the long-head to the short-head form is an integral event in the initiation of infection, however, for this phenomenon is not shared by all of the T-even phages. Although T6 can exist in a long-head form (Cummings, Chapman, and DeLong, 1969), T2H and T4B do not exist in this conformation.

Perhaps the most mystifying aspect of phage head structure is the mode of packing of the DNA. Since this point is discussed at greater length below, suffice it to say here that a T-even DNA molecule is orders of magnitude longer than the head, yet it is packed in such a way that the phage can expel all of its DNA in less than a minute. Models have been proposed to explain DNA packing (cf. discussion in Kozloff, 1968), but no "hard" evidence has yet been obtained.

Another protein component of the head is the so-called internal protein, which is found complexed with DNA inside the head. This protein, which has not been well-characterized biochemically, has been estimated to represent between 3% (Hershey, 1955) and 7% (Minagawa, 1961) of the total protein of the virion. It enters the infected cell along with the DNA, and represents that small proportion of phage protein which was found inside infected cells in the blendor experiment of Hershey and Chase (1953). The biological role of this material is unknown. Because it is basic and can complex with phage DNA *in vitro* (Chaproniere-Rickenberg, Mahler, and Fraser, 1964), functions have been proposed which involve stabilization of the packed structure of

DNA. Kellenberger (1962) proposed that internal protein acts as a "condensation principle," about which nascent DNA can coalesce in the infected cell, during the process of virus assembly. A fragmentary report by Bachrach and Friedmann (1967) raised the possibility that internal protein might act as a regulator of the transcription of the viral genome, since a preparation of this material inhibited an RNA polymerase system *in vitro*. This is hardly surprising in view of the large number of basic materials which bind to DNA and reduce its effectiveness as a template *in vitro*. No evidence for the specificity of the effect of the internal protein was obtained by Bachrach and Friendmann. Whatever biological role is eventually uncovered for this material, it is likely to be more than just a saltlike neutralization function, for the internal proteins of the closely related phage T2 and T4 are immunologically quite distinct (Levine, Barlow, and Van Vunakis, 1958). This implies a higher order of specificity in function than would be expected merely for charge neutralization.

In addition to the internal protein, phage heads contain additional non-DNA material, but their roles as integral structural components of the virion are not yet clear. Ames, Dubin, and Rosenthal (1958) found that T4 contains the polyamines putrescine and spermidine, in amounts sufficient to neutralize about 30% of the negative charge on the viral DNA. However, these materials can be removed by repeated washing of phage preparations without affecting phage viability, which raises some question about their essentiality. As Kozloff (1968) has pointed out, however, they may exert their biological effects during the folding of DNA in the process of virus assembly. T-even phage heads also contain some acid-soluble polypeptide material, amounting to about 1% of the total carbon of phage T2H (Hershey, 1957). More recently, Eddleman and Champe (1966) have shown the presence of two distinct polypeptide components in the heads of T4D and have presented genetic evidence that these compounds play some role in formation of the head. Sternberg and Champe (1969) have extended this to locate the structural gene for one of these peptides and have shown that it maps in a region of the T4 genome which is concerned with formation of the head (cf. Chapter 5).

STRUCTURE OF THE TAIL

The T-even phage tail is a truly remarkable structure, capable of recognizing attachment sites on susceptible bacteria and guiding the viral nucleic acid through the rigid bacterial cell wall, all in an exceedingly short period of time. In this chapter, I would like to present some information on the chemistry and structure of distinct tail components, reserving until Chapter 4 a detailed discussion of how these components interact during the early steps of viral invasion. Much of the information discussed in this chapter has come from studies with purified tail components either after chemical fragmentation of virions or by use of mutants defective in various stages of virus assembly (cf. Chapter 7).

Tail Collar

This structure was first observed in electron micrographs by Anderson (1960). It is a thin, hexagonal-shaped disk, some 360 to 375 Å in diameter. Studies on isolated tail collars have been hampered first by the fact that an isolated collar in a phage preparation greatly resembles an isolated tail plate and second, because the isolated collar appears to be very fragile. Little is known about the function of this structure, although there is some evidence that it may be involved in assembly of the phage head (cf. Chapter 7).

Tail Sheath

The protein of the sheath comprises the bulk of the tail protein, containing some 8 million daltons per virion (N. Sarkar, S. Sarkar, and Kozloff, 1964). The sheath is contractile; during the early stages of infection, it undergoes a shortening from about 800 Å to 350 Å, with concomitant thickening (Kozloff and Lute, 1959; Brenner et al., 1959). Because of this property, the sheath has been studied by a number of workers who wish to use it as a microsystem for understanding muscular contraction.

Brenner et al. have obtained excellent electron micrographs of purified sheaths, which suggest that each sheath contains 144 morphologically distinct subunits. Data from amino acid analyses are in conformance with this, for histidine, the rarest amino acid in the T2L sheath, is present to the extent of 144 residues per sheath, or one per subunit (N. Sarkar, S. Sarkar, and Kozloff, 1964). This would give each subunit a molecular weight of 8 million/144, or 55,000. However, recent studies by Farid and Kozloff (1968) indicate that there are two carboxyterminal amino acids in purified sheaths, namely, serine and glycine. The total number of C-terminal residues per sheath is 288 ± 15, suggesting that each of the morphological subunits may consist of two or more distinct protein molecules.

The T2 sheath has a number of the chemical attributes to be expected of a micromuscle. Kozloff and Lute (1960) found that well-washed T2L preparations contain about 140 molecules of tightly-bound calcium per virion or about one molecule per sheath subunit. Nucleoside triphosphates, primarily ATP and deoxyATP are also bound, again to the extent of about one molecule per sheath subunit (Wahl and Kozloff, 1962). With these and other low-molecular weight components, one has the problem of proving that the material in question is truly an integral structural component, or whether it becomes adventitiously associated with the phage structure during maturation or virus purification. A major difficulty arises in trying to establish criteria for purity of a phage preparation. Granted, one can remove bound calcium or nucleoside triphosphates by prolonged washings or other treatments, but how does one recognize when he has passed the point of obtaining pure phage, and each subsequent treatment represents a step in the degradation of a once-intact structure? In the case of the nucleoside triphosphates, present evidence suggests that they are indeed true phage (as opposed to phage-associated) components. No triphosphates are found in the short-tailed phage T7 or in a number of phages with noncontractile tails. Moreover, some functional role is suggested by the presence in purified phage preparations of an ATPase activity (Dukes and Kozloff, 1959), which Tikhonenko and Polglazov (1963) have reported to be located in the tail sheath. Apparently ATP is required for maintenance of the tail in the relaxed state, but, contrary to what one might expect, ATP hydrolysis is not required for contraction, although it does occur under some conditions. However, liberation of the nucleotides, either intact or as hydrolysis products, always accompanies contraction (Wahl and Kozloff, 1962). Addition of ATP to a purified preparation of contracted sheaths causes them to become elongated (N. Sarkar, S. Sarkar, and Kozloff, 1964). These facts support a model for contraction which involves a change in conformation of the subunits, with the presence of ATP being necessary for maintenance of the conformation associated with the relaxed state.

Further support for the above idea comes from electron microscopy of purified sheaths, both contracted and relaxed (Bradley, 1963; Moody, 1967a, b; Krimm and Anderson, 1967). Because of their structural regularity as seen in electron micrographs, a useful technique in interpreting these pictures has been optical diffraction (Taylor and Lipson, 1964; Klug and Berger, 1964). One places an electron micrograph in a diffractometer and obtains a diffraction pattern by allowing visible light to impinge upon the sample, much as one can obtain an X-ray diffraction pattern by allowing X-rays to impinge upon a crystal. Just as in X-ray crystallography, formation of the pattern depends upon the existence of structural regularities, and it can yield much information about the structure under study. The model favored on the basis of these studies is one in which contraction involves intercalation of adjacent rows of subunits; the relaxed sheath is composed of 24 rows of 6 subunits each, while the contracted sheath is seen as containing 12 rows of 12 subunits each. Further information on the contractile process will undoubtedly require a deeper understanding of the protein chemistry of the sheath than we presently have; in other words, what is the chemical nature of the morphological subunits, how do they interact with one another, and what is the role of the nucleoside triphosphates in stabilizing the extended conformation?

Tail Tube

This component was first clearly visualized in the electron micrographs of Brenner et al. (1959) as a tube-like protein, inside the tail sheath and running almost the full length of the tail. S. Sarkar, N. Sarkar, and Kozloff (1964) isolated T2 tail tubes in pure form and determined that each virion contains 487,000 daltons of tube protein. Because of the presence of five half-cystine residues per tube and because the other are amino acids also occurred in multiples of five residues per tube, a model was proposed which consisted of five identical subunits. The relative stiffness of the tube, and its dimensions, suggest that these subunits exist largely in the α-helical conformation, and wound about one another, leaving in the interior a hole some 25 Å in diameter. If viral DNA entering an infected cell passes through the tube, as seems almost certain (see Chapter 4), it should be noted that this hole is barely wide enough for single-file passage of the nucleic acid.

Tail Plate

This disk-like structure at the distal end of the tail was also clearly seen first in the electron micrographs of Brenner et al. (1959). This is a complex structure, requiring, as we shall see later, the participation of many gene products in its assembly. Kozloff, Lute, and Crosby (unpublished observations cited in Kozloff, 1968) have obtained partially purified preparations of T4 tail plates, using a phage mutant defective in head formation. In infection by this mutant a number of tail parts accumulate and can be separated by sucrose gradient centrifugation of extracts of such cells.

The plate can exist in two conformations, one resembling a hexagon and the other a six-pointed star (Simon and Anderson, 1967a). As we shall see in Chapter 4, there is good evidence that a transition from the hexagon, in intact phage, to the star is a key event in triggering the contraction of the tail.

For some time it has been known that purified phage contain a lysozyme activity (Barrington and Kozloff, 1956; Koch and Weidel, 1956). This appeared to be located near the distal part of the tail (Anfinsen, 1961), and it appeared possible that it is located in the plate, where it could serve to digest a portion of the bacterial cell wall during the earliest phase of viral invasion. Infected cells synthesize a large amount of lysozyme which is responsible for rupture of the infected cell at the end of the latent period (Streisinger et al., 1961). This enzyme has been very well characterized, as we shall see later. Several investigators have wondered whether this lysozyme, which we shall call phage lysozyme, is the same as the lytic activity associated with phage tails (the tail lysozyme). This question has been partly resolved in recent studies utilizing T4 mutants unable to form phage lysozyme (Emrich and Streisinger, 1968; Duckworth, 1969; Yamazaki, 1969). There is a lytic activity associated with mature phage (presumably the tail lysozyme), but it is not the same as the phage lysozyme. Its role in the initiation of infection is unknown. The phage lysozyme itself plays no detectable role in the initiation of infection.

One of the most intriguing components associated with the tail plate is a pteridine compound (Kozloff and Lute, 1965), which has recently been identified as 7, 8-dihydropteroylhexaglutamic acid. This is apparently unique to the phage, for all pteridines found in uninfected cells have either three or seven, but not six, glutamate residues conjugated with the pteridine ring. Analyses of purified T4 preparations show about six molecules of pteridine per phage, and this has been located in the tail plate (Kozloff, Lute, and Crosby, 1966; Verses et al., 1968). Kozloff and his colleagues would like to believe that the pteridine is involved in binding the tail fibers to the tail plate, and the apparent equivalence between pteridine molecules and fibers lends some credence to this idea, although there is no clear idea yet of the nature of the involvement. In recent studies on the nature of the pteridine-binding component which binds the folate molecules to the plate, Verses et al. (1968) have reported that the enzyme dihydrofolate reductase is present in purified T4D, and appears to be specifically associated with the tail plate. These unexpected findings will be discussed later, in connection with the mechanism of viral invasion and with the properties and metabolic roles of phage-induced enzymes.

Tail Fibers

These are attached to the vertices of the base plate, giving a total of six fibers per virion. The fibers are some 1300 Å in length, and are kinked in the middle. Brenner et al. (1959) reported that each fiber consists of four subunits, with a molecular weight of 100,000 per subunit. More recently, Edgar and Lielausis (1966), using immunological tests, have shown that T4 fibers contain at least three distinct components. It has long been known that the first step in T-even phage infection is attachment of the tip of the tail to a site on the bacterial cell wall (Anderson, 1953). The tail fibers form the adsorption organelle, i.e., that part of the phage structure which is responsible for recognition of the attachment site on the cell surface. Isolated fibers will bind to, and clump, susceptible bacteria, but have no effect upon bacteria which have mutated to resistance to the phage from which the fibers were derived (Williams and Fraser, 1956; Wildy and Anderson, 1964). As we shall see in the next chapter, bacterial mutation to resistance leads to an inability of the phage to adsorb.

Electron micrographs of T-even phages show, in addition to the fibers, a number of shorter structures, the tail pins, protruding fom the base plate. These are but 100 Å long. Simon and Anderson (1967b) in an electron microscopic study of early events in T4 infection, observed the presence of short fibers, some 350 to 400 Å long, attaching the base plate to the cell wall. Phage bearing these fibers usually show no tail pins, and Simon and Anderson believe that the short fibers and pins are interconvertible forms of the same structure.

SOME COMMENTS ON CAPSID STRUCTURE

The preceding information on chemical and physical properties of T-even substructures is summarized in Table 3–2. Certainly, a great deal of work remains to be done in this area, including clarification of the shape of the head, the mechanism of sheath contraction, and the detailed structure of the tail plate. However, a closely related, and equally important, area of research deals with the forces holding the viral substructures together. Such work should tell us much about how viruses are assembled. It also may reveal general principles of the architecture of complex subcellular organelles. Two major approachs are being used. The genetic approach, in which conditional lethal mutants are used to determine the roles of viral

TABLE 3–2
COMPONENTS OF T4 VIRIONS

Component	No. of Subunits per Virion	MW of Each Subunit	Daltons per Virion $\times$ 10^{-6}
DNA	1	130,000,000	130
Major head protein	~2,000	42,000	80
Internal protein	?	?	2.5-6
Putrescine	27,000	88	2.4
Spermidine	4,700	145	0.7
Polypeptides	?	?	~2
Collar protein	?	?	?
Sheath protein	144	55,000	8
Nucleoside triphosphates	140	~340	0.05
ATPase	?	?	?
Calcium	140	40	0.006
Tube protein	5	100,000	0.49
Plate protein	?	?	?
Dihydropteroyl-hexaglutamate	6(?)	1,100	0.007
Dihydrofolate reductase	6(?)	31,000[a]	0.19
Tail lysozyme	?	?	?
Tail fiber protein	24	100,000	2.4
Tail pins	?	?	?

Except where indicated, these data are from sources referred to in the text.
[a] Mathews and Sutherland (1965).

gene products in various steps of virus assembly, will be discussed in Chapter 7. To date, this approach has not yielded a great amount of chemical information, but it certainly has the potential to do so. The second approach involves the techniques of quantitative analysis and protein chemistry. It was this approach which led to the discovery of folate compounds in phage tails, as mentioned above (Kozloff and Lute, 1965). It had been long known that certain strains of T4 and T6 require tryptophan in order for adsorption to occur (Anderson, 1945, 1948). Cummings (1964) showed that the presence of the adsorption cofactor causes a conformational change in the virion. Electron microscopic observation showed that in the absence of tryptophan the fibers are oriented proximally, toward the head, but, upon activation, the fibers become extended distally. It was also known that the adsorption of T2H, which does not require tryptophan, is inhibited by indole. The reaction with indole had properties expected of the formation of a change transfer complex (Kanner and Kozloff, 1964), leading Kozloff and Lute to speculate that such a complex may exist in the virion, and play a role in binding fibers to the phage and determining their orientation. Addition of exogenous indole compounds disrupts these complexes, with the results described above. A systematic search for a phage component which could act as a π-electron acceptor in such a complex led to the discovery of the viral folate. In a recent confirmation of the charge transfer hypothesis, Gamow and Kozloff (1968) have treated T4D with an alkylating agent specific for tryptophan residues in proteins. The phage, which normally do not require tryptophan, assume a cofactor requirement after treatment. Presumably the tryptophan which is modified is normally a participant in the postulated charge transfer complex.

In another recent study, Kozloff et al. (1969) have examined the inactivation of T-even phages by arginine. This inactivation is accompanied by changes in tail structure. Kozloff and his colleagues concluded that an arginine residue in a tail protein is involved in maintaining a polar bond with another tail structure. Exogenously added arginine competes with this residue, and disrupts the bond. Further investigation has suggested that the critical bonds involve C-terminal arginyl residues in a tail sheath protein (Shapiro and Kozloff, 1969).

The above studies have brought to light two types of noncovalent bonds involved in maintenance of phage structure. That most, if not all of these bonds are noncovalent is suggested by the efficient dissociation of phage into its components which is caused by dimethyl sulfoxide (Cummings, Chapman, and DeLong, 1968) and by mercaptoethanol (Molina, Pesce, and Schito, 1966), as well as by the study of To et al. (1969), which was cited on p. 21.

We have mentioned various conformational changes which intact phage can undergo, including extension of the head, proximal or distal orientation of tail fibers, and contraction of the sheath, which can be triggered *in vitro* by a variety of agents. These transitions have all been studied in the ultracentrifuge, and there has been considerable confusion over which conformational transitions are responsible for which changes in sedimentation coefficient. The reader is referred to a recent paper by Cummings, Chapman, and DeLong (1969), in which these questions have been clarified by combined electron microscopic and ultracentrifugal analysis.

A final point relates to the functional and evolutionary significance of the tail. The T-even tail is the most complex such structure yet observed. However, the reader should be aware that many other phages also possess tails of quite different structure. Phages T5 and λ, for example, sport very long tails, thinner than those of the T-evens, and noncontractile, while T3 possesses a very short tail. Little is known about the function of these or other non-T-even tails, and more information would certainly be desirable. There also exist a large class of non-tailed phages, including the single-stranded DNA phages and the RNA phages. The structures of these phages are much simpler than those of the T-evens, being either rod-shaped or polyhedral. Little is known at present about the means by which nucleic acid from these viruses leaves the capsid and enters the host cell. Kozloff (1968) has pointed out that all of the known tailed phages contain double-stranded DNA, while all of the known non-tailed phages contain RNA or single-stranded DNA. The possible significance of this interesting generalization remains to be seen.

STRUCTURE OF PHAGE NUCLEIC ACIDS

A study of the chemistry and anatomy of bacteriophage nucleic acid molecules reveals considerably more variety than has been observed in cellular biochemistry. We have already referred to the existence of single-stranded DNA or RNA as the genetic material in certain phages. We have also mentioned the discovery that T-even phage DNAs contain the unique pyrimidine 5-hydroxymethylcytosine, and no cytosine. However, some aspects which were originally thought to be peculiar to phage biology have now been shown to have general relevance. For example, the RNA of the minute phage ϕx174 is circular (Fiers and Sinsheimer, 1962). Although this discovery was somewhat unexpected when it was made, circular DNA molecules have now been shown to exist in several types of subcellular organelles in higher organisms, as well as in other virsuses. One wonders what other "peculiarities" of phage nucleic acid structure will also be shown to have general significance.

Let us first consider the chemical composition of phage DNA molecules. Not only do T-even DNAs contain hydroxymethylcytosine, but the DNAs are glucosylated, with the hydroxymethyl group of HMC forming a glycoside bond through carbon 1 of the glucose (Sinsheimer, 1954; Volkin, 1954; Jesaitis, 1955; Lichtenstein and Cohen, 1960; Lehman and Pratt, 1960). The pattern of glucosylation is species-specific; T2 and T4 contain most of their glucose residues bound in α-configuration, while 72% of the HMC residues in T6 are diglucosylated, with the disaccharide having been identified as gentiobiose (Kuno and Lehman, 1962). T4 contains no unglucosylated HMCs, while some 25% of these residues are not glucosylated in either T2 or T6. As we shall see later, glucosylation is essential for phage survival, inasmuch as it appears to protect intracellular phage DNA against nucleolytic attack.

Although most other viral DNAs are composed of the usual adenine, cytosine, guanine, and thymine, there are a few more examples of unique bases in phage DNAs. Several phages infecting *Bacillus subtilis* contain 5-hydroxymethyluracil (HMU) in place of thymine (Kallen, Simon, and Marmur, 1962; Roscoe and Tucker, 1964), while at least one *subtilis* phage, PBS2, contains uracil substituted for thymine (Kahan, 1963; Takahashi and Marmur, 1963a). There were several reports that some of these DNAs are glucosylated, although the nature of the linkage was not clarified (Takahashi and Marmur, 1963b; Rosenberg, 1965). However, a recent report (Alegria and Kahan, 1968) suggests that the apparent detection of glucose was due to contamination of the DNA with bacterial polysaccharide.

Size of T-Even Phage DNA

For a time, some ten to fifteen years ago, the reported molecular weights of DNA preparations from all sources increased steadily year by year. This upward trend reflected the increasing awareness of investigators of the ease with which giant polynucleotides fragment under hydrodynamic shear forces; the molecular weights of some DNA samples decrease severalfold when one merely sucks a solution into a pipet and blows it out again. Such extreme fragility certainly could not have been predicted on the basis of the chemistry of any other known substances. In the case of viral DNAs there was a theoretical upper limit to any molecular weight which could be reported with any credibility—namely, that molecular weight corresponding to one phage-equivalent unit of DNA. From the molecular weight of a T2 virion, about 215 million, and the DNA-to-protein ratio, about 1.5, Kozloff (1960) had estimated the DNA content of an individual T2 virion to be 115–140 million. Therefore, the maximum possible size for T2 DNA would be this same figure, if each T2 particle contains but a single molecule of DNA.

This, in fact, turned out to be the case. Using a method called P^{32} "star tracking," Rubenstein, Thomas, and Hershey (1961) and Davison et al. (1961) determined that T2 contains only one molecule of DNA per virion. In this method phage with P^{32}-labeled DNA were prepared by growth in radioactive medium. DNA was gently extracted from whole phage, by a method involving very little shear. The sample was imbedded in a photographic emulsion and allowed to sit for a while, for exposure of the film.

A similar sample was prepared from whole phage, similarly labeled with P^{32}. Each P^{32} emission produces a track, which exposes a small area of the emulsion. After a time the emulsion is developed. Each labeled molecule or phage is seen as the center of a many-pointed star, with each point representing a track. From the known specific activity of the phosphorus and the number of tracks in a star, one can compute the number of phosphorus atoms in the molecule or phage and thus, since the composition of DNA is well-known, the molecular weight. Both T2 virions and DNA molecules showed identical numbers of tracks when exposed and developed under identical conditions, leading to the conclusion that the DNA molecule which is released is of the same size as the entire complement of DNA in a single virion. The molecular weight determined by this method was 130 million, a value which has since been confirmed by several other techniques. In fact, this figure is now so well-established that T2 and T4 DNAs are routinely used as sedimentation standards for zone centrifugation studies, particularly in sucrose gradients. Their value as sedimentation markers arose from an empirical study by Burgi and Hershey (1963), which showed that the molecular weight of an unknown DNA can be determined from the rate of sedimentation in a sucrose gradient compared to the same rate for a marker sedimented under the same conditions—preferably, in the same tube. The formula, assuming the densities of the two samples to be equal, is $\frac{D_2}{D_1} = \left[\frac{M_2}{M_1}\right]^{0.35}$, where D refers to distance traveled and M is the molecular weight.

Two of the other methods, namely electron microscopy and radioautography, which have been used to verify the molecular weight of T-even DNA, are of special usefulness, inasmuch as they permit visualization of the nucleic acid molecules under study. In the most common radioautographic technique, DNA is labeled with tritium by growth of the organism in radioactive medium, usually containing tritiated thymidine, and under conditions favoring incorporation of the labeled precursor into DNA. The organisms are very gently lysed under conditions which permit a small proportion of the DNA molecules to unwind in solution and to be deposited on a membrane filter. The filter is then brought into contact with photographic film, and allowed to sit for several weeks or months. The tritium emissions, being of much lower energy than those of P^{32}, do not travel nearly as far in the photographic emulsion, and are ultimately seen, after development of the film, as a series of "grains", or small spots on the film. A DNA molecule is seen as a dotted line whose length, measured microscopically, is equal to the length of the original molecule (Figure 3–4). Although this method has had its greatest success in studies on the replication of bacterial chromosomes (Cairns, 1963), it has also been a valuable tool in the experiments on phage DNA replication, as we shall see later. The value of this technique in studies on replication arises from the fact

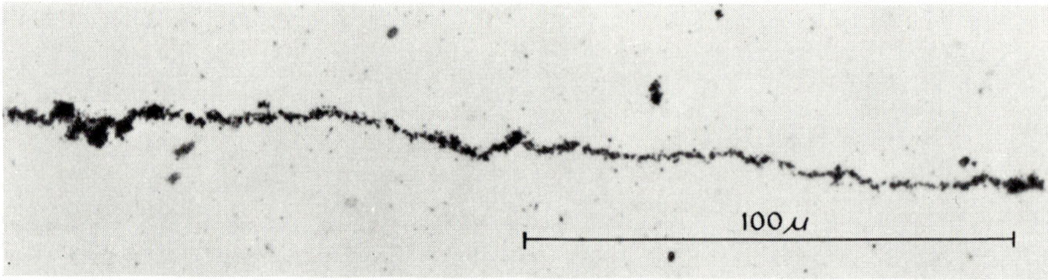

FIGURE 3-4 A radioautograph of a DNA molecule (from Huberman, 1968). This is replicating DNA from a T4-infected cell, visualized by the method of Cairns (1963). Note that the molecule is several times longer than a mature T4 DNA molecule. For further discussion of this point, see Chapter 7. (I am grateful to Dr. Joel Huberman and the Cold Spring Harbor Laboratory of Quantitative Biology for permission to reproduce this photograph.)

that the relative frequency of grains, or "grain density," is directly related to the frequency of thymidylate residues in a given stretch of polynucleotide. Thus, by knowing the specific activity of the thymidine used, the time of exposure, and other details of the experimental regimen, one can draw conclusions about the strandedness of a given region of DNA, whether it is in the process of replication, and so forth. A major limitation of this technique arises from the fact that one is visualizing single molecules, and conclusions drawn may not have validity for the bulk of the molecules in a population.

Electron microscopy has greater resolving power than tritium radioautography, but it is less informative about the state of replication of a molecule. Moreover, radioautography has the advantage of not requiring preliminary purification of the DNA visualized, since only radioactive molecules are seen, and DNA is the only macromolecule which incorporates thymidine to a significant extent. This is especially advantageous in studies on viral DNA replication, since, as we shall see later, some replicative forms are far more fragile than DNA extracted from whole phage. At any rate the relative ease and high resolving power of the electron microscopic methods have led to a large variety of ingenious applications, as we shall see throughout the rest of this book. The most useful method is the so-called Kleinschmidt technique (Kleinschmidt and Zahn, 1959). DNA is complexed with a basic protein, usually cytochrome c, and the mixture is carefully floated on a drop of distilled water. The suspension spreads to form a monolayer, during which period many of the DNA molecules become disentangled. The preparation is then caught on an EM grid, which is "shadowed" with a heavy metal spray and then visualized. A typical picture of a DNA molecule is shown in Figure 3–5. Although it appears that we are actually looking at a nucleic acid molecule, the reader should

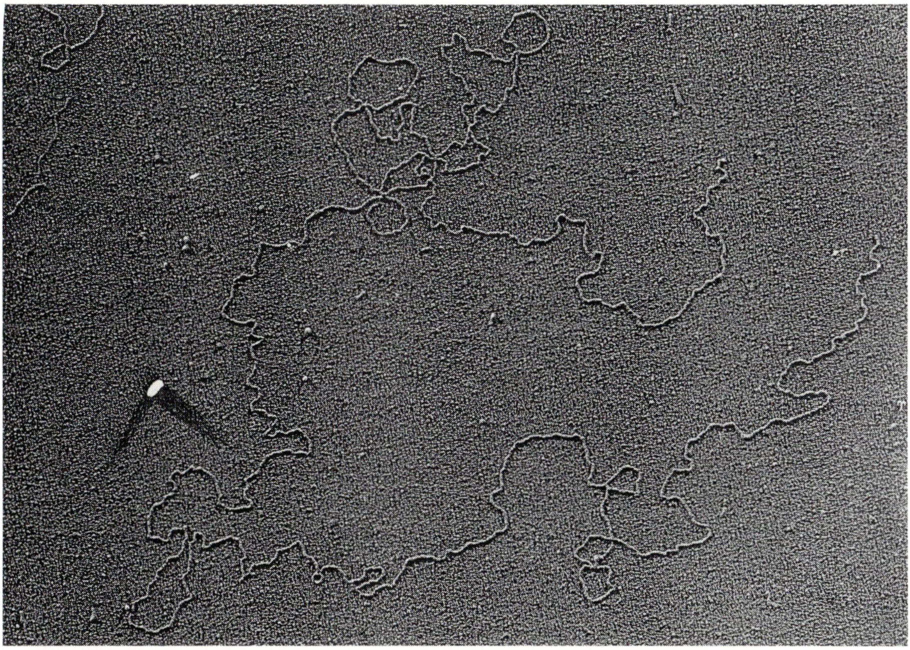

FIGURE 3-5 A T5 phage DNA molecule, visualized by the Kleinschmidt technique (from Bujard, 1969. I am grateful to Dr. Herman Bujard and the National Academy of Sciences for permission to reproduce this photograph).

not become prematurely excited, for most of what is actually seen is the shadowing material. At a recent national meeting, a speaker claimed a resolving power of some 20 to 50 nucleotide pairs for this method, whereupon a knowledgeable member of the audience issued a reminder that the width of a shadowed preparation is 100 to 200 Å, meaning that the width of the polynucleotide chain as visualized is equivalent to the length of 30 or more nucleotide pairs in a Watson-Crick helix.

By measuring the contour length of an electron micrograph of DNA, and knowing the magnification factor, one can calculate the molecular weight. One further assumes that the DNA is double-stranded, according to the Watson-Crick model. Kleinschmidt et al. (1962) determined a contour length of 54 microns for T2 DNA. This yields a molecular weight of 120 million, in close agreement with the value arrived at by other methods. This has been cited, in fact, as a confirmation of the Watson-Crick model. Every investigation of T-even DNA has confirmed that it is a linear, double-stranded structure. Perhaps the most direct demonstration is the recent feat of Guha and Szybalski (1968) in achieving a preparative separation of the two strands of T4 DNA. We shall have occasion to refer to this accomplishment further in Chapter 7.

Circular Permutation and Terminal Redundancy

As stated above, the DNAs of T-even phages have been shown to be linear duplexes. However, genetic data, primarily from Streisinger's laboratory (Streisinger and Bruce, 1960) suggested that the genetic map of T4, generated by determining recombination frequencies in a variety of mixed infections (cf. Chapter 5), is topologically equivalent to a circle. Typically, in a three-factor cross, the outside markers would appear to be closely linked to one another, as well as with the internal marker. These data were extended by Streisinger, Edgar, and Denhardt (1964), who proposed a theory to account for the circularity of the genetic map and the linearity of the molecule. The theory postulated that the nucleotide sequences of the DNA molecules in a T4 population are circular permutations of one another. In other words, if we liken the genetic text to our alphabet, some molecules would start at A, others at G, still others at X, and so on, but the order of genes or nucleotides would always be identical. Thus, A would always be followed by BCD . . . , G by HIJ . . . , and W by XYZ. This proposal was confirmed by two types of physical experiments carried out in the laboratory of Thomas (Thomas and MacHattie, 1964; Thomas and Rubenstein, 1964). In the more direct demonstration, DNA was denatured and allowed to renature. Electron microscopic examination of the produce revealed circular structures. This is expected if random association of individual strands leads one strand to pair with another of complementary nucleotide sequence but different beginning and ending points, as shown in Figure 3–6.

Another unexpected property of the T4 DNA molecule was also suggested from genetic observations, namely, the existence of heterozygotes (Hershey and Chase, 1951). In a cross of two phage types differing only at the so-called *r* locus, one would have expected all of the progeny to be either *r* or r^+. (We shall discuss the biochemistry of the *r* function later; for now, it is sufficient to know that *r* and r^+ phage produce different plaque types, which can be easily distinguished under proper plating conditions.) In fact, about 2% of the plaques resembled neither *r* nor r^+. Rather, they were mottled. Examination of the phage in these mottled plaques revealed that they contained both *r* and r^+ phage. Thus, although T4 is haploid, containing but a single chromosome, a small proportion of the genes are present in two copies. Further investigation revealed that heterozygosity is limited to a small portion of the genome. For example, only 3% of r-r^+ heterozygotes are also heterozygous for the *h* marker, another gene determining host range specificity.

The existence of one class of heterozygotes could be explained by proposing that the T4 genome is *terminally redundant,* i.e., that a small amount of genetic information is repeated at both ends of the genome. Thus, an r-r^+ heterozygote of this type would have the *r* allele at one end,

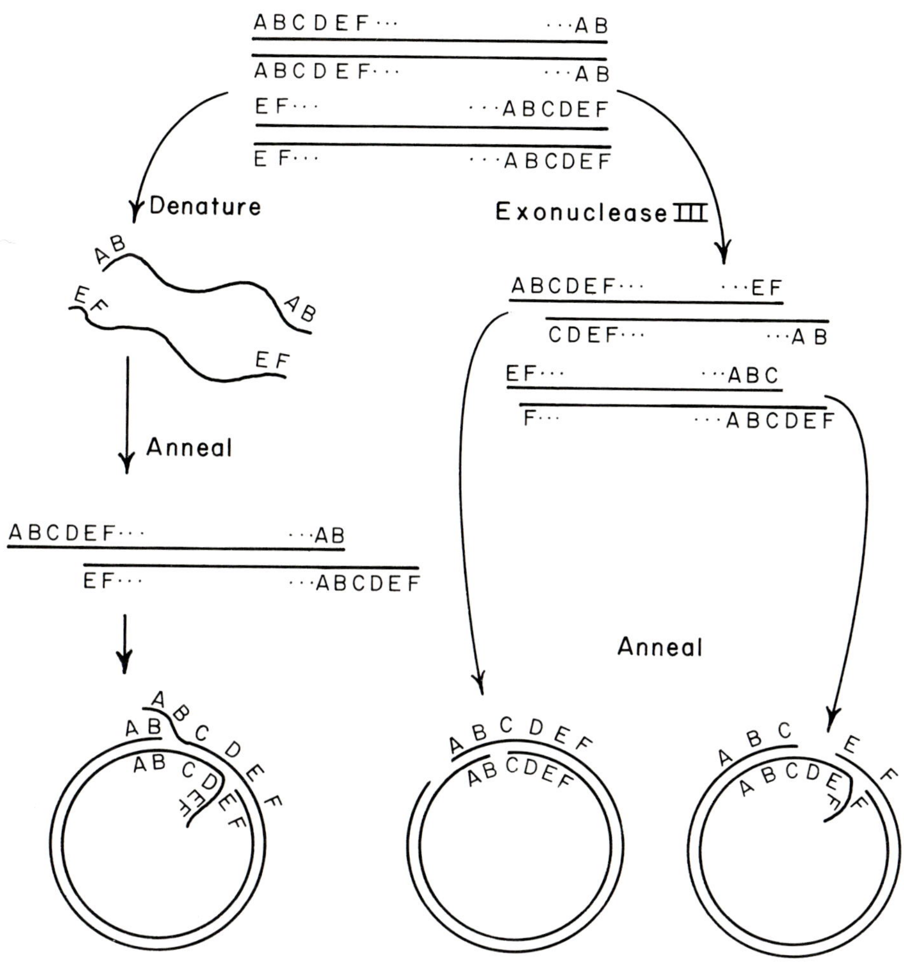

FIGURE 3-6 Experiments designed to prove that T4 phage DNA is circularly permuted with respect to base sequence (left half of picture) and terminally redundant (right half of picture). For further explanation, see text.

and the r^+ allele at the other end. Returning to our analogy between the T4 genome and the alphabet, a chromosome starting with AB . . . would end with . . . YZAB, a chromosome starting with GH . . . would end at . . . EFGH, and so forth. This proposal was put to test by MacHattie et al. (1967). T2 DNA was subjected to limited digestion with exonuclease III from *E. coli,* an enzyme which removed nucleotides stepwise from the 3′ termini of double-stranded polynucleotides. After digestion, the DNA was subject to conditions which promote annealing, and subsequently examined in the electron microscope. Circular structures were seen, as expected if the enzyme treatment exposes complementary 5′-terminated single-stranded regions at opposite ends of each molecule (Figure 3–6). The redundant regions were estimated to be 1 to 3% of the entire length.

Recent studies by Streisinger, Emrich, and Stahl (1967) relate the extent of heterozygosis with the length of the DNA molecule. In crosses involving phage which bore long deletion mutations, the frequency of terminal redundancy heterozygotes was greater than in crosses involving short deletions or point mutations. Thus, when the genome is abnormally short, as with a deletion mutant, some compensating mechanism acts to ensure that essentially all of the packaged DNA molecules are of equal length, so that the redundant ends are made longer. This finding has important implications for the mechanism of DNA replication and maturation, as we shall see later.

Comparative Anatomy of Phage DNA Molecules

A considerable number of phage DNAs have now been subject to the type of analysis described above, particularly by Thomas and his collaborators (cf. Thomas, 1967; Thomas, Ritchie, and MacHattie, 1967). Considerable diversity is seen, some of which is summarized in Table 3–3. Linear and circular molecules are seen, containing either single- or double-stranded DNA, although all linear DNAs are duplexes. Some phage DNAs, as seen above, are circularly permuted with respect to base sequence, while others are unique collections of sequences. However, all linear DNAs so far examined are terminally redundant. In the case of the temperate phage λ, the redundant regions are single-stranded, such that circles can be formed *in vitro* without prior digestion by exonuclease III (Hershey, Burgi, and Ingraham, 1963). An additional distinctive feature, seen with the DNA of T5 and a few other phages, is the existence of single-stranded interruptions, located at specific sites (Abelson and Thomas, 1966; Bujard, 1969). The functional significance of these interruptions is unknown.

One would like to infer some unifying principle from information about the diversity of viral DNA forms, which might explain some aspects of viral origin and evolution. Such unification so far has not been obtained. However, Thomas et al. (1967) have pointed out that all known viral DNA forms, whether isolated from mature virions, as replicative intermediates, or as prophage in the case of temperate phages (see Chapter 9) are terminally repetitious. This, of course, includes both linear forms and circular forms, since, if one proceeds all the way around a circle, he will encounter the beginning sequence again. Thomas et al, have used this generalization as the basis for a speculative scheme which might account for the origin of viruses. However, although the scheme is appealing, there is very little evidence to support it.

TABLE 3–3
COMPARATIVE ANATOMY OF SOME PHAGE DNA MOLECULES

Phage	Host	MW of DNA $\times 10^{-6}$	Nucleotide Pairs (Approx.)	Shape	No. of Strands	Sequence	No. of Nucleotide Pairs in Redundant Region	Remarks
T-even	*E. coli*	130	200,000	Linear	2	Permuted	2000-6000	Glucosylated; contains hydroxymethyl cytosine
λ	*E. coli*	32	49,000	Linear	2	Unique	10	Redundant regions are single-stranded
ϕx174	*E. coli*	1.7	5,500[a]	Circular	1	——	—	——
fd	*E. coli* (male only)	1.7	5,500[a]	Circular	1	——	—	——
T5	*E. coli*	76	117,000	Linear	2	Unique	Not known	Contains specifically located single strand breaks
T3	*E. coli*	24.3	37,000	Linear	2	Unique	220	——
T7	*E. coli*	24.3	37,000	Linear	2	Unique	260	——
P22	*S. typhimurium*	26.5	41,000	Linear	2	Permuted	1000-2000	——
SP82	*B. subtilis*	130	200,000	Linear	2	Unique	Not Known	Contains hydroxymethyl-uracil[b]
SP50[c]	*B. subtilis*	100	154,000	Linear	2	Unique	Not Known	Contains randomly located single strand breaks

Unless otherwise indicated, these data are taken from sources referred to in the text or from Thomas and MacHattie (1967).
[a] Number of nucleotides (total).
[b] Green (1964).
[c] Reznikoff and Thomas (1969).

was the fact that a bacterium sensitive to a particular bacteriophage could mutate to a form which was resistant to attack by virtue of an inability to absorb the phage. The most reasonable explanation of this was that the mutation conferred a structural change in that part of the cell wall structure necessary for adsorption. In addition, a single mutation could lead to resistance toward more than one phage. For example, a single-step mutation in *E. coli* confers resistance simultaneously to phages T3, T4, and T7, suggesting that the three phages share a common receptor. Moreover, the adsorption of any of these three phages is competitively inhibited by either of the other two, lending further credence to the common-receptor theory.

Some progress has been made in identifying the bacterial surface structures which are involved in adsorption. Originally this involved dissection of the *E. coli* cell wall. Weidel and his colleagues (Weidel and Primosigh, 1958; Weidel, Frank, and Martin, 1960) showed that the *E. coli* cell wall is composed of three layers: an outer liproprotein layer, a lipopolysaccharide component, and a rigid inner mucoprotein layer. Phage receptors were localized in the outer two layers by electron microscopy and by phage inactivation studies. In the latter type of experiment, phage are incubated with various cell wall fractions. If a fraction contains the receptor site for the phage under study, the phage is inactivated, for it binds to the site and injects its DNA into the medium. By these means it was learned that phages T2 and T6 are bound to some component of the outer lipoprotein layer, while the receptors for T3, T4, and T7 are located in the lipopolysaccharide layer. None of these receptors have been isolated in purified form, although the receptor for T5 is relatively easily obtained by treatment of the bacterial cell wall with dilute alkali. This receptor turns out to be a small structure containing both lipoprotein and lipopolysaccharide (Weidel and Kellenberger, 1955). Unfortunately, further work on the chemistry of the fiber-wall interaction reached a hiatus with the untimely death of Weidel. This would appear to be a ripe subject for investigation, since fibers can be prepared in nearly pure form (Brenner et al., 1959; Cummings et al., 1968), and, as we have seen, the attachment reaction can be demonstrated *in vitro*.

We have also discussed another aspect of the chemistry of adsorption, namely, the action of adsorption cofactors and the role of charge transfer complexes in determining the orientation of the tail fibers and their ability to adsorb. Brown and Anderson (1969) have recently presented evidence for a role of cell wall material in maintaining newly released phage in an adsorbable state. Phage newly liberated from infected cells contain bacterial "debris" attached to their baseplates. Several lines of evidence indicate that this is cell wall material and that, as it is lost during storage of the phage, the phage lose their ability to adsorb spontaneously. This explains an old observation that cofactor-dependent phage cannot form plaques on minimal agar plates lacking cofactor, but that cells already infected by such phage can form plaques when plated under the same conditions (Anderson, 1948). Presumably the phage released from the first and subsequent rounds of growth on the plate retain their nascent adsorbability long enough to initiate new rounds of infection, whereas free phage plated under the same conditions cannot adsorb at all to initiate the initial cycles of infection. These findings are of interest in connection with the recent finding of Simon (1969) that phage are attached tail-first to the cell membrane during assembly of the tail.

A great deal of morphological information about the early stages of infection has come from electron microscopic investigations, using thin sectioning of infected cells. Cota-Robles and Coffman (1963) demonstrated a local disorganization of cell wall material in the area of attachment of the tip of a phage tail. They also showed that contraction of the tail precedes the ejection of the contents of the phage into the cell. These studies were greatly extended by Simon and Anderson (1967a, b), whose work clarified a number of previously unresolved questions. From a series of spectacular electron micrographs, these authors described the following sequence of events early in infection by T2 or T4: (1) Initial attachment occurs *via* the tips of the tail fibers; (2) Once the phage is attached, the tips of the tail pins become fixed to the cell surface, and the phage appears to "squat down" as the head is brought closer to the surface; (3) The tail plate undergoes a major conformational change, from the hexagonal shape characteristic of mature phage to a six-pointed star. This is a key event, for it appears to (a) loosen the attachment of the plate to the distal end of the tail core, (b) remove a plug from the center of the core, opening up a hole in the core, through which the DNA can slide, and (c) somehow trigger the rearrangement of sheath subunits which seems to be involved in contraction of the tail sheath (Moody, 1967a, b); (4) Contraction of the sheath pulls the baseplate away from the cell wall and drives the core down. At the same time, the tail pins are lengthened and become what has been identified as "short fibers"; (5) The core penetrates the cell wall, and the DNA is ejected through it. Some of the electron micrographs upon which these conclusions were based are shown in Figure 4–1, while a schematic diagram of the process is presented in Figure 4–2.

Simon and Anderson were the first to show that the core actually penetrates the cell wall. Whether this is due entirely to mechanical forces generated by contraction of the sheath is not clear. For some time it had been known that enzymatic digestion of some cell wall material occurs in infection (Barrington and Kozloff, 1956). One explanation of this, which has found relatively wide acceptance, is that the "phage tail lysozyme," of which we spoke in Chapter 3, weakens the site of ultimate core penetration by attacking the cell wall material in that area. However, the status of this "hole-boring" theory must remain unproven, first because the lytic

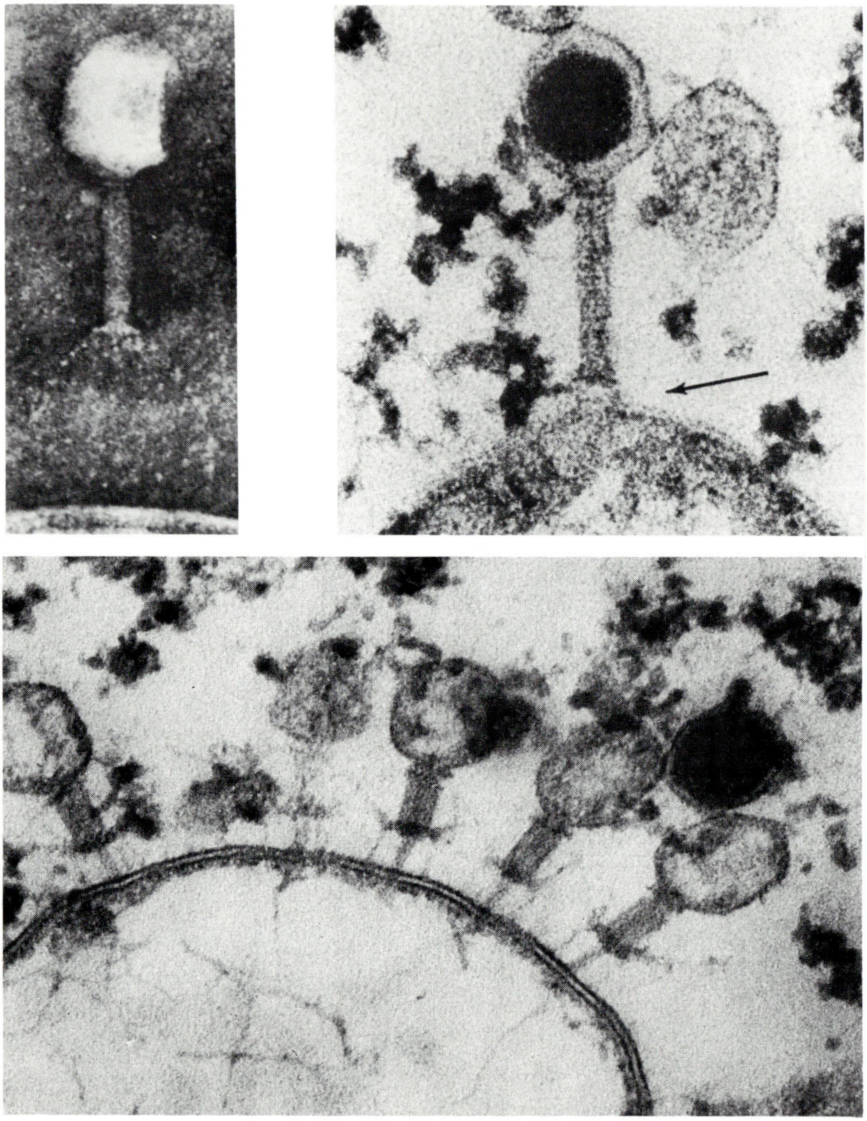

FIGURE 4-1 Early steps in T4 infection, as visualized in the electron microscope. The upper left-hand photograph shows a phage particle making initial contact with a bacterial cell wall through the tips of the long tail fibers. The upper right-hand photograph shows, in thin section, a phage particle having approached the cell wall to the point where the tail pins are in contact with the wall. DNA is still inside the phage head. In the bottom photograph, also seen in thin section, the tails have contracted, the tail cores have penetrated, and DNA is being injected into the cell (from Simon and Anderson, 1967b. I am grateful to Dr. Lee Simon and to Academic Press © for permission to reproduce these photographs).

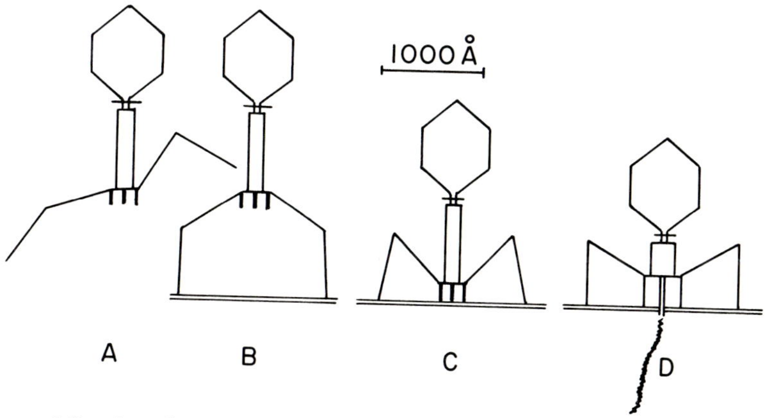

FIGURE 4-2 A schematic diagram illustrating early steps in infection by T-even phages (modified from Simon and Anderson, 1967b).

activity associated with mature phage has not been clearly shown to be associated with the tail tip, and secondly because to my knowledge the digestion of cell wall material which occurs has not been shown to be localized at sites of phage attachment.

Perhaps the most important unresolved question about the processes of adsorption and penetration is the mechanism of DNA transfer from the phage head to the interior of the cell. This process is quite efficient, and expenditure of metabolic energy is not required, for phage will inject DNA into heat-killed or cyanide-treated cells, or into purified cell wall fragments. Yet the DNA molecule, some 500 times longer than the head which it inhabits, somehow unwinds, passes through a hole in the core which is barely wider than the width of a double helix, and traverses the highly charged lipoprotein membrane of the cell, all in less than a minute. This is the type of problem which forces biochemists to turn to physics—or perhaps to metaphysics—for help.

Evidence is beginning to accumulate—and this will be discussed further in Chapter 7—that the parental DNA, rather than penetrating into the cytoplasm of the infected cell, may remain bound to a site near or at the membrane. Jacob, Brenner, and Cuzin (1963) proposed that the *E. coli* chromosome replicates itself through a membrane attachment site. There is now considerable evidence supporting this idea (Smith and Hanawalt, 1967; Schachtele, Anderson, and Rogers, 1968; Tremblay, Daniels, and Schaechter, 1969), and several investigators have begun to explore the involvement of the cell membrane in phage DNA replication. Although the nature of this involvement is quite obscure at the present time, we shall have occasion to return to this theme at several points. Bayer (1968) has recently published a paper which is of some interest in relation to membrane-associated processes in phage metabolism. Bayer studied phage adsorption to plasmolyzed *E. coli* cells. These cells are treated with high concentrations of sucrose, which cause a loss of intracellular water, due to osmosis. Under these conditions, the protoplasm, surrounded by the membrane, shrinks away from the cell wall. However, in some places the membrane is not pulled away from the wall, and these sites are visible as wall-membrane adhesions in thin sections, when viewed in the electron microscope. Bayer showed that phage are adsorbed specifically at these adhesions. Interestingly enough, the total number of adhesions per cell is between 200 and 400, roughly the same as the total number of phage adsorption sites per cell, as determined by Schlesinger (1932). Although the chemical nature of these adhesions is unknown, the fact that phage adsorb at these sites would appear to lead to the introduction of the phage chromosome at those sites providing the most ready access to the protoplasm.

EFFECTS ON THE CELL ENVELOPE

We have briefly alluded to the possible role of the cell membrane in the infection process. A number of workers have explored changes in the bacterial surface in hopes of understanding a number of problems in phage biology. These include lysis inhibition, in which reinfection of an infected cell leads to prolongation of the latent period, with corresponding increase in burst size (Doermann, 1948b); superinfection exclusion, in which superinfecting genomes are prevented from being expressed under certain conditions (Dulbecco, 1952); the biochemistry of *r,* or rapid-lysis mutations, in which the capacity for lysis inhibition is lost; and the mechanism by which virulent phage infection abolishes genetic activity of the host. One can reasonably explain all of these phenomena by postulating various changes in permeability or other properties of the cell envelope. We shall discuss these phenomena *per se* later in the book. Here I would like to summarize work which has been done on changes in cell permeability and on chemical composition of the cell surface which occur as a result of infection by T-even phages.

Perhaps the earliest indication that T-even phage infection causes changes in the cell envelope came from the work of Doermann (1948b), who observed turbidity changes in infected bacterial cultures. Immediately after infection with T2 or T4, there is a 25 to 30 percent drop in turbidity. At about 10 minutes the turbidity starts to increase and by 15 or 20 minutes has regained its original value. At the end of the latent period the turbidity drops irreversibly, as the culture clears due to cell lysis.

The reason for the initial drop in turbidity is not completely clear. Apparently the cells become more transparent. This could be due either to swelling of the cells, or to extensive leakage of material from the cells. Puck and Lee (1954, 1955) examined the permeability of infected cells and showed that extensive leakage of cell contents does indeed occur within the first few minutes after infection by T1 or T2. This leakage is reversible in normal infection; a "sealing" reaction begins at 3 to 5 minutes, and when it is complete, the cells are less permeable than they were before infection. In fact, Puck and Lee used the sealing reaction to explain the phenomenon of superinfection exclusion. They found that when a T2-infected culture was superinfected with T2 at 8 minutes, the DNA of the superinfecting phages failed to enter the cell, and they attributed this to the permeability barrier created by the sealing reaction.

Puck and Lee attributed the initial leakage of cell material to the presence of holes bored by the phages, through which material could exit from the cells. In support of this they found that not only the extent of leakage but the molecular size of leaked material depended upon the multiplicity of infection. Under conditions of single infection acid-soluble phosphorus-containing compounds were lost, whereas at MOIs between 5 and 30, some RNA and the enzyme β-galactosidase leaked out. Calculations of phage size in relation to surface area led to the conclusion that the leakage of macromolecules at high MOI was not due to juxtaposition of phages on a cell surface, which would lead to formation of larger holes. To resolve this apparent discrepancy, they proposed that phage infection has a "spreading effect" on the cell membrane, which causes it to change its permeability in the vicinity of each bound phage. The spreading reaction continues until its effects are overcome by the onset of the sealing reaction, which requires the expenditure of energy. In support of this, it was found that T2-induced leakage is far greater at 4°, where sealing presumably does not occur, than at 37°. The nature of the sealing reaction is not known, but it probably represents, at least in part, the continued synthesis of bacterial cell wall material, and may also represent the synthesis of phage-specific envelope components (cf. below).

The status of the "spreading reaction" is unsettled in light of recent work by Israeli and Artman (1970). These workers showed that, under the conditions employed by Puck and Lee, the release of β-galactosidase represents, not leakage from intact cells, but rupture of some cells by lysis from without. In the nonnutrient medium employed by Puck and Lee, this occurs at rather low multiplicities of infection.

In a recent study Silver, Levine, and Spielman (1968) have confirmed much of the work of Puck and Lee. However, their studies on the effect of MOI also failed to substantiate the idea of a spreading reaction. This is perhaps not surprising, since Silver et al. were using magnesium and potassium efflux as a measure of permeability, and the loss of these small cations may occur under quite different conditions than those allowing leakage of macromolecules. The idea of a spreading reaction, while still unproved, does have a certain appeal, particularly when viewed in the light of recent work on colicins. These are proteinaceous antibiotic substances elaborated by certain bacterial strains. They bind to receptor sites on susceptible bacteria and kill them by as yet unknown mechanisms (for review see Nomura, 1967). Colicins can cause profound metabolic changes within affected cells, yet they never leave the surface of the bacterium. For example, colicin K (which shares a common receptor site with T6) inhibits synthesis of DNA, RNA, and protein in susceptible cells (Nomura, 1963), while binding of colicin E3 causes the irreversible inactivation of all of the 30S ribosomal subunits inside the cell (Konisky and Nomura, 1967). In both cases, the colicin has been shown to remain bound at the cell surface (Maeda and Nomura, 1966). In an attempt to explain these and other results no less strange, Changeux and Thiery (1967) have proposed a general mechanism for colicin action. They suggest that the bacterial membrane is composed of subunits, which can exist in one of at least two conformations. Viewed in this way, the membrane is a gigantic allosteric molecule. Binding of a colicin causes the nearest membrane subunits to undergo a conformational transition to the form which is not normally favored. Next, to use their words, "as a consequence of the mutual cooperative effects between protomers, the conformational transition tends to propagate from one protomer to its neighbor and the complete cell membrane can shift from one conformation to the other." In other words, a spreading reaction. This conformational change, in turn, disrupts a number of sites at which membrane-associated processes, such as DNA replication, or respiration, are occurring. Although this whole model is quite speculative at present, its appeal lies in its ability to account for some most peculiar phenomena. The extension of this line of thinking to the phage-infected cell should be apparent to the reader.

In their studies on cation fluxes Silver et al. (1967) reported that infection by T2 has no effect on the ability of cells to take up potassium or magnesium from the medium and they concluded that infection does not affect the bacterial pumps for these ions. Indirect evidence has been presented (Guttman and Begley, 1968) that T4 infection induces the formation of a new magnesium pump in infected cels. To support this hypothesis, they studied rates of DNA synthesis in cells which had been treated with ethylenediaminetetraacetate, a treatment known to increase permeability and to allow ions to leak out (Leive, 1965). Infection was carried out in high magnesium and in low magnesium, using either wild-type T4 or a mutant presumed to be defective in its ability to form the pump. The authors reported that DNA synthesis in the mutant was more sensitive to the Mg^{++} concentration in the medium, although my own inspection of their data does not bear this out. In their Figure 4, inhibition of DNA synthesis caused by low Mg^{++} is about 50 percent, whether infection is carried out with wild-type T4 or with the mutant. Certainly, there is ample evidence that the ability of a phage-infected cell to regulate its internal cation concentration is critical for success of the replication cycle. We shall explore this evidence further in Chapter 7, when we discuss the biochemistry of the rapid-lysis, or *r* mutants. However, to my knowledge, none of the details of this control have yet been revealed at the enzymatic level.

Another approach which has been used to explore phage-induced changes in the properties of the bacterial membrane is studies on the sensitivity of intracellular phage to acridine dyes (Silver, 1965, 1967; Hessler, Barnes, and Baird, 1967). Although free phage and uninfected cells are not affected by these compounds, infected cells are able to take them up, whence they act to prevent the maturation of vegetative phage. T-even phages contain two genes which determine sensitivity to acridine dyes. Mutation to acridine resistance in either of these genes leads to decreased uptake of the dye by infected cells. Silver (1967) has explored the functioning of one of these genes. Cells infected with wild-type, sensitive phage show a greatly increased permeability to acridines within the first few minutes after infection. Protein synthesis and an energy supply are necessary for the development of increased permeability, but they are not required late in infection for maintenance of increased dye uptake. Silver concluded from this that a viral gene causes an alteration in cell permeability which leads to increased dye uptake. The gene product is not an "acridine pump;" otherwise, the maintenance of uptake would be energy-dependent. Rather, he believes that it is an agent which changes the cell permeability so as to expose intracellular binding substrates for the dye. The nature of the gene product, and its effect on the membrane, are still unknown.

Several investigators are approaching the question of virus-induced changes in the membrane by examining the metabolism of specific membrane of specific membrane components. Initial studies have focused on the phospholipids, for these are located primarily in the membrane (Kates, 1964; Quigley and Cohen, 1969), and good microanalytical techniques for these compounds are available (cf. Wells and Dittmer, 1966). Buller and Astrachan (1968) found that T4 infection causes a slight decrease in the incorporation of P^{32}-orthophosphate into the phosphatidylethanolamine fraction and a corresponding increase into the phosphatidylglycerol fraction. Similar results were noted by Furrow and Pizer (1968), who also observed a slight decrease in the overall rate of P^{32} incorporation into phospholipids. A further phage-associated change was noted in a more recent report by Peterson and Buller (1969), who observed an increase in P^{32} incorporation into the cardiolipin fraction. There is some question whether these changes are truly phage-specific. Furrow and Pizer concluded, from experiments with chloramphenicol added at different times, that protein synthesis between 3 and 6 minutes after infection is necessary in order for these changes to be observed. However, Peterson and Buller reported that addition of cyanide to uninfected cells leads to the same pattern of isotope incorporation as that seen in infected cells, and they concluded that the observed effects on phospholipid metabolism are a consequence of impaired respiration in the infected cell.

Preliminary evidence for a direct interaction between a component of the baseplate and a phospholipid membrane component has been reported by Benz et al. (1970). These workers are studying a system in which urea-treated T4 particles can infect *E. coli* spheroplasts, even when the long tail fibers have been removed by the urea treatment. This infection, however, is inhibited more or less specifically by preincubation of the phage with purified phosphatidylglycerol from *Salmonella* cell membranes. This preincubation causes release of DNA from the phage.

Thus far in this chapter, we have explored the events connected with phage adsorption and injection of DNA, as well as some of the consequences in terms of structure and function of the cell envelope. While much remains to be learned about these processes, it is clear that what is known provides an explanation for the onset of the eclipse period. Once DNA is injected into a cell, it is physically removed from its adsorption organ, namely, the contractile tail. Therefore, if an infected cell is ruptured during the eclipse period, naked viral DNA is released to the medium. This DNA is noninfectious, since it has no way of finding its way back into another cell. A complete understanding of the eclipse now demands that we learn how the injected DNA expresses its genetic information to direct the production of fully formed progeny. In other words, it is now appropriate for us to explore the metabolic consequences of phage infection.

METABOLISM OF PHAGE-INFECTED CELLS

The earliest experiments on the intermediary metabolism of phage-infected bacteria were carried out by Cohen in the late 1940s. In one series of experiments the effect of phage infection on polymer synthesis was observed (Cohen, 1947). The basic experiment was to multiply infect a growing culture with sufficient phage to infect virtually all of the cells simultaneously. Samples were removed at intervals and assayed by colorimetric methods for DNA, RNA, and protein. The results of a typical experiment, summarized in Figure 4–3, showed that the rate of protein synthesis is maintained at the preinfection level, while virtually no RNA accumulation is observed. After a 7-minute hiatus in DNA synthesis, this process begins at a rate some 5- to 10-fold higher than that characterizing the cells before infection. The overall level of metabolic activity of the cells did not change upon infection; no change, for example, was seen in the rate of respiration as a result of infection (Cohen and Anderson, 1946). Later it was found that infection halts cell division and that there is no further increase in cell mass. Thus, phage infection allows cells to maintain the total metabolic effort characteristic of the immediate pre-infection period. Yet even these initial studies indicated that profound metabolic changes do occur, leading to the diversion of metabolites into new pathways of utilization. Elucidation of the nature of these diversions has formed several chapters in the history of phage biology, some of which are still being written.

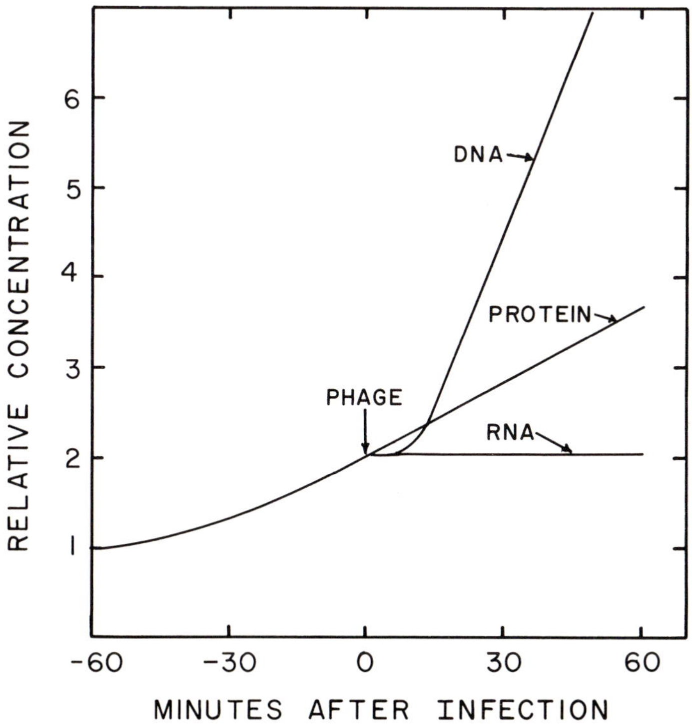

FIGURE 4-3 Patterns of polymer synthesis in T-even phage-infected *E. coli,* as determined by colorimetric assays performed on aliquots of cells sampled before and after infection.

The availability of radioisotope methodology allowed some clarification of Cohen's early findings on polymer synthesis. Cohen (1948) used P^{32} labeling to explore the origins of phosphorus in the DNA of progeny phage. In his studies about two thirds of the phosphorus came from the medium, and about one third from the cells themselves—mostly from bacterial DNA, which is degraded after infection (Kozloff, 1953). As we shall see later, another source of DNA in progeny phages is the parental DNA itself.

The discovery of 5-hydroxymethylcytosine as a specific T-even phage DNA component (Wyatt and Cohen, 1952) allowed a distinction to be made between synthesis of bacterial and viral DNA. Hershey, Dixon, and Chase (1953) grasped this biochemical handle, and showed that all of the DNA formed in T2-infected cells contains HMC and is, therefore, virus-specific. In addition, Hershey et al. were able to present their results in meaningful quantitative terms. From the known DNA content of phage and its base composition and from the specific activity of the isotopic pyrimidine precursors used in their experiments, they were able to express their results as phage-equivalent units of DNA per infected cell. These data

1956 on the mechanism of protein synthesis for them to suggest its now well-known role as a viral messenger, although they did mention that its appearance might be directly related to the synthesis of virus-specific proteins. That phage-specific RNA is a distinct physical entity was shown by Nomura, Hall, and Spiegelman (1960), who were able to separate pulse-labeled RNA from unlabeled cellular RNA by zone electrophoresis and by sucrose gradient centrifugation. Hall and Spiegelman (1961) showed that labeled phage RNA could form specific hybrid molecules with denatured phage DNA but not *E. coli* DNA, under conditions promoting denaturation of polynucleotide strands. Thus, the phage RNA was shown to have base sequences corresponding to sequences in phage DNA. This plus the finding that RNA labeled after phage infection becomes associated with ribosomes which were assembled before infection (Brenner, Jacob, and Meselson, 1961) greatly increased the attractiveness of the idea that the viral RNA might be acting as some sort of informed genetic intermediary. This idea was formally stated by Jacob and Monod (1961) in their classic paper on genetic regulation of protein synthesis. Their postulation of the existence of messenger RNA and its subsequent demonstration in many viral and cellular systems forms a basic chapter in the intertwining histories of biochemistry and genetics. Although a detailed discussion of these developments is beyond the scope of this book, the mechanisms controlling the formation and utilization of phage messengers will be discussed at some length in Chapter 7. We should observe here, however, that the properties just described for virus-specific RNA, and the proposal that it is a messenger, did not automatically make it so. One had to prove that genetic information in DNA, which was apparently conserved in the formation of viral RNA, was subsequently utilized in the synthesis of proteins. The first evidence for this in viral systems came from genetic experiments by Benzer and Champe (1961). These workers found that the pyrimidine analog 5-fluorouracil (FU) could cause phenotypic reversion of certain *r*II mutants of T4 (we shall discuss the *r*II mutations further in Chapters 5 and 7). In other words, the mutants could behave like wild-type in the presence of FU, but the mutant genotype was retained. Benzer and Champe proposed that FU was incorporated into phage messenger RNA (mRNA) in place of uracil and was occasionally translated as cytosine, leading to a corresponding change in the amino acid sequence of the protein specified by that mRNA. This certainly appears to be the most plausible interpretation, since FU can be incorporated into RNA (Shalitin and Ben-Ishai, 1966), but not into DNA (Cohen et al., 1958). However, the most direct demonstration of the messenger role of phage RNA is the synthesis of phage lysozyme *in vitro,* in an amino acid-incorporating system primed specifically by RNA extracted from phage-infected cells (Salser, Gesteland, and Bolle, 1967). We shall have more to say about this *tour de force* later in the book.

Arrest of Gene Expression in the Host

The early studies on nucleic acid metabolism discussed above indicated the existence of some virus-directed mechanism for turning off the synthesis of bacterial DNA and RNA rapidly and quite specifically. Furthermore, although the overall rate of protein synthesis did not change as a consequence of infection, it soon became apparent that all proteins made after infection are virus-specific. This realization initially came through the discovery that induced bacterial enzymes cannot be formed after T-even phage infection (Monod and Wollman, 1947) and later through a series of experiments showing that the synthesis of every known bacterial enzyme, whether induced or constitutive, ceases after infection (for partial list see Cohen, 1968, p. 11). At any rate, it became apparent that T-even phages have some mechanism for rapidly and completely arresting all bacterial gene expression, i.e., synthesis of DNA, RNA, and protein. The nature of this arrest has received much less attention than it deserved until fairly recently. Certainly, an understanding of the arrest mechanism is one of the key problems in phage biology today, for it should help to explain the extreme virulence of the T-even phages and should provide a clue to the actual lethal event in T-even phage infection. It should be observed that the number of phages which, like the T-evens, completely abolish host cell genetic activity, is relatively small. Since the existence of the arrest mechanism has enormously simplified the dissection of virus-specific metabolic events, this would appear to provide further evidence for the statement, made in Chapter 1, that the early choice of T-even phages as objects of intensive study, was a lucky accident. Consider, for example, the difficulties which Volkin and Astrachan would have encountered in interpreting their data, had cellular RNA synthesis been occurring after infection.

The earliest proposal advanced to explain the arrest of genetic activity in the host was the physical destruction of the bacterial chromosome. Cytological examination showed that the bacterial "nucleus" is destroyed after infection (Luria and Human, 1950), and isotope-labeling studies showed that the cellular DNA is degraded (Kozloff, 1953). The idea that this physical destruction is responsible for the functional inactivation of the chromosome was rigorously tested by Nomura et al. (1962). These workers found that host DNA remains physically intact as long as five minutes after infection, by which time host-cell macromolecular synthesis has ceased. Moreover, the chromosome was functionally intact as well. This was shown by transferring by conjugation the chromosomes of infected male bacteria at five minutes to phage-resistant, β-galactosidase-negative female bacteria. β-galactosidase was synthesized in the recipient, indicating that the transferred chromosomes were undamaged, even though they were not being expressed in the infected donor cells.

Other early work by Nomura and his colleagues (Nomura, Okamoto, and Asano, 1962; Okamoto, Sugino, and Nomura, 1962) suggested the involvement of a viral gene in the host-cell arrest. Infection in the presence of chloramphenicol led to continued synthesis of ribosomal and transfer RNAs (rRNA and tRNA) in the host. Phage mRNA was also produced, but, interestingly, host mRNA and DNA were not formed under these circumstances. In addition, phage mRNA was turned over as in normal infection, but nucleotides released from degradation of this material ultimately appeared in the cellular rRNA and tRNA. Nomura and his coworkers concluded that synthesis of a virus-specific protein, which is blocked in chloramphenicol, is necessary for shutoff of bacterial rRNA and tRNA synthesis. These findings suggest the possibility, which to my knowledge has not been explored, that a separate mechanism, independent of protein synthesis, exists for arresting host-cell mRNA and DNA synthesis.

In a recent study, Ennis (1970) has presented evidence that a phage-coded protein is involved in the arrest of host-cell mRNA synthesis. In a mutant *E. coli* strain which cannot accumulate potassium from the medium, protein synthesis is dependent upon exogenous potassium concentration. At low potassium levels, complete shutoff of host mRNA is not observed after T4 infection, leading Ennis to conclude that a phage protein is responsible for shutoff. However, Ennis has not established that inhibition of protein synthesis is the only metabolic consequence of potassium limitation in this system, or even the primary consequence.

The conclusion that a viral protein is solely responsible for host-cell arrest was not satisfying, inasmuch as phage ghosts, which are unable to direct phage protein synthesis, could still shut off the synthesis of bacterial polymers (French and Siminovitch, 1955; Duckworth and Bessman, 1965; Duckworth, 1970). This apparent paradox was partially resolved by Nomura et al. (1966), who demonstrated the existence of at least two mechanisms in T4 phage-directed arrest. In the presence of chloramphenicol a shutoff of cellular biosynthesis was seen, but the extent of shutoff was dependent upon multiplicity of infection (MOI). Some of these data are reproduced in Fig. 4–5. Nomura *et al.* concluded that there is some second effect, not dependent upon phage gene action, which acts at the bacterial surface, in a manner analogous to some colicins, whose effects had also been shown to be multiplicity-dependent. It is clear, however, from the data in Figure 4–5, that this mechanism cannot be responsible for normal arrest of host-cell mRNA and DNA synthesis, since at an MOI of 6, within the range normally used in biochemical experiments, the inhibitions observed in the presence of chloramphenicol are considerably less than 50%. Thus, there would appear to be at least three separate mechanisms involved in host-cell arrest: an MOI-dependent effect at the cell surface, a chloramphenicol-sensitive step which blocks host rRNA and tRNA synthesis, and a chloramphenicol-insensitive effect which inhibits host mRNA and DNA synthesis.

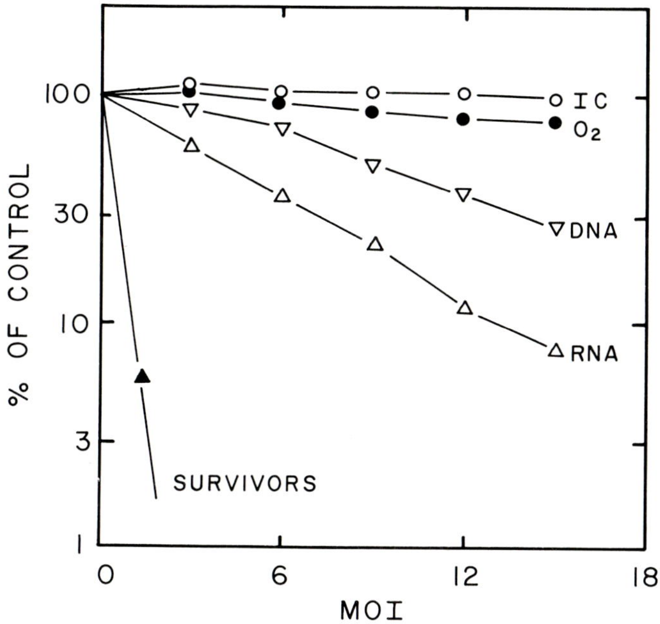

FIGURE 4-5 Multiplicity-dependent shutoff of bacterial functions in infection of *E. coli* by T4 in the presence of chloramphenicol (modified from Nomura et al., 1966).

Duckworth (1970) has reinvestigated the effect of T4 ghosts upon cellular metabolism, and she has confirmed the MOI-dependent, "colicin-like" effect noted by Nomura et al. However, in her hands complete arrest of macromolecular biosynthesis is effected by multiplicities as low as 4 ghosts per bacterium. This arrest is not chloramphenicol-sensitive. As Duckworth points out, however, one cannot conclude that the mechanism of ghost-induced shutoff is the same as that of the arrest caused by intact phage, and it is impossible at present to ascertain the precise relationship between these two phenomena.

The possibility of an arrest mechanism dealing with host protein synthesis at the level of translation must also be seriously considered. A great deal of work has been done in Sueoka's laboratory (cf. Kano-Sueoka, and Sueoka, 1968) on structural alterations in transfer RNA molecules occurring after T-even phage infection. The possibility has been raised that selective inactivation of one or more cellular transfer RNAs, possibly coupled with synthesis of new species of tRNA, might lead to failure of recognition of certain triplet codons in the host mRNA, while permitting the translation of all of the phage mRNA triplets. Direct evidence for the involvement of such a mechanism has recently been obtained (Kano-

Sueoka and Sueoka, 1969) and will be discussed at greater length in Chapter 7.

Kaempfer and Magasanik (1967) and Kaempfer and Sarkar (1967) have examined the transcription of host cell genes after infection, using β-galactosidase-forming capacity as an index of mRNA synthesis. These authors find that inhibition of enzyme induction does not require protein synthesis. They did not investigate whether control in their system was at the level of transcription or translation. It is possible that the effect they were observing is the chloramphenicol-insensitive step affecting host mRNA formation, which was mentioned above. Kaempfer and Magasanik reported that the decay of enzyme-forming capacity after phage infection of strains lacking a sex factor (F^- strains) was twice as great as that seen in normal cells after removal of inducer. Curiously, this accelerated rate of decay was not seen in cells bearing a sex factor (F^+ or Hfr). The basis for this unexpected result was not explained and indeed is at variance with more recent results of Rouvière et al. (1968). These workers found a twofold acceleration of decay of β-galactosidase-forming capacity in an Hfr strain. However, since they did not discuss their results in light of the discrepancy with the results of Kaempfer and Magasanik, the possible role of the sex factor in arrest of host-cell protein synthesis must remain an open question.

Terzi (1967) compared the effects of T4 infection on *E. coli* and on *Shigella dysenteriae.* In *Shigella* the arrest was not complete and depended upon the MOI, reminiscent of the situation seen with chloramphenicol-treated *E. coli.* Terzi suggested that the phage protein which is active in arresting *E. coli* metabolism is inactive in *Shigella,* and that only the surface effect is seen. Although numerous studies, including Terzi's, point to the existence of a viral protein as a key agent in the arrest of host-cell functions, the protein has not been isolated, nor its precise mode of action ascertained. There is now good reason to believe that at least some of the requirement for phage protein synthesis relates to alterations in the subunit structure of the host-cell RNA polymerase, which is used for transcription of all phage genes (see p. 174). These alterations, in turn, affect the ability of the enzyme to recognize different DNA templates.

One way to seek more information on the role of phage proteins in host-cell arrest is to try to obtain mutants which are defective in this process. However, it is difficult to design a screening process, since such mutants may or may not be lethal, although the fact that this function is apparently dispensable in *Shigella* suggests that they would not. One way to seek such mutants is to look for phage which will permit induced enzyme synthesis under conditions of single infection. Several years ago the author tried without success to use this approach. However, the problem is an important one, and the reader is invited to try to apply his imagination toward its solution.

The virulence of T-even phages as it relates to the expression of heterologous viral genomes has been the object of two recent studies. Hattman and Hofschneider (1967) found that T4 infection blocks replication of the RNA phage M12, and they suggested that the inhibition was directed toward translation of the viral RNA message. Hattman (1970) has recently confirmed this suggestion. Hayward and Green (1965) found that T4 infection blocks the replication of the temperate DNA phage, lambda. In this case the block seemed to be directed toward transcription of the viral DNA. Protein synthesis was required in order for this inhibition to be effective, in contrast to the results of Kaempfer and Magasanik. At any rate, T4 replication appears to be "dominant" over the replication of two other viruses, with controls operating at the levels both of transcription and translation.

Kinetics of shutoff of host synthesis have been explored by three groups, who explored the synthesis of cellular mRNA after infection. Rouvière et al. (1968) pulse-labeled RNA immediately before infection and measured the post-infection loss of radioactivity hybridizable with *E. coli* DNA. They found a half-life of 40–60 seconds for this decay process, in general agreement with conclusions drawn about mRNA decay from studies with induced enzymes. Rouvière et al. also pulse-labeled cells at various times after infection and measured radioactivity hybridizable with *E. coli* DNA. They reported that shutoff of host mRNA is extremely rapid, since essentially no such hybridizable label was found in a culture pulsed at one minute after infection. However, Landy and Spiegelman (1968) were able to detect significant host mRNA synthesis as late as three minutes after infection, while Kennell (1968) reported, rather unexpectedly, that about half of the RNA made as late as four minutes after infection is host-specific. This discrepancy in experimental results can be comprehended if we realize that Kennell's hybridizations were carried out at very low RNA:DNA inputs—10^{-3} or less—in order that all of the possible DNA binding sites would be present in excess so that essentially all of the hybridizable RNA would be bound. The studies of Rouvière et al. were carried out at RNA:DNA input ratios of 0.2, and, if we accept Kennell's reasoning and his data, use of these experimental conditions might lead to an underestimation of true *E. coli* mRNA synthesis by a factor of 10, while the results of Landy and Spiegelman might be low by a factor of two to three. Kennell has suggested that this continued synthesis of host-cell mRNA after infection might represent completion of polynucleotide chains which had been initiated prior to infiction. However, several studies on rates of RNA chain elongation in *E. coli* (cf. Bremer and Yuan, 1968a; Winslow and Lazzarini, 1969) give values between 20 and 60 nucleotides per second. Thus, the complete synthesis of even a polycistronic RNA molecule would not require more than two or three minutes. Therefore, Kennell's report would appear to indicate either that infection slows the rate of transcription of host genes, or that initiation of host mRNA molecules occurs after infection.

As Cohen (1968) has pointed out, most of the evidence implicating a viral protein in host cell arrest has come from experiments involving treatment of host cells with chloramphenicol. Since the precise mode of action of this compound is not known, it is possible that it has effects other than inhibition of protein synthesis. Cohen has proposed a mechanism for host-cell arrest which does not require protein synthesis. This model postulates that there is a rapid influx of liquid into the cells occurring as a result of puncture of the cell wall by the phage tail. This causes the cells to swell and disrupts critical membrane attachment sites, necessary for replication and transcription of the bacterial genome (cf. Jacob, Brenner, and Cuzin, 1963). The findings of Doermann (1948b) relative to turbidity changes in infection are cited as evidence for swelling. This, of course, is not the only possible interpretation of Doermann's data. Moreover, although Cohen's hypothesis is appealing in its simplicity, it does raise a number of questions. For example, not all of the tailed phages arrest gene expression in the cells they infect, yet some opening of the cell surface must occur in order for the phage genome to reach the cytoplasm. Why do these cells not take in liquid from the medium, with a resulting disruption of sites of gene action? Moreover, this model does not account for the differential abolition of host-cell genetic capability which is seen under some conditions. As stated above, infection of chloramphenicol-treated cells blocks host DNA and mRNA synthesis, but allows continued synthesis of rRNA and tRNA. Regardless of the actual effect of chloramphenicol, this result is difficult to explain in terms of the Cohen model unless one puts forth *ad hoc* hypotheses such as the existence of different types of sites for gene expression which are differently susceptible to disruption. Finally this model does not account for the element of translation level control of protein synthesis, as indicated from the inhibition of replication of the RNA phage M12 (Hattman and Hofschneider, 1967). It is, of course, highly possible, even probable, that the Cohen effect is the same as the MOI-dependent "surface" effect noted by Nomura and his collaborators. However, the evidence for more than one different mechanism involved in shutoff strikes me as too great to accept any single explanation for this mystifying phenomenon.

The Early Proteins; Phage-Induced Enzymes

As stated above, T-even phage infection blocks the synthesis of every specific bacterial protein which has been examined. Although it has long been believed that all cellular protein synthesis ceases upon infection, it has been shown only recently (Levinthal, Hosoda, and Shub, 1967) that all proteins made after infection are virus-specific. Since the rate of protein synthesis does not decrease following infection, the formation of viral proteins must commence immediately. Hershey, Garen, Fraser, and Hudis (1954) carried out pulse-labeling studies of protein synthesis at various times after infection, using S^{35}-sulfate as a protein precursor. These workers found that the proteins formed earliest in infection are not incorporated into progeny phage. For example, only 10% of the label incorporated into protein during the first 5 minutes of infection was eventually found in mature virions. By contrast, when the isotope was added later in infection, this value rose to 60%. Thus, these early studies suggested the existence of two classes of phage proteins: the "early" proteins, which in general, are not incorporated into virions, and the "late" proteins, which are. Although this statement is a gross oversimplification, it has provided a convenient nucleus for the crystallization of thought relative to the control of phage gene expression.

The demonstration that the early proteins are not structural components of virions immediately raised the question of their actual function. A role for phage proteins in DNA synthesis had already been indicated by Cohen and Fowler (1947), who found that addition of the amino acid analog 5-methyltryptophan to T2-infected cultures not only blocks protein synthesis, as expected, but also DNA synthesis. These findings were extended by Tomizawa and Sunakawa (1956), who found that the time at which protein synthesis is blocked by chloramphenicol is critical in terms of the subsequent effect on DNA synthesis. When the inhibitor was present from the time of infection onward, no DNA synthesis was observed; some synthesis was permitted in cells to which chloramphenicol was added at 5 minutes, while the addition of the agent at 10 minutes or later had virtually no effect on DNA synthesis. Thus, it appeared that the early proteins are specifically required for DNA synthesis, but that their continued synthesis past 10 minutes is not necessary.

Cohen and his colleagues were primarily responsible for the clarification of the metabolic roles of the early proteins—a milestone in biochemical virology. In 1954 Barner and Cohen had found that infection of a thymine-requiring bacterial mutant, *E. coli* $15T^-$, with T2 overcame the nutritional deficiency: DNA was formed at normal rates even in the absence of thymine. Since the utilization of thymine nucleotides for DNA synthesis is considerably greater in infected cells than in uninfected cells, it appeared that the phage must have somehow induced the formation in the infected cell of a metabolic system for synthesis of thymine nucleotides—a phenomenon which Cohen called "virus-induced acquisition of metabolic function." At the same time, other workers in Cohen's laboratory were investigating the biosynthesis of hydroxymethylcytosine in phage DNA by administering various labeled pyrimidine precursors to infected cultures (Cohen and Weed, 1954; Green and Cohen, 1957). Both of these lines of investigation culminated in the discovery by Flaks and Cohen (1957) of two new enzyme activities, specifically present in infected cells, which could account for the formation of both thymine- and HMC-containing nucleotides. The enzymes, thymidylate synthetase and deoxycytidylate hydroxymethylase, respectively, catalyze the following reactions:

$$\text{dUMP} + \text{methylenetetrahydrofolate} \xrightarrow[\text{synthetase}]{\text{dTMP}} \text{dTMP} + \text{dihydrofolate}$$

$$\text{dCMP} + \text{methylenetetrahydrofolate} \xrightarrow[\text{HMase}]{\text{dCMP}} \text{HMdCMP} + \text{tetrahydrofolate}$$

Thymidylate synthetase had already been shown to be present in uninfected *E. coli* (Friedkin and Kornberg, 1957), but Flaks and Cohen found that the amount of enzyme increased severalfold after infection. Moreover, the enzyme, which is absent in *E. coli* 15T$^-$ and other thymine auxotrophs, is abundantly present in infected cells of these strains (Barner and Cohen, 1959). This fact, plus the complete absence of dCMP hydroxymethylase in uninfected cells of all strains (Cohen, 1961), suggested that the phage carries genetic information for the synthesis of enzymes necessary for various stages of virus production. A major role of the phage early proteins in DNA synthesis was shown through these contributions. Since dCMP hydroxymethylase represents the only biosynthetic route leading to the HMC nucleotides, its absence due to blockage of protein synthesis would lead to a deficiency of an essential phage DNA constituent.

At about the same time as the above work was being done, Kornberg and his colleagues were beginning to achieve success in their studies on the enzymatic synthesis of DNA (cf. Kornberg, 1961). Turning their attention to T2-infected *E. coli,* they were quickly able to detect the presence of several new enzymes associated with infection, some of which, like dCMP hydroxymethylase, are undetectable in uninfected cells and others which, like dTMP synthetase, are present at levels severalfold higher in infected than in uninfected bacteria (Kornberg, Zimmerman, Kornberg, and Josse, 1959). The search was joined by several other laboratories, and there are now about two dozen known enzymes associated with T-even phage infection and which play direct or indirect roles in phage DNA synthesis. A more detailed discussion of these enzymes and their metabolic functions is presented in Chapter 6.

When phage-induced enzymes were first discovered, it seemed reasonable to propose that they were true phage gene products, synthesized from the instructions encoded in the vital genome. In agreement with this was the fact that the phage-induced enzymes are not formed in the presence of chloramphenicol (Flaks, Lichtenstein, and Cohen, 1959). However, it was impossible *a priori* to exclude the possibility that these enzymes could exist in a preformed state in uninfected cells and that infection might activate them by addition or deletion of a polypeptide fragment, as in zymogen activation, or by making available some essential cofactor. Since some of the phage-induced enzymes, such as dTMP synthetase, were present in uninfected cells, albeit in low amounts, this was a real possibility which had to be ruled out if biochemists and geneticists were to be able to fully exploit the opportunities afforded by the discovery of these enzymes. In the case of dTMP synthetase, Greenberg, Somerville, and DeWolf (1962) showed that the phage-induced enzyme was physically distinct from the host-cell enzyme, since they could be separated chromatographically. This type of evidence, while suggestive, is not conclusive, and unambiguous routes were sought to prove that phage-induced enzymes are synthesized entirely after infection, under genetic control of the phage. One of these routes is essentially biochemical, and the other is genetic. The biochemical route was traversed in Cohen's laboratory, in an experiment which I carried out with Fred Brown (Mathews, Brown, and Cohen,

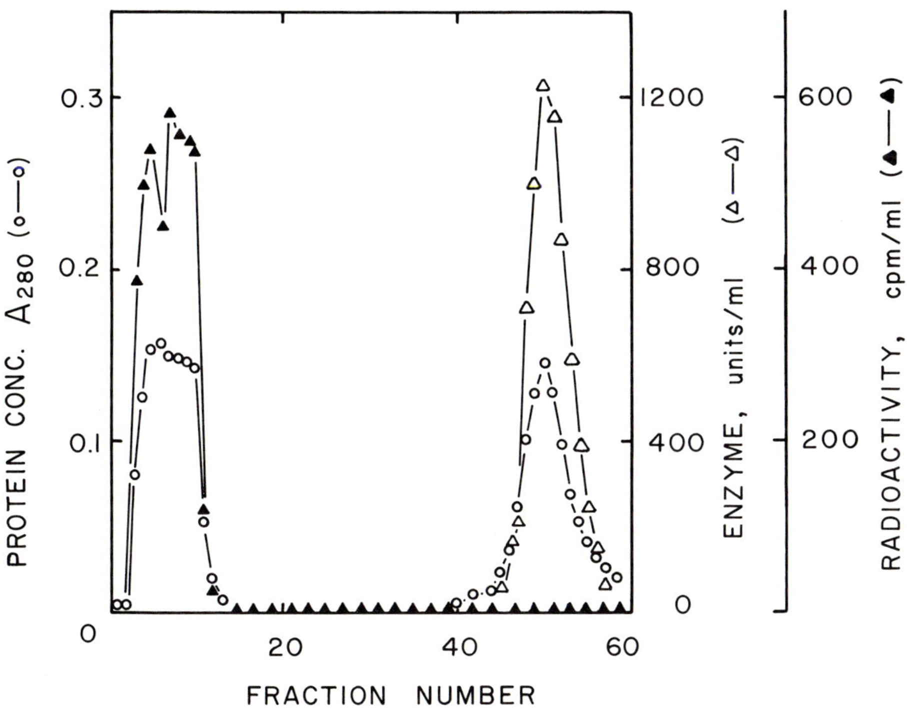

FIGURE 4-6 An experiment demonstrating the synthesis *de novo* of dCMP hydroxymethylase following infection by phage T6. The drawing shows the elution profile from CM-Sephadex of a fraction containing enzyme which was formed in non-radioactive medium by cells whose proteins were labeled before infection with C^{14}. For further details see text (redrawn from Mathews, Brown, and Cohen, 1964).

1964). We developed a method for purification of dCMP hydroxymethylase, to the point where it was virtually free of contamination with host-cell protein. A large culture of *E. coli* B45, a methionine auxotroph, was grown in the presence of C^{14}-methionine, in order to label bacterial proteins. The cells were placed in nonradioactive medium and infected with T6. The hydroxymethylase was purified from this culture and shown to be nonradioactive, as shown in Figure 4–6, indicating that it must have been formed entirely from amino acids present in the cell after infection.

Although the above experiment clearly proved that a phage-induced enzyme is synthesized *de novo* after infection, it did not rule out the possibility that this results from derepression of a bacterial gene. Mathews and Cohen (1963a) showed that T2-induced dTMP synthetase differs in its kinetic properties from the T6-induced enzyme, strongly suggesting that the enzyme is formed under genetic control of the phage. However, the most direct experiments on this point involved the use of conditional lethal mutants of phage T4, and we shall postpone discussion of these experiments until we have had an opportunity to briefly consider the biochemical genetics of T-even phages in the next chapter.

5

Biochemical Genetics of T-Even Phages

The purpose of this chapter is not to review the entire field of phage genetics, but to discuss those aspects which are most relevant to a comprehension of biochemical events in the phage-infected bacterium. As we have already emphasized, biochemists could not possibly have exploited phage systems as successfully as they have without the background of information provided by phage genetics. In particular, the development of conditional lethal mutants of bacteriophage has provided the biochemist with invaluable genetic tools. The skeptic need only examine the current literature of T-even phages to convince himself of the truth of this statement. Nearly every paper includes an acknowledgment to R. S. Edgar, whose laboratory is the chief source of supply for these mutants. In order to understand how these tools are being used, one must appreciate the nature of conditional lethal mutations at the molecular level. In discussing selected aspects of phage genetics from the standpoint of a biochemist, we shall be concerned, for example, not so much with the mechanism of constructing a genetic map, but with questions of its physical significance: what is the relationship of the genetic map to the phage chromosome? How does one identify the product of a given gene and determine its metabolic function? What does the genetic map tell us about regulation of gene expression? For timely reviews of the general field of phage genetics, the reader is referred to the second edition of Hayes' excellent textbook (Hayes, 1968) and to the book by Luria and Darnell (1967). A highly readable, yet quite detailed, account of the status of the field some seven years ago is presented in Stent's book (Stent, 1963).

PREBIOCHEMICAL PHAGE GENETICS

Stent's book appeared at what might be described as a watershed period in the history of phage genetics and, as such, it gives an excellent picture of this field in the prebiochemical era. Until 1963 the geneticists had used markers which were recognized primarily on the basis of plaque morphology or resistance to various agents. As such, there was little understanding of the biochemical basis of most known mutations. In addition, most of the mutants in use were altered in what must be called nonessential functions, since the mutants had to be able to replicate and form plaques in order to be detected. At the same time, bacterial genetics was in full flower. By means of selective enrichment and plating techniques, one could isolate mutants defective in many metabolic functions, both essential and nonessential. A central feature of most techniques for isolation of bacterial mutants was *conditional lethality:* one would devise a procedure for recognizing a mutant by virtue of its ability to grow under one set of conditions but not another. An early application of this concept was the isolation of auxotrophic mutants. For example, a leucine-requiring mutant was recognized as a bacterial strain which could not grow on minimal medium, but could grow on minimal medium supplemented with leucine. By and large, selection techniques for nutritional deficiencies could not be applied directly to phage genetics; the phage gets its raw materials for biosynthesis from the host and thus depends upon the nutritional state of the host. One could not, for example, isolate a "leucine-requiring" mutant of phage T4. Production of leucine is an essential function of the host bacterium, both for growth (in minimal medium) and for support of phage infection. However, the general concept was one which could be applied to phage. In order to be able to study the growth of any organism, one must be able to recognize the steps essential for growth. If one was to apply genetics to this end in phage, one needed a means to isolate mutant phage defective in these functions. But in order to study phage, one must be able to grow them. Hence, the need for conditional lethal mutants: strains selected for ability to grow under one set of conditions but not another.

The reason for identifying 1963 as a "watershed year" in phage biology is that this year saw the first comprehensive publication describing conditional lethal mutations in phage T4 (Epstein et al., 1963). These included temperature-sensitive, or *ts* mutations, developed in Edgar's laboratory, and amber, or *am,* mutations, developed by Epstein and his colleagues. The *ts* mutants are recognized by their ability to form plaques at 30°

but not at 43°. The amber mutants were originally isolated as strains which could replicate in a particular strain of *E. coli* K, called CR63, but could not grow in *E. coli* B. As we shall see later, this differential growth response is due to the presence of suppressor genes in certain host strains. These allow for translation of a part of the genetic message which is meaningless in the strains lacking these genes. A large number of *ts* and *am* mutants were described in the report by Epstein et al. (1963). The new mutations were distributed over the entire genetic map, many of them being located in hitherto unrecognized genes. Thus, the geneticist was provided with many new markers and simple methods for detecting them. Also, since the new mutations were all located in genes whose expression was necessary for replication, the biochemist was handed an invaluable tool for physiological-genetic analysis. He could now ask, for example, which genes were responsible for the formation of the phage-induced enzymes, of which about a dozen were known in 1963. Moreover, one could investigate the metabolic consequences of the failure of a given gene to be expressed, by experimentation on cells infected under non-permissive conditions. A good deal of preliminary work along these lines was reported in the original communication by Epstein et al.

Actually, it is inaccurate to identify 1963 as the year in which conditional lethal phage mutants became available. Campbell had, several years previously, described host-defective mutants of the temperate phage λ (Campbell and Balbinder, 1958; Campbell, 1959), and had at least partially recognized the basis for these mutations when he named them suppressor-sensitive, or *sus* (Campbell, 1961). (Later it was found that the *sus* mutations in λ are the same as the *am* mutations in T4. See p. 223). However, Campbell's work does not appear to have had the immediate and dramatic results which followed the development of the *am* and *ts* systems in T4, in terms of the development of a physiological phage genetics. I am not sure of the reason for this, but it is probably related to the fact that in the early 1960s there were far fewer biochemists working on temperate phages than on the T-even phages. There were simply fewer people available who could exploit Campbell's discoveries. It is only within the past few years that the special advantages of λ and the other temperate phages as experimental material have been generally recognized, as we shall discuss in Chapter 9.

Edgar (1966) has raised the question of why the conditional lethals in phage were not discovered much earlier than they were. Certainly, conditional lethal mutations had been recognized in other organisms for many yars prior to 1963. Temperature-sensitive mutations in *Neurospora* and *E. coli* had been studied in the early '50s (Horowitz and Leopold, 1951). Why did not phage workers immediately apply these principles to their own field? There seems to be no ready answer.

Before proceeding further with a discussion of conditional lethal mutations, let us discuss the principal genetic markers available to T-even phage geneticists in the pre-conditional lethal area. Although not a great deal was known about the biochemistry of these genes, some knowledge has accumulated within the past few years, and it is appropriate to present some of it here.

The "Classical" Markers

Rapid lysis mutants. These mutants are recognized by virtue of their ability to form plaques much larger than those of wild-type phage, with sharp, as opposed to fuzzy edges. These mutants occur in fairly high frequency in natural T-even phage populations, to the extent of about 1 in 10^3 or 10^4. Early studies on rapid-lysis, or *r* mutants, showed that they are unable to establish lysis inhibition (Doermann, 1948b). It will be recalled from Chapter 4 that lysis inhibition results when an infected cell is superinfected by a homologous phage. This leads to prolongation of the latent period for as much as several hours. Since phage development continues for most of this period, the burst sizes for lysis-inhibited cultures are much larger than those seen in the absence of this effect. The absence of lysis inhibition explains the aberrant plaque morphology seen with *r* mutants. In development of the plaque, *r* phage can undergo more cycles of infection and lysis than wild-type, or r^+, phages, since the latent period is shorter. Therefore, a larger area can be cleared before growth of the bacteria, and their ability to support infection, ceases. Also, since bacteria in a developing r^+ plaque are lysis-inhibited, many of the cells will not have ruptured by the time bacterial growth ceases, and this factor will limit the size of the cleared area in the plaque.

r mutants map in three separate regions of the genetic map in T4, the *r*I, *r*II, and *r*III genes. The *r*II system has been used extensively in analysis of genetic fine structure, as we shall see later in this chapter. These mutants are conditionally lethal, a feature which established their attractiveness for genetics long before the *am* and *ts* mutations came into existence. Although able to replicate on a wild-type *E. coli* strain, such as the B strain, they cannot grow on strains lysogenic for phage λ. This has simplified the task of measuring mutation frequencies quantitatively, or of selecting for recombinants in $r \times r$ crosses. One simply determines the infectious titer on a host bearing λ. In this way revertants from *r* to wild-type can be detected, as can wild-type recombinants in mixed infections.

Little is known about the physiology of lysis inhibition, nor of the restriction of *r*II growth by lysogenized λ, in spite of many man-years of effort directed toward both problems. We shall consider these questions further in Chapter 7.

Host range mutants. We earlier mentioned that bacteria can mutate to become resistant to infection by certain phages. Such mutants are usually selected by seeding an agar plate with bacteria and with a large excess of phage—of the order of 10^5 particles. After development of the plate, we see that it is essentially clear because the many plaques formed overlap each other and cover the entire plate *(confluent lysis).* However, a few distinct bacterial colonies are seen, and most of these are cells which have mutated to resistance to the phage on the plate. Now if one of these phage-resistant mutants is grown up and plated along with an excess of the same phage as used in isolation of the mutant, one will find a few plaques on the plate. Analysis of the phages in these plaques shows that they have mutated to a form which can now infect the previously resistant bacterial mutant. We call this a *host-range* mutation, since it has extended the host range of the phage to include a bacterium resistant to infection by the wild-type virus. If we call a T2-resistant bacterial strain *E. coli* B/2, then a host range mutant of T2 which can infect B/2 is called T2*h*.

As we mentioned earlier, the mutation to phage resistance involves structural modification in that part of the cell wall responsible for interacting with the phage tail fibers. The phage mutation, which extends its host range, must occur in that part of the phage responsible for recognition of the cell surface, namely the tail fibers. All *h* mutants of T2 map in a single gene (Streisinger and Franklin, 1956), indicating that only a single fiber subunit is involved.

Bacteria can undergo secondary mutations which make them resistant to attack to host-range mutants, and the *h* mutants can undergo further mutation, allowing them to infect these new bacterial mutants, and so forth. It is clear that some type of genetic equilibrium must have been established in natural environments, in order to ensure the survival of both phage and bacteria. It is also clear that the term "wild-type," as applied to host-range properties, is illusory. A wild-type strain is one which is withdrawn from a natural environment, and stored under conditions preserving the original genotype. A T-even phage strain is called wild-type if it will infect *E. coli* B but not its phage-resistant mutants. However, we see that members of a natural phage population must show considerable variation in host-range properties.

Osmotic shock-resistant mutants. We stated earlier that when T-even phages are diluted rapidly from a medium of high ionic strength into distilled water, the heads rupture, with release of DNA into the medium. This is the usual procedure for preparing phage ghosts. However, some phage mutants, which are designated osmotic shock-resistant, or *os,* survive this treatment. It turns out (Brenner and Barnett, 1959), that *os* mutations all map in a single gene, that determining the structure of the head protein. The heads of *os* mutants are more permeable than those of wild-type phage, such that ionic equilibrium is reached sufficiently rapidly under shock conditions that the heads do not burst. These mutants are also more permeable to the entry of many materials; for example, *os* mutants are much more sensitive to the lethal effects of nitrogen mustards than wild-type. Thus, the *os*$^+$, or shock-sensitive, character probably provides a selective advantage to phage, in rendering them more impervious to the action of noxious agents.

Cofactor-requiring mutants. We have already discussed at some length the requirement of certain strains of T4 and T6 for an adsorption cofactor, such as tryptophan. Cofactor-requiring (*c*) strains can undergo single-step mutations to give cofactor-independent (c^+) strains. The property of cofactor independence or dependence appears to be determined by one of the tail fiber components.

Acridine dye-resistant mutants. These mutants have also been discussed, in Chapter 4. They are distinguished by virtue of their ability to form plaques when plated in the presence of an acridine dye, such as acriflavin. It will be recalled that these mutations appear to affect the permeability of the cell surface in such a way that the dye can be taken up into infected cells, where it exerts its lethal effect at some stage of phage assembly.

Ultraviolet-sensitive mutants. Phage T4 is more resistant to inactivation by ultraviolet light than is T2. Resistance is determined by the *v* gene in T4 (Harm, 1963), and to a lesser extent by the unlinked *x* gene. T4*v* mutants show uv sensitivity similar to that of T2. Indirect evidence has been presented (Haber, 1966) that the product of the *v* gene is an enzyme which can repair damage to DNA, which is the target of UV irradiation. Thus, *v* mutants are less able to repair damage to their DNA than wild-type T4. Repair enzymes have been well-characterized in bacteria (Setlow and Carrier, 1964; Boyce and Howard-Flanders, 1964). One action of such enzymes is to excise, from damaged DNA, cyclobutane-type thymine dimers which are formed as a consequence of radiation (Setlow, 1966). Until very recently, however, the T4-specific uv repair enzyme had not been purified, nor its precise mode of action determined. Recent work of Ebisuzaki (1969) implicates the T4-specific enzymes, DNA ligase and DNA polymerase (see Chapter 6), in repair. However, one cannot conclude that either of these enzymes is the product of the *v* gene, for T2 probably can induce the ligase and can certainly induce the polymerase. Subsequently, Friedberg and King (1969) suggested that the *v* gene product is an endonuclease specific for uv-damaged DNA. This has been greatly extended in very recent work of Sekiguchi and his colleagues, which we shall discuss in Chapter 6. To my knowledge mapping studies on this gene have not been published, although one published genetic map of T4 (King, 1968) does indicate, without comment, a position for the *v* gene.

Lysozymeless mutants. As we mentioned earlier, phage-infected bacteria rupture when treated with chloroform. This is due partially to dissolution of the membrane by the organic solvent and partially to degradation of the cell wall by phage lysozyme once the cells have been killed. Phage mutants which are unable to form lysozyme cannot lyse their host cells, nor do cells infected by these mutants lyse in the presence of chloroform. However, upon treatment with crystalline egg white lysozyme, they do rupture. A number of lysozymeless mutants have been isolated on this basis (Streisinger et al., 1961).

All such mutants map in the *e* gene (*e* is an abbreviation for endolysin, another name for phage lysozyme). This appears to be the first gene in phage for which the actual gene product was identified and isolated. Because lysozymeless mutants can be easily recognized on plates, and because lysozyme itself is a small protein whose complete amino acid sequence is now known, both in T2 (Inouye and Tsugita, 1968) and T4 (Tsugita and Inouye, 1968), this system has proven eminently satisfactory for verification of a number of genetic postulates. We shall discuss these later in the chapter.

Other mutants. Other plaque-type mutants in T-even phages include minute-plaque (*mi*) mutants, turbid-plaque (*tu*) mutants, and star (*s*) mutants. The bases for naming these mutants is evident from the size, clarity, and shape, respectively, of the plaques formed by them, but little is known about the biochemical changes involved. *Mi* strains may bear leaky mutations in essential genes.

MUTAGENESIS

Before proceeding to a consideration of the biochemical bases of conditional lethal mutations, let us briefly consider how mutants come into existence, both naturally and in the laboratory. This treatment will be somewhat cursory, and the reader is again referred to Hayes (1968) for a complete discussion of this topic. Our primary concern here will be with the mutagenic agents used in phage genetics, with some consideration of the types of mutation induced by each agent. Although chemical mutagenesis has been used as the basis of many studies on the nature of the genetic code, a detailed treatment of this work is beyond the scope of the present book.

Any phage preparation will contain spontaneously-arising mutants, with frequencies for mutations in specific genes varying from about 10^{-9}, for host-range mutants, to nearly 10^{-3}, for *r* mutants. The frequency of appearance of a particular type of mutant appears to depend both upon the size of the gene (the larger the gene, the greater the probability of its being affected by a mutational event) and the extent to which it is selected against by natural processes. Rapid-lysis mutants have lower burst sizes than wild-type phage in most hosts. However, these mutants are fully viable under most conditions, so that they are not as strongly selected against as many other types of mutants. Some types of mutants can be easily isolated from untreated phage stocks. For example, *r* mutants are easily spotted by virtue of their distinctive plaque morphology. If one prepares 10 plates, each containing about 500 infectious units, from a laboratory phage stock, one has an excellent chance of picking up at least one *r* mutant. Isolation of an *h* mutant is even simpler. One plates a very large number of phage with a plating strain resistant to that phage. Essentially every plaque which appears is due to the spontaneous formation of a mutant whose host range extends to the plating bacteria. However, most mutants which an investigator might wish to isolate are either not present in sufficiently high frequency nor can be so easily recognized by selective plating techniques. For these reasons, various agents capable of increasing the mutation frequency have come into use.

Before we discuss the mechanism of action of various mutagenic agents, let us consider the types of mutations which can occur. In this discussion it is assumed that the reader is familiar with the principles of DNA structure, as proposed by Watson and Crick (1953), and with the nature of the genetic code, as determined primarily from *in vitro* studies on amino acid incorporation (cf. Nirenberg et al., 1965; Khorana et al., 1966). Mutations which involve a change of one base pair to another, without a change in the number of nucleotide pairs in a gene, are called *point mutations.* These include *transitions,* in which a purine is replaced by a purine, or a pyrimidine by a pyrimidine, and *transversions,* in which a purine is replaced by a pyrimidine, or vice versa. Some mutations involve removal of nucleotides from a polynucleotide chain, such that the mutant gene is shorter than the wild-type gene. These are called *deletion* mutations. Similarly, mutations which result in the lengthening of a polynucleotide chain are called *additions* or *insertions.*

Types of Mutagenic Agents and Their Effects

There are five major classes of mutagens used in phage genetics. Elucidation of the modes of action of these agents has involved a combination of inference from the chemical structure of a particular mutagen, coupled with some knowledge of its interaction with nucleic acid components *in vitro,* and genetic analysis of the mutants formed. The most powerful genetic techniques have involved studies on the revertibility of induced mutations. Many induced mutations can undergo reverse mutation to wild-type under the influence of the same agent used for the original mutagenesis. Now, if mutagen A can revert mutations induced by mutagen B, and if something is known about the mechanism of mutagenesis by B, then one can draw inferences about the types of mutation induced by A. The *r*II system in T4 has been of particular utility in these studies because of the ease of scoring wild-type revertants. One simply plates on *E. coli* K12 (λ), which restricts the growth of *r*II mutants. Any plaque seen under these conditions represents a revertant.

Let us now consider the major types of mutagens and their effects.

Base analogs. These are structurally related to normal DNA constituents, such that they can be incorporated into DNA during replication. Presumably, the electron distribution about the incorporated analog is sufficiently different from that of the base which it replaces that there is an increased frequency of mispairing during replication. Among the most potent members of this class of mutagens are 5-bromouracil (BU) and 5-bromodeoxyuridine (BUDR), analogs of thymine and thymidine, respectively. These analogs can be incorporated nearly quantitatively into DNA in place of thymine, such that some adenine-thymine (A-T) base pairs become A-BU base pairs. Freese (1959) proposed that, because of the altered electron density caused by substitution of the highly electronegative bromine atom for the methyl group of thymine, the enolic structure of BU is stabilized, such that it occasionally "looks like" cytosine, and pairs with guanine during replication. Thus, what was originally an A-T pair has undergone mutation to become a G-C pair, as shown in Figure 5–1.

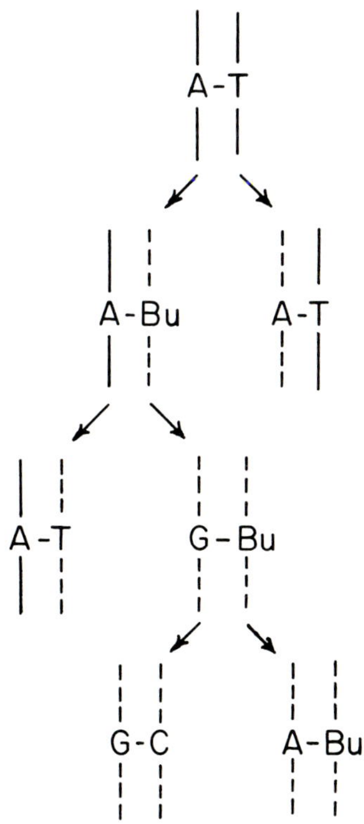

FIGURE 5-1 A mechanism for mutagenesis by 5-bromouracil following incorporation into DNA. The solid lines represent parental DNA, and the dashed lines represent newly-synthesized DNA. The mutation shown here is a transition from an AT base pair to a GC pair.

Another thymidine analog which is mutagenic in T4 is trifluorothymidine, which has a —CF_3 group in place of the methyl group (Gottschling and Heidelberger, 1963). The mutagenic specificity of this compound has not been precisely determined, but it appears to be similar to BUDR in its effects.

Another base analog in common use is 2-aminopurine (AP), which is incorporated into DNA (although much less efficiently than BU) in place of adenine. As such, it can occasionally interact with cytosine, inducing a transition of an A-T pair to a G-C pair. AP can also revert AP-caused mutations, so that it seems capable of inducing transitions in either direction.

Agents which interact directly with DNA. The base analogs can function as mutagens only after their incorporation into DNA. Therefore, they are only active when administered during phage growth. Members of the second class of mutagens can interact with resting DNA, or DNA in mature phage. These include substances such as nitrous acid, which is known to deaminate a wide variety of compounds, including nucleic acid bases, *in vitro*. Thus, nitrous acid would be expected to deaminate adenine in nucleic acids to hypoxanthine, which would pair with cytosine, and cytosine would be deaminated to give uracil, which would be read as thymine. Guanine is also subject to deamination, but the product, xanthine, would probably also be read as guanine.

Direct evidence for the presumed action of nitrous acid has come from studies with tobacco mosaic virus (cf. Fraenkel-Conrat, 1968). Comparison of the structures of mutant coat proteins with the known amino acid sequence of the wild-type coat protein has shown that nearly all amino acid substitutions can be accounted for by using known codon assignments and the assumption that nitrous acid is acting as a deaminating agent. For example, several mutants have been examined in which isoleucine is found in a position normally occupied by threonine. Examination of codon assignments for these amino acids reveals that a base triplet coding for threonine (ACU, ACC, ACG, or ACA) can change to one coding for isoleucine (AUU or AUC) only by change of a C to a U. Note that, of the four threonine codons, only the first two can be changed to isoleucine codons by single-base changes. Thus, the replacement of threonine by isoleucine indicates that the codon used *in vivo* for threonine at that site is either ACU or ACC.

Another important member of this class of mutagens is hydroxylamine (HA), whose effect is more specific than that of nitrous acid. Although the precise nature of the reaction is not known, hydroxylamine reacts only with cytosine among the nucleic acid bases. Thus hydroxylamine is expected to yield G-C to A-T transitions. Since HA mutagenesis is not revertible by HA (Freese, Bautz, and Bautz-Freese, 1961), but is highly revertible by AP (Freese, Bautz-Freese, and Bautz, 1961), it is concluded that HA-induced transitions occur only in the G-C to A-T direction.

Depurinating agents. These include alkylating agents, such as nitrogen mustards, which appear to interact rather specifically with DNA at the 7-position of guanine residues. The presence of an alkyl group at this position labilizes the deoxyribosyl linkage, such that the purine is released as 7-alkylguanine. A gap is left in the nucleotide sequence, although the sugar-phosphate chain remains intact. Since any of the four bases can be inserted into this gap with equal probability, one can expect these agents to yield a variety of transitions and transversions.

In more common use than the nitrogen mustards are the less toxic agents ethyl ethanesulfonate (EES) and ethylmethanesulfonate (EMS) (Loveless and Howarth, 1959). These mutagens also act as alkylating agents for guanine residues, with 7-ethyl guanine being released at affected sites. In agreement with the expectations regarding the mode of action of alkylating agents, it is found that EES is capable of inducing either transitions or transversions (Bautz-Freese, 1961).

Another potent member of this class of mutagens, which has recently come into wide use, is N-methyl-N′-nitro-N-nitrosoguanidine, more commonly called nitrosoguanidine, or NG. This agent is so efficient that treatment of *E. coli* under conditions resulting in 50 percent survival gives an average of one mutation per surviving cell (Adelberg, Mandel, and Chen, 1965). The active alkylating species formed from NG appears to be diazomethane (Magee and Farber, 1962). Nitrosoguanidine has the interesting property of mutagenizing selectively at the replication point (Cerdá-Olmedo, Hanawalt, and Guerola, 1968). The basis for this is not known, but may involve structural distortions occurring as DNA strands are unwound at the replication point, which could make the target bases more accessible to the mutagen. NG is ineffective when applied to resting phage or to nongrowing bacteria. Baker and Tessman (1968) have found that NG differs in its mutagenic specificity for T4 and for the single-stranded DNA phage S13. As previously shown for *Salmonella typhimurium* (Eisenstark, Eisenstark, and Van Sickle, 1965), NG induces transitions in both directions in S13 wth equal frequency. However, almost all of the transitions induced in T4 are of the G-C to A-T type. This difference is ascribed to differing environments in the area of the replicating points for the two phages. Whatever the nature of the effect of NG, it is clear that this mutagen can provide a valuable tool for probing the nature of replicating DNA.

Acridine dyes. These agents, including proflavin and 5-amino acridine, have long been known to be mutagenic for phage, but not bacteria (DeMars, 1953). Early studies on the mutations induced by these agents showed that they could not be reverted by base analogs, but they could be reverted by other acridine dyes. Thus, they form a special class of mutants. Crick et al. (1961) proposed that acridine dyes, which are planar molecules, can intercalate themselves between adjacent bases in one strand of a DNA double helix. This can result in insertion or deletion of a single base, once the error is perpetuated by a further round of replication. Thus, these mutations are either additions or deletions, explaining their inability to be reverted by base analogs. These are called phase shift mutations, since the "reading frame" is shifted out of the normal phase. Consider a region of messenger RNA, and the polypeptide sequence which it specifies, based upon current codon assignments. Consider also the effect of the deletion of a single nucleotide pair in the corresponding stretch of DNA, which would result in removal of the nucleotide underlined in the following diagram.

						deletion						
. . ACG	<u>U</u>AC	UUU	ACG	CAC	. . .	→	. . . ACG	ACU	UUA	CGC	AC .	. . .
. . thr	tyr	phe	thr	his	. . .		. . . thr	thr	leu	arg	thr	. . . .

Since the code is read in units of three, the removal of one nucleotide shifts the initial nucleotide of each triplet one position to the right, so that the entire message past the site of mutation is changed. The insertion of a single nucleotide would have a similar effect, with the "reading frame" being shifted one nucleotide in the opposite direction.

Phase shift mutations can revert spontaneously or under the influence of acridine dyes. One mechanism proposed to allow for reversion of a deletion of this type is a compensatory insertion of a single base at a nearby site, such that the original number of nucleotides, and thus, the original reading frame, is restored. Similarly, a compensatory deletion can correct for the presence of an extra base in an insertion mutation. This and several other predictions of Crick et al. have been verified in a series of brilliant experiments, originating in the laboratories of Streisinger and Tsugita, on T4*e* mutants, altered in the structural gene for phage lysozyme. We shall consider this work later in the chapter.

It is of interest that the majority of spontaneous mutations in phage are not point mutants, as evidenced by their inability to be reverted by base analogs. A large proportion can be reverted by acridines. However, some spontaneous mutations are long deletions and cannot be reverted to a measurable extent under any known conditions.

Ultraviolet radiation. Various forms of radiation, including x-rays and gamma rays, have been known for many years to possess mutagenic effects. The most convenient source of mutagenizing radiation for most phage workers is ultraviolet light. Although uv light is widely used in preparing phage mutants, the precise nature of its mutagenic action is not known. Uv irradiation of DNA results in a number of chemical transformations. Precise identification of the specific transformations responsible for lethal or mutagenic effects is not a simple task. One possible mechanism of uv mutagenesis involves transformation of cytosine in DNA either into uracil or a cytosine hydrate, either of which would be read as thymine. A prediction of this hypothesis is that most uv-induced mutations would be G-C to A-T transitions. This prediction has been partially borne out in studies on phages S13 (Howard and Tessman, 1964) and T4 (Drake, 1963). However, in both of these studies a significant proportion of the mutations studied were not transitions at all.

Most of the above treatments are effective only when the mutagen is applied to vegetative phage, i.e., to phage-infected bacteria. Most induced mutations appear as a result of misreading, or intercalation of material, during DNA replication or recombination. An exception is nitrous acid, which can deaminate cytosine residues in resting DNA. At any rate, it was long assumed that phage DNA *in situ* is sufficiently stable that mutations would not occur unless the DNA were replicating. However, Drake (1966) showed that mutants can accumulate even when a phage lysate is stored under conditions favorable for survival, with no opportunity for replication. Analysis of spontaneous reversion of *r*II mutants of T4 showed that this was observed only with mutants containing a G-C pair at the altered site. Phase shift mutations did not revert spontaneously. In a later study of forward mutations, Drake and McGuire (1967) found that many of the mutations occurring were transversions, and they suggested that these might occur following alteration of purine residues, or depurination. At any rate, it is well for the phage worker to bear these phenomena in mind. Many biochemical experiments, particularly those involving amber mutants, require genetically pure phage preparations. For such work, it is wise to prepare new stocks regularly, starting from single plaques each time.

CONDITIONAL LETHAL MUTATIONS; THE T4 GENETIC MAP

We have already referred to the development of conditional lethal mutations and their importance to bacterial virology. The *ts* mutations were developed by Edgar and his colleagues (Edgar and Lielausis, 1964; Edgar, Denhardt, and Epstein, 1964), at about the same time that Epstein and his co-workers were isolating a large number of *am* mutants (Epstein, Bolle, and Steinberg, unpublished work). Several hundred mutants of both classes had been isolated by 1963, and at the present time, well over one thousand conditional lethals of T4 are in existence. Once a large number of mutants was at hand, Edgar and Epstein and their colleagues set about the task of determining how many genes were represented among the mutants, and how the new mutants were distributed about the linkage map of T4. It should be recalled that at this time there was little clear information about the chemical nature of these mutations. The best explanation for *ts* mutations was that each one led to a structural change in a protein which rendered it more temperature-sensitive than the wild-type protein. It was known that the phage particle itself does not become more heat-sensitive; rather, the mutation affects some aspect of the phage developmental cycle. The nature of *am* mutations was somewhat less clear. It was inferred that permissive bacterial strains contain "suppressor genes" which permit expression of the altered genes which are inactive in the nonpermissive hosts. However, the term suppression was ill-defined in biochemical terms at that time. We shall discuss the biochemistry of *am* and *ts* mutations a little later.

The first step in constructing a genetic map was the assignment of each mutant to a gene. This was done by complementation tests. Two mutants are used to mixedly infect a bacterial culture under restrictive conditions. If the mutations in the two phage strains occur in different genes, then each phage can supply the genetic function which is defective in the other, and a crop of viable progeny is produced. On the other hand, if both mutations fall in the same gene, then neither parent can provide the missing gene function, and no progeny phage are formed. By carrying out a large number of mixed infections of this type, using different mutant pairs, it was possible to assign each mutant to a complementation group, or gene. Actually, this task was complicated somewhat by the phenomenon of "intragenic complementation" (Edgar, Denhardt, and Epstein, 1964; Bernstein, Edgar, and Denhardt, 1965). Two mutants which fall into the same complementation group, as judged by complementation tests with other mutants, will sometimes complement each other. This can occur, for example, if an enzyme normally functions as a "multimer," containing two or more polypeptide subunits in each molecule of completed enzyme. It is possible, in some cases, for a defective subunit from one mutant to combine with a subunit from another mutant which bears a different defect, in such a way that the final product has partial or full activity. However, it was found that intragenic complementation in the T4 conditional lethals was only seen in *ts* × *ts* crosses. Amber mutants failed to complement, either with other *am* mutants, or with *ts* mutants in the same gene. The reason for this will become clear later. At any rate, this failure of the ambers to complement within the same gene allowed unambiguous assignment of each mutant to a gene. Moreover, the existence of intragenic complementation among the *ts* mutants was later found to be useful for the ordering of mutant sites within a gene (Bernstein et al., 1965). All of the *ts* mutants in a given gene can be tested pairwise for complementation, and the resulting pattern used to prepare a complementation map of the gene. Such maps are reasonably similar to corresponding maps determined by measuring recombination frequencies in pairwise matings (see below).

The conditional lethal mutations isolated in the early work of Epstein and Edgar and their colleagues fell into 47 complementation groups. They were then ordered and arranged on the T4 linkage map from recombination frequencies observed in a large number of genetic crosses. In performing a cross, one mixedly infects a bacterium with phage bearing genetic differences in two or more sites, under conditions permitting replication of all parental genomes. One then examines the progeny to see what proportion contains some genetic characteristics of both parents. Consider, for example, a mixed infection with two phages, one of which bears a mutant allele in gene A and is wild-type with respect to linked gene B, while the other is wild-type at A and bears a mutation at B. The parental genotypes can be represented thusly: a + and + b. Most of the phage in the progeny will have either one or the other parental genotype. However, a fraction will bear either the complete wild-type genotype, + +, or the double mutant genotype, a b. It is beyond the scope of this book to discuss recombination in detail. However, it probably involves a process of breaking and joining of DNA strands, such that the resultant molecules contain regions derived originally from each parent. This is called a crossover event. In general, the farther apart are two

genetic markers, the higher is the probability that a crossover will occur between them during vegetative growth, and the higher will be the recombination frequency. The relationship between recombination frequency and physical distance on the phage chromosome will be examined in a later section.

Crosses involving conditional lethal mutants were carried out under conditions permissive for the growth of each mutant. The total progeny resulting from each mixed infection were measured by plating under permissive conditions. The wild-type recombinants were enumerated by plating under conditions restrictive for both parental strains. In determining the proportion of all recombinants in the progeny, both wild-type and doubly mutant, it was assumed that recombination is reciprocal, i.e., that for each event creating a wild-type genome, a corresponding event creates a double mutant genome. Thus, the total proportion of recombinants was arrived at by multiplying the fraction of wild-type recombinants by two. (It should be observed that recent evidence argues against the long-held assumption of reciprocal recombination, at least in mass lysates of T4 (Doermann and Parma, 1967; Mosig, Ehring, and Duerr, 1968)).

From a large number of two- and three-factor crosses, it was possible to construct a genetic map of T4 (Edgar, Denhardt, and Epstein, 1964; Stahl, Edgar, and Steinberg, 1964; Streisinger, Edgar, and Denhardt, 1964). As we have already observed, the map is topologically equivalent to a circle. This had already been inferred from results of three-factor crosses

(crosses involving three different mutant sites). Outside markers had been shown to be linked to each other, as well as to an inner marker (Streisinger and Bruce, 1960). It was also found that the newly discovered genes, represented by the conditional lethal mutations, were distributed all over the genetic map, rather than being clustered in a few parts. Preliminary experiments (Epstein et al., 1963) showed that the distribution of genes is nonrandom; there was a definite tendency for different genes concerned with the same function to be clustered.

Since the initial studies, a number of new genes have been discovered, on the basis of the isolation and characterization of many more conditional

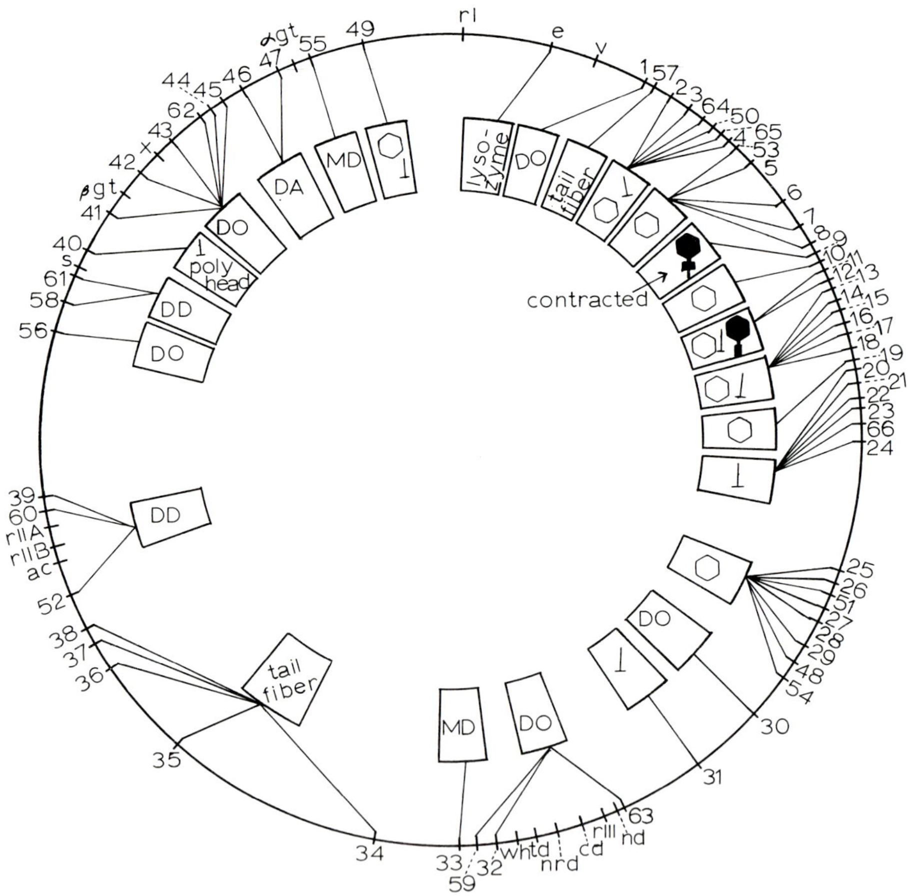

FIGURE 5-2 The T4 linkage map. Most of this map is derived from Edgar (1968), but some of it is from sources described in the text. The inner circle indicates the gross phenotype associated with defective functioning of each of the indicated genes.

lethal mutants. In 1966 a total of 66 such genes had been recognized, each of which was assigned a number. A map showing the positions of these genes is presented in Figure 5–2. This map also includes several of the classical markers, along with some other genes whose function we shall discuss in the next chapter.

Once the conditional lethal mutations of T4 had been arranged on a genetic map, there came the task of identifying the function of each of the newly discovered genes. In other words, what is the product of each gene, and what is the metabolic role of each gene product? Because of the large number of genes and gene products to identify, this has been a major undertaking, involving the efforts of many laboratories in the years since 1963. Initial studies were directed toward identification of the gross phenotype associated with each genetic defect (Epstein et al., 1963). Cells infected under nonpermissive conditions with a particular mutant were examined for their ability to carry out various reactions associated with normal development, including DNA synthesis, spontaneous lysis, and synthesis of phage components. The latter was detected by means of electron microscopic examination of lysates of the infected cells. From these screening studies, it was possible to identify the general function associated with each genetic defect. These are also indicated in Figure 5–2. Some mutants were unable to initiate DNA synthesis and are called "DNA-negative," or DO. Others, called "DNA arrest," or DA mutants, could start DNA synthesis in normal fashion, but it ceased soon after initiation. Another class was delayed by several minutes in its onset of DNA synthesis, and these are called "DNA delay," or DD mutants. Mutants in genes 33 and 55 are called "maturation-defective," or MD. These mutants can synthesize DNA, but none of the viral structural components. A large class of mutants, comprising over 40 genes, are able to carry out early functions, such as DNA synthesis, but are defective in some stage of virus assembly, such as formation of a particular viral substructure, or the joining together of two or more such substructures. For purposes of the initial work, as indicated in Figure 5–2, the defective phenotypes of each of these genes are indicated merely by schematic diagrams of the phage parts which were detected by electron microscopic examination. For example, phage bearing mutations in any of the genes 34 through 38, or gene 57, can form phage particles which are complete except for the absence of tail fibers. Because of this absence, of course, these incomplete particles are noninfective. In other instances, grossly abnormal phage structures, such as "polyheads" or "polysheaths" were formed. We shall have more to say about the functions and interrelations of the genes governing virus assembly in Chapter 7.

It seemed reasonable to propose that some of the genes involved in DNA synthesis are structural genes for the phage-induced enzymes known to be necessary for phage DNA synthesis. Indeed, it had already been reported (Wiberg et al., 1962), that a T4 amber mutant, *am* N122, was unable to induce the formation of dCMP hydroxymethylase. Later, Epstein et al. (1963) reported that this mutant maps in gene 42. This, of course, does not prove that gene 42 is the structural gene for the hydroxymethylase. However, as we shall see, such evidence was supplied later. In addition, several other structural genes for early enzymes have been identified, as we shall see in the next chapter.

None of the gene products associated with DNA delay genes have yet been identified. However, evidence is accumulating (C. Yegian, personal communication) that these gene products act to regulate the intracellular ionic concentration. A failure of one of these gene products to function can, presumably, increase the length of time necessary for the cell to establish the proper intracellular environment necessary for maintenance of DNA synthesis at the normal rate.

The clustering of genes with related functions is worthy of comment. Note that most of the genes involved in DNA synthesis are located between 10 and 12 o'clock on the circular map as it is drawn in Figure 5–2, while a few more such genes are located at 5 o'clock. Similarly, most of the genes involved in head formation are located near three o'clock (these mutants make completed tails, but no heads), and the genes for tail fiber assembly fall largely between six and eight o'clock. However, the clustering is not complete. Gene 1, which is essential for DNA synthesis, and gene 57, which is involved in tail fiber assembly, are both located quite far from genes whose products perform closely related functions. However, the organization of genes on this map is clearly nonrandom, and it would be exceedingly interesting to know how such a map evolved (Stahl and Murray, 1966).

Let us now consider the biochemical nature of *am* and *ts* mutations. We have already mentioned the probability that *ts* mutations involve the generation of "missense," which leads to formation of altered proteins which are more heat-sensitive than the corresponding native proteins. This was first verified by Wiberg and Buchanan (1964) in studies on *ts* mutants of T4 which bore defects in gene 42 (DO phenotype) and were unable to synthesize DNA in cells infected at 42°. We have already mentioned that a gene 42 amber mutant was unable to induce dCMP hydroxymethylase. Wiberg and Buchanan found that the hydroxymethylase induced by their *ts* mutants at 30° was inactivated at 40°, in contrast to the wild-type enzyme. Moreover, the hydroxymethylase induced by two different mutants showed differences in a range of properties, including temperature-sensitivity. Thus, mutations in gene 42 led to clear alterations in the properties of a particular enzyme, which was taken as proof that gene 42 is the structural gene for that enzyme. This study also confirmed the generally accepted concept of a *ts* mutation, and it provided the first evidence that a phage-induced enzyme is a gene product of the virus, not the host (see p. 64).

A number of structural genes for other phage-induced enzymes have been identified by the same criteria as those used for the hydroxymethylase by Wiberg and Buchanan. However, to my knowledge, no one has yet performed the kind of protein chemistry necessary to make this type of evidence ironclad for a phage system, i.e., identification of the actual amino acid change resulting from a particular *ts* mutation. This is perhaps a suitable time to observe that the phage-induced enzymes present excellent systems for studying the relationship between protein structure and function. Purification of phage-induced enzymes, where it has been accomplished is a relatively straightforward task (cf. Mathews, Brown, and Cohen, 1964; Fleming and Bessman, 1967; Duckworth and Bessman, 1967; Weiss et al., 1968). A large number of *ts* and other types of mutants are available, and more can be prepared without difficulty. Thus, one can have available a series of protein molecules, most of them differing from each other only at a single amino acid residue, which can be purified and compared as catalysts. To date, however, this type of approach has not received the attention which it deserves.

The biochemistry of the amber mutation was not as readily apparent as that of the temperature-sensitive mutation. It was clear relatively early (cf. Campbell, 1961) that the ability of a bacterial strain to support growth of an amber mutant depended upon the presence of a suppressor gene. The position of a suppressor gene on the bacterial chromosome could be determined by standard techniques in bacterial genetics, but little was known about the actual nature of suppression. An early clue was provided from genetic studies of the *r*II region (Benzer and Champe, 1962). Some *r*II mutants were "ambivalent" in their behavior, i.e., they could grow on some strains of *E. coli* which are normally restrictive for *r*II growth, but not on others. Genetic analysis led to the conclusion that these mutations "give rise to coding units that do not correspond to an amino acid, i.e., they are nonsense mutations." The presence of a suppressor mutation in some way allowed this nonsense to be translated. As one possible explanation of this phenomenon, Champe and Benzer suggested that the suppressor strain might contain an altered transfer RNA which could respond to a nonsense coding unit and specify the insertion of a particular amino acid at the site of the mutation. We now know that the ambivalent *r*II mutations represent amber mutations in the *r*II gene, and that altered tRNA molecules are indeed responsible for suppression. The elucidation of these facts and identification of the nonsense coding units in molecular terms has provided an exciting chapter in molecular genetics. Space does not permit a detailed coverage of this work, but the reader is referred to a recent article by Garen (1968), which provides such a review. However, let us briefly cover the major points as they relate to phage systems.

Dirksen, Hutson, and Buchanan (1963) provided confirmation of the conclusions of Benzer and Champe, in studies carried out at the enzymatic level. A gene 42 *am* mutant, unable to induce dCMP hydroxymethylase in a restrictive host, was used to infect three different permissive hosts. Hydroxymethylase was extracted from each infected cell culture, and the three preparations were compared. Clear differences were seen in heat-sensitivity of the three preparations, suggesting that different suppressor genes allow the nonsense coding unit to be translated in different ways.

Sarabhai et al. (1964) exploited a series of amber mutants in gene 23, which is the structural gene for the head protein, to show that the presence of a nonsense coding unit, or triplet, in messenger RNA, causes termination of polypeptide chain elongation. The choice of gene 23 was particularly felicitous, since the majority of the protein synthesized late in infection by T-even phages is the head protein. Analysis of the polypeptide fragments synthesized by the mutants under restrictive conditions showed a close correlation between length of each fragment produced and the position of the corresponding mutation, as determined by genetic mapping. Thus, the amber triplet is seen as a true nonsense triplet, beyond which translation cannot occur. Moreover, this study provided the first proof of the long-held assumption that genes and the proteins they specify are colinear.

The nucleotide sequence of the nonsense triplet corresponding to the amber mutation was determined by structural analysis of the alkaline phosphatase formed by revertants of *E. coli* strains bearing amber mutations in the structural gene for this enzyme (Weigert and Garen, 1965). *In vitro* studies on the genetic code (cf. Nirenberg et al., 1965; Khorana et al., 1966) were showing that only three out of the possible 64 trinucleotide sequences in RNA failed to direct the incorporation of any amino acid into polypeptide in a cell-free system from *E. coli:* UAG, UAA, and UGA. Weigert and Garen identified the amino acid inserted at the site of the amber mutation in several alkaline phosphatase revertants, compared these with the amino acid at the same site in the wild-type enzyme, and concluded that an assignment of UAG as the amber triplet was the only one consistent with all of their data. Similar conclusions were reached in studies on chemical mutagenesis in gene 23 of T4 (Brenner, Stretton, and Kaplan, 1965). These studies also revealed the elistence of a second class of nonsense mutants, in which the mutant triplet was identified as UAA (Brenner et al., 1965; Weigert, Lanka, and Garen, 1967). In keeping with the trivial basis used for naming amber mutations (Edgar, 1966), the term "ochre" was applied to this latter class of mutants. Since that time, a class of UGA nonsense mutations has also been discovered (Zipser, 1967) and given the equally meaningless name of "umber" mutants, although some laboratories prefer to call them "opal" mutants. It should be observed that the term "azure" has also come into use, to describe certain mutants of the RNA phage f2 which are actually, in all probability, a type of amber mutants (Horiuchi and Zinder, 1967). Fortunately, the code in *E. coli* contains no more nonsense triplets, for it would appear that the application of this type of nomenclature is well-nigh exhausted.

The chain-terminating nature of the amber mutation explains why intragenic complementation is not observed with strains bearing such mutations. Such complementation requires the formation of complete, albeit imperfect polypeptide chains. Since complete chains are not formed by amber mutations, there are no opportunities for forming the partially active polypeptide aggregates which can be formed by interaction of the products of genes with *ts* mutations.

In light of our discussion of the nature of the amber mutation, we can now ask whether the clustering of genes with related functions on the T4 map has any significance for the functional expression of these genes. Recall the regulatory model of Jacob and Monod (1961), which postulated the existence of the operon: a group of functionally related and physically contiguous genes whose expression is coordinately controlled, partially through the formation of a multigenic mRNA molecule, which codes for all of the structural genes products in the operon. Do operons exist in T4? One way to answer this question is to look for *polarity mutations.* If a nonsense mutation is located in one gene of an operon, then all of the genes transcribed after this gene also fail to be expressed, even though they are structurally intact. This is because the nonsense triplet blocks translation of the mRNA past the point of the mutation, and this includes the sequences specifying the structures of the other gene products which are not formed. (Actually, this block is not always complete, for reinitiation of mRNA translation past the point of the nonsense triplet can occur under some conditions. See Sarabhai and Brenner, 1967). At any rate, polarity mutations were sought in complementation tests involving double amber mutants of T4 (Stahl et al., 1966; Nakata and Stahl, 1967). The data suggest that amber mutations in genes 51 and 34 abolish expression of the respective adjacent genes, 27 and 35. Thus, there appear to be at least two regions of co-transcription, each containing but two genes. Apparently such units are relatively rare in T4. These data of Stahl and his colleagues also indicate that all four of these genes, which determine late functions, are transcribed in a clockwise direction as the map is usually drawn. The reason for this should be apparent from an examination of Figure 5–2.

Recently Stahl et al., (1970) have completed a more extensive study of units of co-transcription in T4. There are now at least eight such units, each involving between two and four genes. Of these eight units, only one includes known early-function genes (44 and 45), and this unit, as expected, is transcribed counterclockwise.

We have not yet discussed the mechanism of suppression of amber mutations, although we mentioned the hypothesis of Benzer and Champe that altered tRNA molecules are involved. This is indeed the case, as shown by Capecchi and Gussin (1965) and by Engelhardt et al. (1965). Both groups were studying the synthesis *in vitro* of the coat protein of the RNA phage f2, primed by viral RNA. When RNA from a phage amber mutant was added to the system, no coat protein was made unless seryl-tRNA isolated from a suppressor strain of *E. coli* was also added. When this was done, the *in vitro* protein product had a serine residue in place of a specific glutamine present at the same site in the wild-type coat protein. This experiment showed that suppression involves the synthesis of an altered tRNA, with serine acceptor and transfer activity, which can recognize the amber triplet. At least five genes capable of suppressing amber and/or ochre mutations are now known in *E. coli* (Garen, Garen, and Wilhelm, 1965; Gallucci and Garen, 1966). Each suppressor gene determines the insertion of a specific amino acid at the site of the nonsense mutation. In each case an altered tRNA is responsible for this, although it has not been established that all of the suppressor genes are indeed structural genes for tRNA. Such a gene may be a regulatory gene or a gene coding for an enzyme which modifies tRNA. However, there is now good evidence that the *su*3 gene, which causes tyrosine insertion, is a structural gene for a tyrosyl tRNA (Smith et al., 1966).

PHYSICAL SIGNIFICANCE OF THE T4 GENETIC MAP

So far, the T4 linkage map which we have been discussing exists as a strictly genetic construct, which has been pieced together by measuring recombination frequencies and assuming that the greater the extent of recombination between two markers, the farther they are apart. To what extent does this map represent the physical structure of the T4 chromosome, both with respect to gene order and the distance between markers? Let us first compare recombination frequencies over an entire T4 genome with the known length of the T4 DNA molecule. If two markers recombine with a frequency of 1 percent, we say that they are 1 map unit apart. Now if we add up the map distances between pairs of markers spanning the entire genome, we find that the entire T4 genome is some 800 map units in length. This of course does not mean that recombination frequencies of 800 percent will be seen between distantly located markers, since this type of result would have no significance with respect to genetic methods presently in use. It does mean, however, that there is a high probability of multiple recombinational events occurring between distantly linked markers; they are, for all practical purposes, unlinked.

From the molecular weight of T4 DNA, we know that the entire chromosome contains some 200,000 nucleotide pairs, which means that there should be 200,000/800, or about 200 nucleotide pairs between two markers which are one map unit apart, if one assumes that recombination frequency is a measure of physical length. Let us now refer to the brilliant fine-structure genetic analysis of the *r*II region by Benzer (1957, 1961).

This region consists of two cistrons *r*IIA and *r*IIB, each cistron being defined by complementation tests of the type later used to order genes bearing conditional lethal mutations (see p. 80). Cistron A is about 5 map units long, and cistron B is some 3 map units in length. Thus, the entire *r*II region spans about 1 percent of the T4 genome. Benzer isolated and mapped several thousand spontaneous and induced *r*II mutants. In no case where recombination was observed was the frequency less than 0.01 percent, although the sensitivity of Benzer's techniques would have allowed detection of a considerably lower percentage than that (some pairs of mutants failed to recombine altogether, and such pairs were assumed to be altered at the same site). If one map unit corresponds to 200 nucleotide pairs, then 0.01 map unit corresponds to a distance of the order of a single nucleotide pair. Thus, it would appear that genetics is capable of distinguishing mutations occurring in adjacent nucleotide pairs.

In extending recombination data to form a map of the entire chromosome, certain assumptions must be made, particularly the assumption that the probability of recombination is the same in all regions of the genome. In addition, one assumes that recombination is reciprocal, a presumption we have already seen to be not always correct. How accurate are genetic maps in physical terms? Recently, two approaches have been used for generation of physical maps in T4. Goldberg (1966) has made use of a transformation system, first developed by van de Pol, Veldhuisen, and Cohen (1961), and subsequently refined in the same laboratory (Veldhuisen, Poelman, and Cohen, 1968a,b). *E. coli* cells are converted to spheroplasts by treatment with lysozyme and EDTA. This removes the cell wall, but does not block cellular metabolism. Next, one adds urea-disrupted phage, which are known to be infectious for spheroplasts. Finally one adds genetically marked DNA, which is taken up into the cells, where it can recombine with the phage genomes already present. One further incubates the spheroplasts and determines the frequency with which markers on the added DNA are found in progeny phages. Goldberg's approach was to prepare DNA from phage bearing two linked markers and to fragment the DNA to various known sizes. The minimum length which could still accommodate both markers, as determined by cotransformation of both genes, was determined and compared with the distance between the markers as known in map units. From this he calculated a ratio of 140 nucleotide pairs per map unit, in close agreement with the corresponding figure reached from purely genetic considerations.

The other approach, used by Mosig (1966, 1967) involves her discovery that there exists in T4 stocks a class of small-headed phages which contain, on the average, two thirds as much DNA as normal phage (Mosig, 1963). These were originally discovered on the basis of their lower buoyant density. These phage are unable to replicate under conditions of single infection, but they can contribute markers in mixed infection, such that a cell infected by two or more light phages usually yields a burst. The best available evidence indicates that each light phage contains a single DNA molecule, two thirds the length of a complete T4 DNA molecule. These two-thirds-length segments are apparently cut at random from the entire genome, for all phage genes are represented with equal frequency in a population of light phage. Mosig has measured the frequency with which marker pairs in light phage are contributed in mixed infections with whole phage. Obviously, the farther apart are two genes on the chromosome, the lower is the probability that a random fragment of DNA, two-thirds as long as normal, will bear both genes, and vice versa. From a large number of such experiments, Mosig has constructed a physical map which corresponds closely to the genetic map except that it indicates a different gene order between genes 5 and 7, and it indicates that recombination frequencies are abnormally high between genes 34 and 35.

How complete is the T4 map in its present state? From the fact that T4 DNA contains 2×10^5 nucleotide pairs and that three nucleotide pairs bear the information for insertion of a single amino acid in protein synthesis, we can conclude that T4 DNA contains sufficient information to code for between 150 and 200 average-sized protein molecules, containing about 300 to 500 amino acid residues each. The precise number of known genes is difficult to determine precisely, because many of the genes originally recognized from the existence of minute-plaque or turbid-plaque mutations might be the same as corresponding genes later discovered in mapping conditional lethal mutations. If this is so, then the *tu* and *mi* mutants would in all probability bear "leaky" mutations in essential genes, which allow for partial function of these genes, but not sufficient to allow formation of normal plaques. Probably the total number of known genes is about 80, representing about half of the genetic potential of the virus. Of course, we have no way at present of determining whether all of the phage DNA is really genetically functional. A number of T4 genes appear to be nonessential, and mutations in these genes are difficult to detect without special selection techniques (the term "nonessential" is operationally defined as nonessential for plaque formation under usual laboratory conditions. The extension of these conditions to the natural environment for the phage is not always made. As we shall see in Chapter 6, a number of "nonessential" genes do have positive value in terms of survival of the virus). There are also between 10 and 20 T4-specific enzymes whose existence is known or inferred, but for which structural genes have not yet been located. Even so, it is clear that about 30 or 40 percent of the phage genome remains unaccounted for in terms of functional expression. One cannot propose that certain essential genes are present in two or more copies, since, as we have seen, point mutations can abolish the expression of many essential functions.

The *e* GENE AND THE GENETIC CODE

We earlier mentioned that phage lysozyme and its structural gene present an unparalleled system for examining the relationships between genes and their products. Lysozyme is a small protein, containing only 164 amino acids. The enzyme is relatively easy to purify (Tsugita et al., 1968), and mutants in the lysozyme, or *e,* gene are easily detected (Streisinger et al., 1961). The latter involves treatment of a plate with chloroform vapor. Cells in the vicinity of a wild-type plaque, become partially lysed by lysozyme which has diffused outward from the plaque, creating a halo around the plaque. Mutants with less active lysozymes have smaller halos, and double mutants, or mutants with inactive lysozyme, make no halos at all. Exploitation of this system, carried out in the laboratories of Streisinger and Tsugita, has involved the isolation of mutants induced by proflavin, an acridine dye. It will be recalled that such mutants are expected to represent additions or deletions. Double mutants with partial activity were prepared, the lysozyme purified and digested with trypsin, and the peptides separated on an ion-exchange column. The elution patterns for double-mutant enzymes were compared with that for the wild-type enzyme, and altered peptides were analyzed structurally. In one study, Terzaghi et al. (1966) found that the enzyme from one double mutant, *e*J42*e*J44 differed from the wild-type enzyme only in a single sequence of five amino acid residues. Using the codon assignments coming from *in vitro* studies on the genetic code (Nirenberg et al., 1965; Khorana et al., 1966) Terzaghi et al. could conclude unambiguously that the altered lysozyme could have arisen only through deletion of a single nucleotide at one site and addition of a single nucleotide at a nearby site, as shown below. Thus, this study confirmed

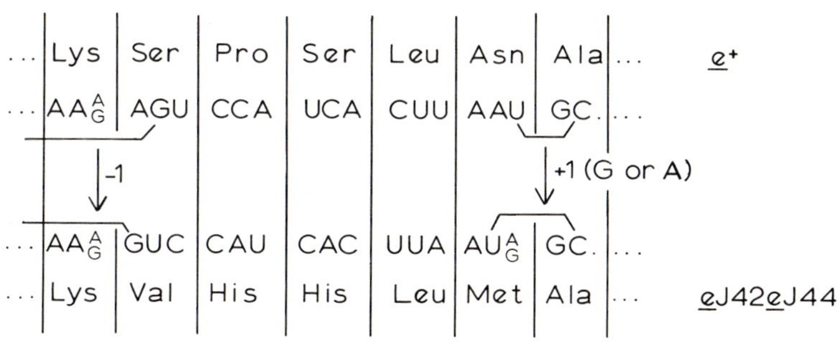

the postulates of Crick et al. (1961), namely that acriflavin dyes can cause additions or deletions, each of a single base, and that a nearby deletion or addition, each of a single base, can be self-correcting. In the case of the *e*J42*e*J44 lysozyme, the deletion in the above diagram alters the reading frame, such that the initial nucleotide in each subsequent coding triplet is shifted one position to the right. If this situation is not corrected, then the rest of the genetic message is "gibberish," and the amino acid sequence bears no relation to that of the wild-type protein. However, in the case of the double mutant, the addition of a nucleotide shifts the reading frame back to normal, and the message past the point of the second mutation is identical to that used in the synthesis of wild-type enzyme. Apparently, the sequence of five amino acids which is altered does not affect the structure of the protein sufficiently to completely abolish enzymatic activity. These studies also established unambiguously that messenger RNA is translated in the 5′ to 3′ direction.

Analysis of another double mutant, *e*J17*e*J44, showed an altered sequence of five amino acids and an additional amino acid inserted next to the changed sequence (Okada et al., 1966). Since *e*J44 had already been shown to be an addition or deletion of a single base, these results could be explained only by postulating that *e*J44 is an addition and that the *e*J17 mutation results in addition of two bases, as shown below. In a similar

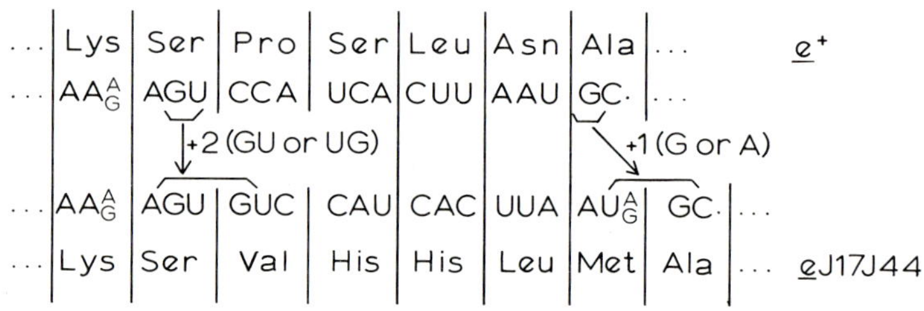

study, Inouye et al. (1967) showed that another frameshift mutation involves the deletion of two base pairs. More recently Okada et al. (1968) analyzed a double mutant containing mutations at the left-hand end of the *e* gene. The wild-type N-terminal sequence Met-Asn . . . was changed in the double mutant to Met-Glu-Tyr . . . , indicating first that the two mutations represent additions of one and two bases, and secondly, that a sequence of bases longer than three (specifying the N-terminal Met) is not necessary for initiation of translation of the lysozyme messenger. However, this study does not take into account the possibility that a unique sequence to the left of the N-terminal codon may be required, or that nascent lysozyme might contain additional material to the left of the Met residue, which is subsequently removed.

In another recent study, Okada et al. (1969) prepared a triple mutant strain and analyzed its lysozyme. The expectation was that a single amino acid would be added or deleted. However, although there was a short altered sequence of amino acids, there was no change in the total number of residues. The most logical conclusion from this was that one of the mutations involves addition of two base pairs, while the other two result in the deletion of one base pair each.

To add to the attractiveness of the *e* gene and its product as a system for biochemical-genetic analysis, the complete amino acid sequence of T4 lysozyme has recently been determined (Tsugita and Inouye, 1968). Interestingly enough, there is no apparent structural relationship to egg white lysozyme. However, as one might expect, the structure is very similar to that of T2 lysozyme (Inouye and Tsugita, 1968). The two enzymes differ at only three amino acid residues, and these differences can be accounted for by three single-base transitions. It is apparent from these investigations that structural analysis of phage lysozymes should provide an important tool in classifying the relationships between various bacteriophages.

GENETIC AND RADIOBIOLOGICAL STUDIES OF PHAGE DEVELOPMENT

It is, I believe, fair to say that genetics and radiobiology provide us with indirect tools for analysis of phage replication, while biochemistry and electron microscopy provide us, basically, with direct methods. In the case of genetics, for example, one draws inferences about molecular events by examining the distribution of phenotypes in the progeny of an infection set up in a particular way. Similarly, in radiobiology we draw inferences from the shapes of survival curves resulting when phage or infected bacteria are subjected to the lethal effects of high-energy radiation, even when the actual effects of such radiation are not known in physical terms. Certainly, the results of such investigations can often be interpreted unambiguously, but it is important to keep these studies in proper perspective. These were among the few methods available for probing molecular events until the development of powerful analytical methods such as zone centrifugation, equilibrium density gradient centrifugation, dual isotope analysis, and high-resolution electron microscopy and radioautography. Radiobiology and genetics allowed the development of models for various processes in phage replication, and with the proliferation of laboratory instrumentation, these models became subject to direct test. For example, the existence of terminal-redundancy heterozygotes in T4 (Streisinger, Edgar, and Denhardt, 1964; Séchaud et al., 1965) suggested a model for phage DNA replication which is now being tested by isolation and analysis of replicative intermediates (Frankel, 1966a, 1968a). These studies are discussed at some length in Chapter 7. Because the focus of this book is primarily biochemical, we shall concentrate upon the latter type of study. However, because genetics and radiobiology have provided the biochemists and biophysicists with starting points for their investigations, it is appropriate to briefly review some of this work here. For more extended treatment, the reader is again referred to the books of Stent (1963), Hayes (1968), and Luria and Darnell (1967).

Mechanism of Reproduction of the Viral Genome

When this question was initially raised there were, *a priori,* three possible answers: (1) the genome could replicate geometrically, with each newly replicated genome spawning two copies, which in turn replicate; (2) the parental genome could be used as the template for formation of all new genomes, and the progeny genomes would not replicate; and (3) each genome would replicate itself once, with the newly formed genome subsequently replicating once, and so on. Luria (1951) answered this question by considering the frequencies with which mutations arise during infections of single cells. If we consider that there is a certain probability of a mutation occurring during any replication event, then each of the above models for genome replication makes a distinct prediction about the frequency distribution of mutants in single-cell bursts, in other words the relative frequency with which bursts containing 0,1,2,3, ... mutants will be seen. For his analysis, Luria examined the production of *r* mutants in several thousand single-cell bursts of bacteria infected with T2. The data most closely approximated the prediction of the geometrical replication model. The data fit even more closely when allowance was made for the fact that not all of the newly formed genomes are represented in the burst, since not all of the newly replicated DNA is packaged into mature virus (Steinberg and Stahl, 1961).

The geometric form of replication is that which would have been predicted had the Watson-Crick model been available in 1951, when Luria's experiments were reported. One might ask why, if genome replication is geometric, the kinetics of phage DNA synthesis are linear (Cohen, 1947; Hershey, Dixon, and Chase, 1953). This is because, after the end of the eclipse period, DNA is withdrawn from the replicating pool at a constant rate for virus maturation, to a form which is unavailable to participate in replication. Actually, most published data on rates of phage DNA synthesis show what appears to be an exponential rate of DNA synthesis between about seven and twelve minutes, before maturation has commenced (cf. Murray and Mathews, 1969a). As might be expected, the rate of DNA synthesis is considerably higher in infection by a maturation-defective mutant, where packaging does not occur, than in wild-type (Mathews, unpublished results.) However, replication is not maintained at an exponential rate, presumably because at some point the total biosynthetic capacity of the cell becomes rate-limiting.

Ultraviolet Sensitivity of Infected Cells

Luria and Latarjet (1947) devised a method which they hoped would allow them to enumerate the phage genomes within a cell at any time during the latent period. Cells were singly infected with phage, and, at various times, portions of the infected cells were irradiated with ultraviolet light. At various times of irradiation, samples of the infected complexes were plated on sensitive bacteria and the titer of infective centers determined. Since we know that one phage particle can give rise to a plaque, it is clear that what is measured in this type of experiment is the ability of the irradiated complexes to produce at least one viable progeny phage. In such an experiment we are measuring the functional survival of the intracellular phage genome, not the capacity of the cell to support phage replication, for it is known that phage will replicate successfully in cells previously irradiated with uv doses much heavier than those used in a Luria-Latarjet experiment.

Luria and Latarjet expected that uv inactivation data obtained at various times after infection would give a family of single-hit survival curves, with shoulders increasing as a function of time after infection (see Figure 5–3, part A). It was already well-known that phage are inactivated by uv light with single-hit kinetics. The expectation was that at a given time, a cell would contain *n* phage genomes, each of identical uv sensitivity to the genome in a virion. The survival curve for such a complex would be multi-hit, i.e., the slope would be identical to that of the survival curve of free phage, but there would be a shoulder which, when extrapolated to zero

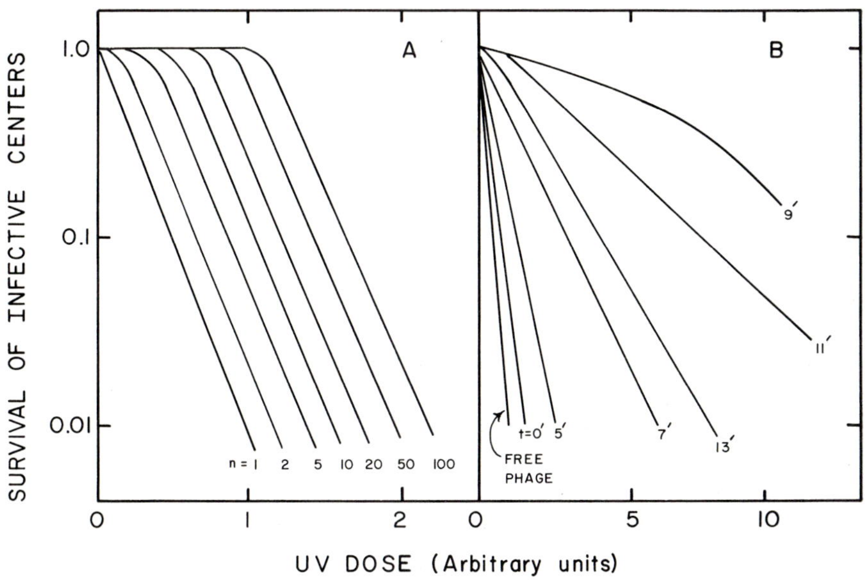

FIGURE 5-3 The Luria-Latarjet experiment. Panel A shows hypothetical uv inactivation curves for infected cells containing the indicated numbers of non-interacting inactivation targets, all of equal sensitivities and all of which must be inactivated in order to inactivate the complex. Panel B shows experimental curves obtained for singly-infected *E. coli* cells harvested at the indicated times after infection with phage T2 (modified from Benzer, 1952).

dose, would given *n*, the number of genomes per cell. When the experiment was done, however, all of the curves were found to be straight lines (Figure 5–3, part B). Beginning at six minutes, the slope decreases greatly, indicating an increased uv resistance of the complexes with time. The best explanation for this phenomenon is a form of multiplicity reactivation (Luria, 1947). It is known that cells multiply infected with uv-inactivated phage are more likely to yield progeny than cells singly infected with phage which have suffered the same dose. The reason for this is that damaged genomes can interact, by recombination, within the infected cell such that the undamaged portions of each combine to give an intact chromosome. Such opportunities, of course, are unavailable when only one chromosome has been introduced into each cell. Similarly, when one irradiates a cell containing a large number of genomes arising from a single genome, as in a Luria-Latarjet experiment, there are many opportunities for multiplicity reactivation, and the capacity of the cell to produce at least one infective phage is greatly increased relative to the cell containing but one or a very few genomes. Consistent with this interpretation is the fact that the development of uv resistance begins at about the same time as the onset of DNA replication. Moreover, at late times in infection, the complexes partially regain their uv sensitivity (Figure 5–3, part C). This may also be expected if some of the genomes are being incorporated into mature virus, in which state they are unable to interact with other genomes (although some *ad hoc* assumptions may be necessary if the vegetative pool remains constant once maturation has commenced). Finally, in the T-odd phages T1, T3, and T7, which undergo fewer rounds of mating during replication than do the T-evens, the pattern of survival curves is very much closer to that predicted by Luria and Latarjet (Benzer, 1952). T5, on the other hand, behaves similarly to the T-even phages.

Sensitivity to P^{32} Decay

Another technique for irradiation of intracellular bacteriophage, and one which was expected to circumvent the effects of multiplicity reactivation, was a "P^{32} suicide" experiment. It had been known for some time that incorporation of P^{32} is a lethal event for T-even phages (Hershey, Kamen, Kennedy, and Gest, 1951), if one prepares the phage in medium of high specific activity and if one stores the phage long enough for a significant fraction of the phosphorous atoms to decay. At least two effects appear to be associated with P^{32} decay. The emission of a β-particle transmutes phosphorus to sulfur, and this disrupts the phosphodiester linkage involving the transmutation, such that a single-strand break, or "nick, results. This alone is not lethal in a double-stranded DNA virus (although it is lethal in phages with single-stranded DNA), since the remaining strand of DNA can serve as a template for replication, and also because the nick might be repaired. However, P^{32} decays occasionally induce double-strand breaks, or "cuts" (Stent and Fuerst, 1960; Thomas, 1959; Tomizawa and Ogawa, 1967; Miller, 1970a). Since the two resulting double-stranded fragments presumably drift away from each other in the cellular environment, there is no way they can be repaired or function as complete genomes.

The design of a P^{32} suicide experiment is as follows (Stent, 1955). One prepares phage with heavily labeled DNA by growth in medium containing P^{32} at high specific activity. These phage are used to infect bacteria, and, at various times, the infected complexes are rapidly frozen and stored for periods allowing fractions of decay of the incorporated radioactivity. (It is known that this treatment does not affect the viability of nonradioactive infected cells). In order to prevent multiplicity reactivation with unlabeled genomes, the bacteria and the medium are both labeled with P^{32} at the same specific activity as the phage. At various times, usually between zero and thirty days, portions of the frozen culture are removed and plated, to see what proportion of the complexes are still capable of releasing at least one infective phage particle. As with the uv survival curves, one notes an increasing resistance to the effects of P^{32} decay, until at about 17 minutes the complexes are almost totally resistant. A distinctive feature of this type of experiment is that considerable stabilization has occurred by five minutes, before DNA replication has begun. In another study (Kozinski and Bessman, 1961), some stabilization was seen as early as three minutes and was essentially complete by 10 minutes. Various *ad hoc* hypotheses have been advanced to explain these observations, including transfer of the information in parental DNA to some nonDNA structure at a time before five minutes (Kozinski and Bessman, 1961; Stent and Fuerst, 1960). However, to my knowledge no completely satisfactory explanation of the phenomenon of P^{32} decay stabilization has been advanced. It would appear to this author that stabilization might well involve the induction of enzymes capable of repairing P^{32} damage. Several enzymes potentially capable of repairing damage to DNA have been found to be induced in T-even phage infection (see Chapter 6). Many of the early enzymes are present in cells in significant amounts as early as five minutes after infection. Presumably, once infected and stored complexes were plated, the repair enzymes could repair the damage, using the nonradioactive ingredients of the plating medium. Even double-stranded breaks could presumably be repaired, since there is ample evidence that incomplete fragments of DNA can replicate (cf. Kozinski and Kozinski, 1968), and presumably these fragments could undergo a process akin to multiplicity reactivation, in order to reform full-length molecules.

Another attractive hypothesis for P^{32} decay stabilization has recently been advanced by Miller (1970b). Using an improved method for detection of DNA replication, Miller has noted a better correspondence than was previously observed between the onset of stabilization and the beginning of replication. Evidence for replication is seen as early as six minutes, about the same time as the most dramatic increase in stabilization occurs. To explain the significant stabilization which is seen earlier, at three minutes, Miller invokes some recent results obtained by himself and Kozinski (Miller and Kozinski, 1970), which show that parental phage DNA becomes complexed with phage-coded proteins as early as three minutes after infection. Miller reasons that such a complex might hold the severed DNA ends together after a double-strand scission had occurred, thereby increasing the probability that successful repair could take place.

Limited Genome Expression

Recently this phenomenon, originally discovered by Krieg (1959) has been subjected to intensive investigation by Snustad (1966a,b). In one study, an *E. coli* strain lysogenic for phage λ and therefore restrictive for growth of *r*II mutants was mixedly infected with a constant low multiplicity of wild-type T4 and varying multiplicities of an *r*II mutant. Snustad found that the higher the MOI of the *r*II mutant, the lower the fraction of infected cells which could be scored as infective centers. Similar results were noted in mixed infections of a suppressor-negative strain with wild-type T4 (constant MOI) and several amber mutants (varying MOI). Snustad interpreted his data in terms of a limited number of parental genomes which can express themselves inside an infected cell, with those genomes which do express being drawn at random from the parental genomes. For example, when the number of *r*II genomes present inside the restrictive strain is much higher than the number of r^+ genomes, then the latter genomes fail to express, and expression of the *r*II genomes leads to failure of phage development. Snustad analyzed his data to estimate the actual number of phage genomes allowed to express; the number was four in one strain and six in another. These results were interpreted in terms of competition of input genomes for sites of expression, or "slots." Snustad proposed that these slots might be membrane attachment sites to which phage DNA must bind in order to be expressed and to replicate. Certainly evidence is accumulating from many sources (see for example p. 47, p. 140, and p. 276) that both bacterial and viral chromosomes are expressed through some type of membrane-bound site, although the nature of the site remains ill-defined.

It is well-known that the number of replicating bacterial chromosomes, and presumably the membrane attachment sites as well, depend strongly upon the nutritional state of the bacterium: the more rapid the bacterial growth rate, the greater the number of simultaneously replicating chromosomes per cell (Lark, 1966; Cooper and Helmstetter, 1968). Snustad (1969) attempted to test his "slot" hypothesis by determining the number of slots in the same bacterial strain under widely varying physiological conditions. This appeared to be identical under all conditions tested, which would appear to necessitate a reevaluation of the slot model. Snustad proposed that the slots might be located near the cell ends and thus be relatively independent of growth rate, since cell length appears to be the principally varying parameter. However, in view of the evidence on relation of replicating bacterial chromosomes to growth rate, this proposal does not appear to be tenable, unless one invokes further *ad hoc* hypotheses concerning the differential availability of existing slots for bacterial and viral genome expression (presumably a slowly growing bacterium would need some mechanism to prevent utilization of all of its two to eight slots, if it contains only one replicating chromosome).

As a by-product of his experimental approach, Snustad (1968) was able to tell if the product of an essential gene is required in "catalytic" or "stoichiometric" amounts. In testing a large number of T4 amber mutants, he found two distinctly different forms of behavior. One class of mutants was co-dominant with wild-type: the burst size in unequal-input mixed infections decreased sharply with increasing MOI of the mutant. The other class was recessive to wild-type, with the burst size showing relatively little dependence upon MOI of the amber mutant. Snustad suggested that this latter class of mutants bears defects in catalytic functions, i.e., in genes coding for products such as enzymes involved in DNA replication. Here the expression of a wild-type genome can be severely limited, as at high mutant MOI, but the functioning of the gene product is affected only slightly, since it can function catalytically. By contrast, if a gene product is, for example, a structural component, then limitation of gene expression is expected to have a direct influence on burst size, and this will then be a very sensitive function of mutant MOI. With this as a basis for classification, Snustad tested at least one amber mutant in each of the genes known to bear such mutations. As expected, those genes dealing primarily with early functions, such as DNA synthesis, were largely of the "catalytic" type, while most of the late-functioning genes were "stoichiometric." However, there were exceptions and some intermediate cases, and these should be helpful in probing the biological functions of these genes.

6

Enzymology of T-Even Phage-Infected Bacteria

In Chapter 4 we discussed the discovery of deoxycytidylate hydroxymethylase and thymidylate synthetase, the first known phage-induced enzymes,* and in Chapter 5 we alluded to the type of evidence used in identifying structural genes for virus-specific enzymes. Within a few years after the earliest paper of Flaks and Cohen (1957) on phage-induced enzymes, several other enzyme activities were found to be either increased or apparently created in T-even phage-infected bacteria. Most of these were enzymes whose role in phage DNA metabolism was more-or-less obvious: for example, a kinase responsible for phosphorylation of the virus-specific compound, hydroxymethyldeoxycytidylate; a pyrophosphatase capable of cleaving deoxycytidine triphosphate, apparently accounting for its absence from viral DNA; and enzymes capable of transferring glucose residues from uridine diphosphate glucose to newly synthesized polynucleotide chains. However, the metabolic roles of a number of more recently discovered enzymes are less obvious. In considering the functions of some of

* It should be observed that the appearance of another phage-specific enzyme, lysozyme, had been strongly inferred nearly thirty years previously (Sertic, 1929). However, at that time lysozyme had not been identified as an enzyme, capable of catalyzing a known reaction. For further discussion of this point, see Cohen (1968) p. 68.

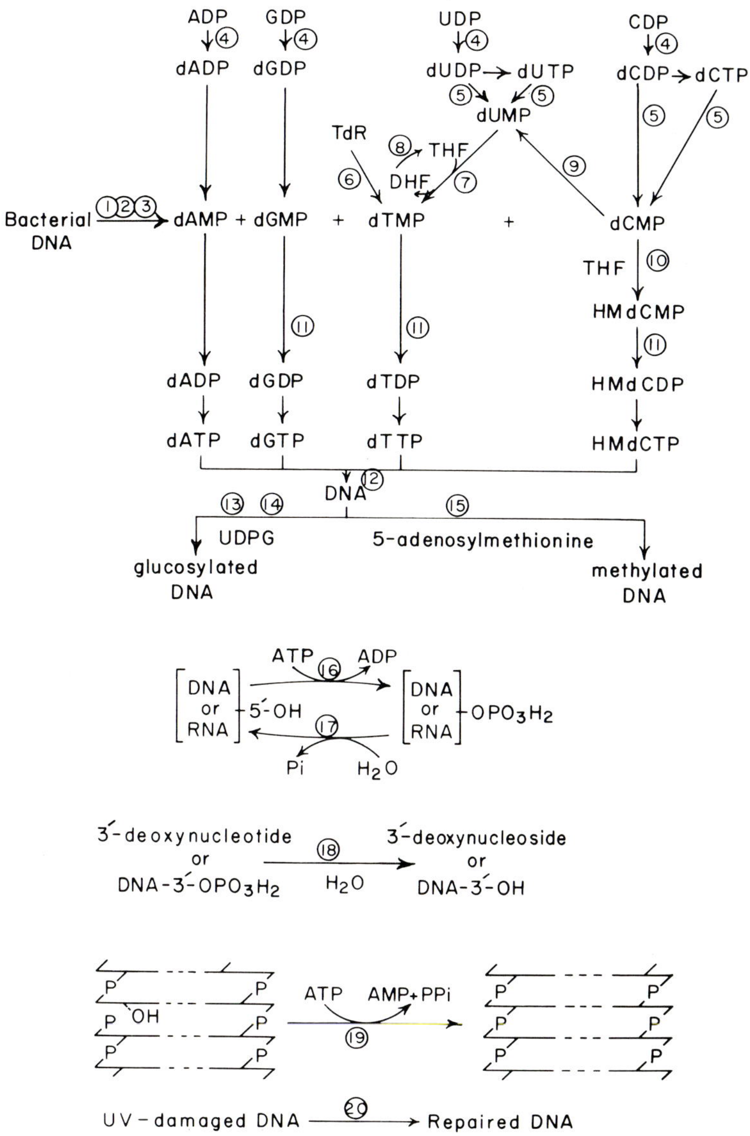

FIGURE 6-1 Enzymes of DNA metabolism which are induced in infection by phage T4. 1, 2, and 3, deoxyribonucleases; 4, ribonucleoside diphosphate reductase (there are at least two phage-specific components to this enzyme system); 5, deoxyuridine and deoxycytidine di- and triphosphatase; 6, thymidine kinase; 7, thymidylate synthetase; 8, dihydrofolate reductase; 9, deoxycytidylate deaminase; 10, deoxycytidylate hydroxymethylase; 11, deoxyribonucleotide kinase; 12, DNA polymerase; 13 and 14, glucosyl transferases; 15, DNA methylase; 16, polynucleotide kinase; 17, 5′-polynucleotide phosphatase; 18, 3′-deoxynucleotidase; 19, polynucleotide ligase; 20, DNA repair enzyme (probably containing more than one component, one of which is an endonuclease).

these enzymes, one is faced very directly with the task of extrapolating from *in vitro* studies with partially purified enzymes and trying to decide what duties an enzyme performs in the intact cell.

Figure 6–1 presents in diagrammatic fashion all of the known reactions catalyzed by virus-specific enzymes in T4-infected bacteria. About half of the enzymes are concerned more-or-less directly with the biosynthesis of nucleotides, particularly pyrimidine deoxyribonucleotides, while the remainder appear to participate in DNA metabolism at the polynucleotide level. It is in this latter group of enzymes that most questions arise with regard to biological function. Theoretically, an enzyme whose substrate is a polydeoxynucleotide can be participating in replication, repair of radiation damage, genetic recombination, or other functions which we have not yet discussed, such as host-induced modification (see p. 124). Since our knowledge of most of these processes at the enzymatic level is far from complete, the task of assigning a metabolic function to a newly-discovered enzyme is a difficult one. The task is eased somewhat if one can isolate a phage mutant deficient in its ability to induce a particular enzyme. However, some of these enzymes do not appear to be "essential" for replication, and screening of conditional lethal mutant stocks does not always

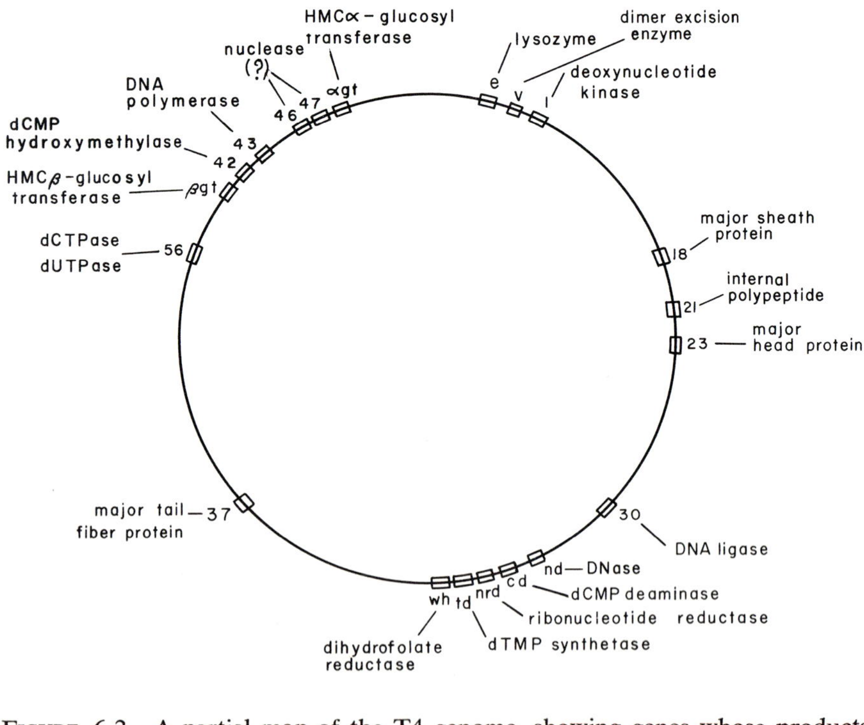

FIGURE 6-2 A partial map of the T4 genome, showing genes whose products have been identified.

turn up the desired mutants. One is then forced to turn to more difficult or ingenious screening and selection techniques.

It should be observed that all of the enzymes catalyzing the reactions listed in Figure 6–1 are synthesized primarily during the first 10 to 15 minutes after infection and are known collectively as the "early enzymes" (although, as we shall see in Chapter 7, they are not all synthesized within exactly the same portion of the latent period). A biochemical-genetic map of T4 is presented in Figure 6–2. This gives the map position of each gene in T4 which has been identified or inferred as a structural gene for a recognized early enzyme. Table 6–1 presents a list of the known early enzymes in T4 for which structural genes have not been found. We shall have occasion to refer to the information in this table and in Figures 6–1 and 6–2 as we proceed through our discussion of viral enzymology.

TABLE 6—1
T4-INDUCED ENZYMES FOR WHICH CONTROLLING GENES HAVE NOT BEEN DISCOVERED

Endonuclease IV
Exonuclease
Thymidine kinase
DNA methylase
Polynucleotide kinase
3′ Deoxynucleotidase
5′ Polynucleotide phosphatase

A comparison of Figure 6–2 with the complete genetic map of T4 (Figure 5–2) reveals the existence of a number of genes whose products are essential for DNA synthesis but which have not been characterized biochemically. Here is a clear opportunity for phage biochemists: if one can identify the biochemical function associated with one of these genes, then one has in all probability identified a new step in DNA replication, a process whose complexity appears to increase with the publication of each new paper (see, for example, contributions to Cold Spring Harbor Symposia on Quantitative Biology, vol. 33, 1968).

ENZYMES OF NUCLEOTIDE METABOLISM

Deoxycytidylate Hydroxymethylase

We have already discussed the discovery and metabolic role of this enzyme at some length. Although the enzyme has been highly purified, probably to near-homogeneity (Mathews, Brown, and Cohen, 1964), relatively little is known of its physical and chemical properties. Pizer and Cohen (1962) determined some kinetic constants for the T6 hydroxymethylase and estimated a molecular weight of 68,000 from sedimentation and diffusion studies of partially purified enzyme. Yeh and Greenberg (1967) have carried out some experiments on the mechanism of the reaction. The overall reaction requires tetrahydrofolate and formaldehyde, in addition to dCMP, and it had been assumed that the reactive species in transfer of the one-carbon group is 5,10-methylenetetrahydrofolate. However, Yeh and Greenberg showed that the first step is a tetrahydrofolate-dependent labilization of the hydrogen at position 5 of dCMP, as shown by an enzyme-catalyzed exchange reaction between dCMP-5-H^3 and water (see Figure 6–3). Whether the subsequent transfer of the single-carbon group involves formaldehyde or methylenetetrahydrofolate is unknown (presumably the true reactant *in vivo* is not formaldehyde *per se* but serine, or some other biological one-carbon donor).

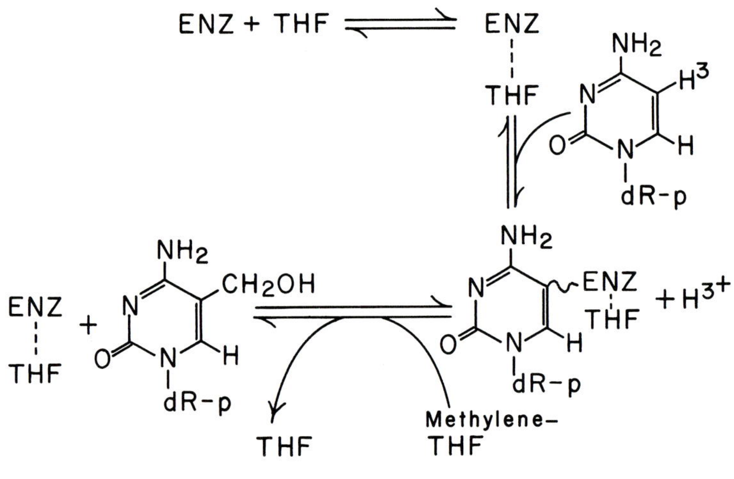

FIGURE 6-3 Steps in the deoxycytidylate hydroxymethylase reaction, as suggested by Yeh and Greenberg (1967).

As we have seen, the discovery of dCMP hydroxymethylase provided an answer to the question of the biogenesis of HMC in T-even phage DNA. T4 strains bearing *am* or *ts* mutations in gene 42 are unable to induce the hydroxymethylase and are also unable to synthesize DNA under restrictive conditions. Although it was assumed that the failure in DNA replication followed directly from the failure to accumulate a pool of hydroxymethyldeoxycytidylate (HMdCMP), Chiu and Greenberg (1968) have recently presented evidence suggesting that dCMP hydroxymethylase might play an additional role in DNA synthesis, as a component of some multienzyme replication complex. This evidence comes from work with T4 *ts*L13, a gene 42 mutant. Mathews and Kessin (1967) had shown that enzymatically active hydroxymethylase is formed in infection at 43°, a nonpermissive temperature, but they suggested that the enzyme is not active *in vivo* at that temperature. However, Chiu and Greenberg presented evidence that the enzyme can catalyze HMdCMP synthesis *in vivo* at 42°, suggesting that the mutant enzyme is defective in an additional, hitherto unrecognized function, possibly as a component of a replication complex. Although their evidence is persuasive, it is mostly quite indirect, involving measurements of DNA synthesis in various temperature-shift experiments. One hopes that they will present direct evidence, from measurements of nucleotide pool sizes, to substantiate their conclusion that the L13 hydroxymethylase is active *in vivo* at 42°.

Deoxynucleotide Kinase

The discovery of dCMP hydroxymethylase, an enzyme capable of generating an HMC nucleotide at the monophosphate level, led to a search for a kinase which could convert the monophosphate to a triphosphate, the proximal precursor of DNA-HMC. Such an activity was detected in T2-infected cells both by Kornberg et al. (1959) and by Somerville, Ebisuzaki, and Greenberg (1959). In addition, both Kornberg et al. and Bessman (1959) noted that the levels of dGMP kinase and dTMP kinase, both present in uninfected cells, increase severalfold after infection. The level of dAMP kinase, which is quite high in uninfected *E. coli,* does not increase after T-even phage infection. Bessman and Bello (1961, 1963a) provided strong evidence that the three phage-induced activities, namely, the phosphorylation of HMdCMP, dTMP, and dGMP, are catalyzed by a single enzyme. The partially purified T2-specific kinase converts all three nucleoside monophosphates to the diphosphate level, in contrast to the situation in uninfected *E. coli,* where a separate enzyme is responsible for the phosphorylation of each of the four common deoxyribonucleoside monophosphates, and there is no route for phosphorylating HMdCMP. The diphosphates are converted to the triphosphate level by an active and nonspecific kinase which is already present in uninfected cells and whose activity does not increase after infection (Bello and Bessman, 1963b).

Wiberg (1966) and Duckworth and Bessman (1967) found that T4 amber mutants defective in gene 1 failed to induce any of the three new kinase activities, and Duckworth and Bessman provided evidence that gene 1 is indeed the structural gene for a trifunctional deoxyribonucleotide kinase. Spontaneous revertants of gene 1 mutants regain all three kinase activities simultaneously. Moreover, the kinase formed by a gene 1 amber mutant in a permissive strain differed from the wild-type enzyme in heat stability and in pH optimum, indicating that suppression of the nonsense mutation has altered the structure of the enzyme.

Deoxynucleoside di- and triphosphatase (dCTPase-dUTPase)

The absence of cytosine from T-even phage DNA raised the question of how cytosine-containing DNA precursors are diverted to alternate metabolic pathways. Since the deoxynucleoside triphosphate, dCTP, is the most direct precursor, Kornberg et al. (1959) sought a phage-specific enzyme capable of cleaving this compound, and found in T2-infected cells a pyrophosphatase activity, which could cleave dCTP to give dCMP plus pyrophosphate. A similar activity was discovered by Koerner, Smith, and Buchanan (1959). Later Zimmerman and Kornberg (1961) found that a purified dCTPase preparation could also attack dCDP, giving dCMP plus orthophosphate. At about the same time, Greenberg and Somerville (1962) were studying an enzyme in uninfected *E. coli* which splits deoxyuridine triphosphate, dUTP, to the corresponding monophosphate, dUMP. Later Greenberg (1966) showed that the activity of this enzyme rises some tenfold after infection by T2. The new enzyme differed from the preexisting dUTPase in being able to cleave dUDP as well as dUTP. Simultaneously, Warner and Barnes (1966a) presented evidence that in T4 a single gene codes for an enzyme with activity against all four substrates: dCTP, dCDP, dUTP, and dUDP. dCTPase and dUTPase activities remained together through extensive purification. Moreover, amber mutants defective in gene 56 failed to induce the new dCDPase, dCTPase, and dUDPase activities, while there was no increase in the preexisting level of dUTPase. This suggests that gene 56 controls the synthesis of another polyfunctional enzyme, one capable of cleaving dCTP, dCDP, dUTP, and dUDP to the corresponding monophosphates. Direct confirmation of this point has recently been presented (Price and Warner, 1968; Munro and Wiberg, 1968). As with previous research identifying the products of genes 1 and 42, this work involved a comparison of the properties of the enzyme induced by gene 56 *ts* mutants or suppressed *am* mutants with the wild-type enzyme.

From the variety of reactions catalyzed by dCTPase-dUTPase, one can suggest a number of metabolic roles: exclusion of cytosine and uracil from phage DNA, or production of dCMP and dUMP as substrates for dCMP hydroxymethylase and dTMP synthetase, respectively. Since pyrimidine deoxyribonucleotides can also be generated *via* reduction of ribonucleoside diphosphates (see below), it is important to know whether this enzyme does play a significant role in biosynthesis. Gene 56 mutants are classified as DO, or DNA-negative, although they are capable of directing a very small amount of DNA synthesis (Warner and Hobbs, 1967; Wiberg, 1967). However, since dCTP is not destroyed in infection with these mutants, the small amount of DNA formed contains some cytosine and is subject to degradation by phage-specific nucleases (see below). Moreover, as we shall see presently, this DNA is nonfunctional even under conditions preventing its breakdown. Warner and Hobbs (1968) have investigated the metabolic roles of dCTPase-dUTPase directly, through measurement of deoxyribonucleotide pool sizes. As expected, dCTP and/or dCDP accumulate to a large extent in the absence of an enzyme to degrade them. However, no accumulation of dUTP was seen, suggesting that the host cell dUTPase suffices to exclude uracil from DNA. Moreover, HMC nucleotides were undetectable, suggesting that dCTPase-dUTPase provides the major pathway in T4-infected cells for biosynthesis of dCMP, essential as a substrate for dCMP hydroxymethylase. Thus, the enzyme appears to be essential both for cytosine exclusion and for dCMP synthesis.

Price and Warner (1969) have extended their studies on the dCTP-dUTP-ase to all three T-even phages. The three preparations all have the same catalytic specificity. However, significant physical differences were noted. The T4 enzyme has a molecular weight of approximately 59,000 and is strongly acidic, whereas the T2 and T6 enzymes are less acidic and have molecular weights close to 51,000.

Deoxycytidylate Deaminase

This enzyme, which is not found in uninfected *E. coli,* appears after infection with T-even phages (Keck, Mahler, and Fraser, 1959). The enzyme catalyzes the hydrolytic deamination of dCMP, to give dUMP and ammonia. Thus, it appears to impart to the phage-infected cell a new pathway for synthesis of dUMP, needed for thymidylate biosynthesis. However, the enzyme is nonessential for phage replication, as shown by the isolation of T4 mutants unable to induce the enzyme (Hall and Tessman, 1966). The isolation procedure is nonspecific, and is suitable for isolation of mutants defective in any function which is nonessential for plaque formation. A T4 stock was heavily mutagenized with hydroxylamine, plaques were picked at random from survivors and used to prepare stocks, and their ability to induce the deaminase measured. Obviously, this procedure is cumbersome if one has to examine a large number of stocks before finding the desired mutant. However, the mutagenic dose was sufficiently heavy that most if not all of the survivors were expected to bear mutations in one or more nonessential genes. As it happens, three out of the first eight plaques picked by Hall and Tessman gave stocks which were deficient in deaminase induction. Mapping studies showed that the mutations in these three strains are located in the same portion of the genetic map, the so-called *cd* gene. It should be emphasized that mutants isolated by this type of procedure are likely to contain several mutations, and that, if one wishes to study the function of a single gene, one must eliminate the unwanted mutations by several cycles of backcrossing against wild-type.

Although it appears likely that the *cd* gene is the structural gene for dCMP deaminase, direct evidence bearing on this point has not yet been presented. Moreover, the metabolic function of the gene has not yet been explored in detail. Hall and Tessman did report that *cd* mutants give a burst size about half that of wild-type T4, suggesting that the presence of dCMP deaminase does allow an increased supply of dUMP, needed as a DNA precursor. However, the relatively small reduction in burst size would not be expected to alter the ability of the mutant phage to form plaques.

Deoxycytidylate deaminase from animal tissues is subject to complex regulatory effects, notably inhibition by thymidine triphosphate and activation by deoxycytidine triphosphate (Geraci, Rossi, and Scarano, 1967; Maley and Maley, 1964). Consideration of these effects suggests that the enzyme forms a central branch point in the biosynthesis of pyrimidine deoxynucleotides, and that control is necessary in order to ensure balanced production of dCTP and dTTP. It is of interest that the phage-specific dCMP deaminase is subject to similar feedback effects *in vitro* (Fleming and Bessman, 1965, 1967; Maley, Guarino, and Maley, 1967; Scocca, Panny, and Bessman, 1969). Coincident with activation or inhibition are substantial changes in sedimentation coefficient, suggesting that the enzyme can exist in two or more states of aggregation (Maley et al., 1967). dCMP deaminase is classified as a "nonessential" enzyme, since *cd* mutants are able to form plaques under standard laboratory conditions. However, the fact that complex control properties have been maintained in this enzyme throughout the course of evolution would suggest that the ability to synthesize the enzyme has positive survival value for the virus. In light of this, it would appear that authors should specify the context in which they apply the terms "essential" or "nonessential" to viral genes or enzymes.

Thymidylate Synthetase

We have already discussed the discovery of this enzyme in phage-infected bacteria by Flaks and Cohen (1957), and enzymological evidence that the enzyme is a viral gene product (Mathews and Cohen, 1963a). Genetic evidence was brought to bear upon this point when Simon and Tessman (1963) isolated "thymidine-requiring," or *td,* mutants of T4. These were selected as minute-plaque formers when a thymidylate synthetase-negative strain, *E. coli* B3, was used as plating bacteria, and the plating medium contained thymidine in growth-limiting amounts. Wild-type phage are able to overcome the limitation by inducing dTMP synthetase, whereas *td* mutants, being unable to induce the enzyme (Shapiro, Eigner, and Greenberg, 1965; Mathews, 1965), must rely for growth upon the small amount of thymidine available and, therefore, they form smaller plaques. Direct evidence has not yet been presented to show that *td* is a structural gene for dTMP synthetase, but this seems quite likely.

The question of the "essentiality" of phage-induced thymidylate synthetase is of some interest, inasmuch as the enzyme is present in uninfected *E. coli.* Simon and Tessman reported that one mutant, *td*8, gave the same burst size as wild-type T4 in infection of *E. coli,* even in the absence of thymidine, suggesting that the phage enzyme is dispensable for growth. However, Mathews (1965) found a two- to threefold reduction in burst size of *td*8 relative to wild-type when cells were diluted several thousand-fold shortly after infection. It was suggested that at high cell concentrations, cells can "cross-feed" each other, to compensate for a nutritional insufficiency of thymidine. Thus, dTMP synthetase presents another example of a phage-induced enzyme which is "nonessential" under one set of criteria, but which does permit optimal phage growth under other conditions. Or, to use the words of Simon and Tessman, "possibly its true worth cannot be fully appreciated outside such environments as colons and sewers."

The mechanism of the dTMP synthetase reaction is of some interest, inasmuch as this enzyme is the principal target for the antitumor effects of 5-fluorouracil and 5-fluorodeoxyuridine (Cohen et al., 1958). These fluorinated pyrimidines are converted to the corresponding 5′ deoxyribonucleoside monophosphate, which is a potent inhibitor of dTMP synthetase. Inhibition of the phage-induced enzyme is so effective as to be essentially stoichiometric (Mathews and Cohen, 1963a). This plus the relatively high level of the synthetase in infected cells makes it an attractive system for studying the effects of fluorinated pyrimidines at the enzymological level. The enzyme has been purified to near-homogeneity in the author's laboratory (unpublished work of J. Krupp and G. Lincoln). It has an approximate molecular weight of 55,000. Preliminary studies show that inhibition by 5-fluorodeoxyuridylate, while kinetically stoichiometric, is not physically irreversible.

Thymidylate synthetase from T2-infected cells catalyzes an exchange between the H at the 5 position of dUMP and water, similar to the exchange catalyzed by dCMP hydroxymethylase (Lomax and Greenberg, 1967). However, in contrast to the results with the hydroxymethylase, this exchange shows a strong dependence upon the presence of formaldehyde, suggesting that methylenetetrahydrofolate, rather than tetrahydrofolate, is the reactive species.

Another metabolic study of *td*8 was carried out by Mathews (1966a), who measured phage DNA synthesis in infected cultures of *E. coli* B3, a dTMP synthetase-negative host. In the presence of added thymine, DNA synthesis proceeds at a linear rate for over one hour (the cultures are lysis-inhibited), while in the absence of thymine, DNA synthesis continues until about twenty phage-equivalent units of DNA have accumulated per cell, and then ceases. Under the conditions of these experiments, about thirty phage-equivalent units of acid-soluble thymine-containing compounds are released per cell *via* degradation of the host-cell DNA. In the presence of FUDR, which inhibits breakdown of host-cell DNA, the extent of phage DNA synthesis is about ten phage-equivalents per cell, while about twenty phage-equivalents of thymine-containing compounds are released *via* bacterial DNA breakdown. Thus, within the limits of sensitivity of these experiments, we can conclude that both the phage and bacterial mutations are "nonleaky," and that the extent of DNA synthesis is partially dependent upon the extent of breakdown of bacterial DNA.

Dihydrofolate Reductase

Since the transfer of a one-carbon group in the dTMP synthetase reaction entails its reduction at the expense of tetrahydrofolate, Seymour Cohen and I wondered whether the increased rate of thymidylate synthesis in infected cells might lead to a requirement for increased regeneration of tetrahydrofolate from dihydrofolate. Accordingly, we examined the level of dihydrofolate reductase, which catalyzes the TPNH-dependent reduction of dihydrofolate to tetrahydrofolate, in T6-infected cells (Mathews and Cohen, 1963b). The level was some twentyfold higher than in uninfected *E. coli,* suggesting the induction of a virus-specific dihydrofolate reductase. In line with this idea was the observation that crude preparations of reductase from phage-infected cells could utilize either TPNH or, to a lesser extent, DPNH. The *E. coli* reductase, by contrast, has a strict requirement for TPNH. Kerstin Sutherland and I examined partially purified preparations of *E. coli* and T6-induced reductase (Mathews and Sutherland, 1965). Many differences were seen, notably in pH optimum, stability, and sedimentation rate in sucrose gradients. The bacterial enzyme has an estimated molecular weight of 22,000, while that of the T6 enzyme is about 31,000. Further biochemical evidence that the phage reductase is a viral gene product was presented by the author (Mathews, 1967a), who showed significant differences in stability and catalytic efficiency among the reductases induced by T2, T4, T5, and T6. The catalytic efficiencies of these preparations were inferred from turnover numbers calculated from aminopterin titration data. The phage reductase preparations are extremely sensitive to inhibition by aminopterin. Since dihydrofolate reductase is the target enzyme for the antileukemic action of aminopterin and the other folate analogs (cf. Bertino, Hillcoat, and Johns, 1967), it is important to understand the mechanism of aminopterin inhibition. The phage systems appear to present excellent material for this type of work. To this end, J. Erickson in the author's laboratory has purified the T4 reductase to homogeneity, and physical studies of aminopterin binding are in progress.

Genetic evidence on the viral origin of dihydrofolate reductase was provided through isolation of T4 mutants presumed to be defective in reductase induction (Hall, Tessman, and Karlstrom, 1967). The selection procedure used is rather involved and quite ingenious. It involves the use of a plating host which requires pyrimidines for growth and which lacks cytidine-deoxycytidine deaminase. The cells are plated, along with a portion of a mutagenized phage stock, in the presence of cytidine as the sole pyrimidine. Mutants are recognized by virtue of the formation of plaques surrounded by a thick white halo. Because of this, the mutants are called white, or *wh*. As expected, *wh* mutants are unable to induce dihydrofolate reductase (Mathews, 1967a,b; Hall, 1967). An explanation for this plating phenotype is as follows. The plating conditions impose a uracil deficiency, since the cells are defective in their ability to deaminate cytidine compounds, and cannot synthesize uracil nucleotides endogenously. The absence of phage-induced reductase in cells infected with *wh* mutants leads to a deficiency in thymidylate synthesis, leading to an accumulation of dUMP. This is excreted in some form which can "feed" the surrounding cells. Thus, the halo is formed by more luxuriant growth of uninfected cells in the vicinity of a *wh* plaque than elsewhere on the plate. In support of this conclusion is the following: (1) *td* mutants also give a halo under these plating conditions, although it is less thick than that of the *wh* strains; (2) *cd* × *wh* double mutants do not possess the white-halo phenotype, presumably because the presence of the *cd* mutation prevents accumulation of dUMP; and (3) *E. coli* B infected with a *wh* mutant does have a partial thymidylate deficiency, which can be circumvented by addition of exogenous thymine (Mathews, 1967b). Of course, it would be desirable to verify this conclusion by means of direct analysis of nucleotide pool sizes.

Hall (1967) has provided direct evidence that *wh* is a structural gene for dihydrofolate reductase. Among several *wh* mutants isolated was one which was partially suppressed in *E. coli* strains bearing amber suppressor genes. Dihydrofolate reductase was induced in *E. coli* CR63 following infection by this mutant, and it differed in its properties from the wild-type T4 enzyme.

In chapter 3 we mentioned the report from Kozloff's laboratory (Verses et al., 1968) that dihydrofolate reductase is a structural component of phage T4D. The precise structural role of this enzyme is not known, but Kozloff believes that it binds the viral dihydrofolate compound which has been localized in tail plates. In support of this, Kozloff (personal communication) has found that T4D can be reversibly inactivated by TPNH, and, to a lesser extent, by DPNH. Kozloff believes that this involves reduction of the viral dihydrofolate *in situ,* which results in loss of phage adsorbability. Unexpectedly a *wh* mutant is also inactivated under these conditions.

The above results would suggest that synthesis of dihydrofolate reductase is an essential function for formation of infective phage. Presumably, most *wh* mutations are of the missense type in which the enzyme structure is altered only slightly and can bind the phage dihydrofolate, even though enzymatic activity is lost. However, one would predict that an amber mutation in the *wh* gene would be lethal, inasmuch as a complete polypeptide chain is not made. This is at variance with Hall's isolation of such a mutant which appears to be as viable as the other *wh* mutants. Perhaps these opposing findings can be reconciled by proposing that the mutation in Hall's strain is near the end of the gene corresponding to the carboxyl terminus, such that a nearly-complete polypeptide chain is made, one which can play the essential structural role but is enzymatically inactive. On the other hand, Kozloff (personal communication) presently believes, improbable as it sounds, that phage particles can incorporate the *E. coli* reductase into their tail plates if the phage-coded enzyme is unavailable.

In a recently-published study of the phage-bound reductase, Kozloff et al., (1970) reported data in support of the above assertion. In one experiment, the T4 amber *wh* mutant, presumably containing the *E. coli* reductase, was found to be far more heat-stable than wild-type T4 strains. Since the activity of the *E. coli* reductase itself is more heat-stable than those of the phage-induced reductases (Mathews and Sutherland, 1965), this suggests that the virion-bound enzyme is the sensitive target for heat inactivation of these phages. In line with this, we have recently found that the heat stabilities of phages T2, T4, T5, and T6 are in the same relative order as those of their induced reductases (P. Gegenheimer and C. K. Mathews, unpublished observations).

Another role for the phage-induced reductase was suggested by Mathews and Cohen (1963b), on the basis of its ability to use either DPNH or TPNH. It was thought that the level of TPNH might decrease after infection, owing to decreased activity of the hexose monophosphate shunt (Cohen, 1951). Thus, the cell would require an enzyme which could utilize DPNH instead. This idea was tested directly by Mathews (1966b), who measured the levels of reduced pyridine nucleotides in T6-infected cells. Somewhat unexpectedly, the levels of both TPNH and DPNH were seen to rise briefly, and then return to the preinfection level. When infection was carried out in the presence of FUDR, the TPNH level did not return to the preinfection value, although the DPNH level did. Under these conditions the block in dTMP synthesis presumably leads to concomitant inactivity of dihydrofolate reductase. This suggests that TPNH is the true biological reductant for dihydrofolate, since it is not consumed when the reductase is turned off, while DPNH is used up normally.

In a discussion of these experiments, Cohen (1968) has pointed out that they were performed with a glucose-ammonia-salts medium, and that the use of a medium containing amino acids might more closely approximate natural conditions. Specifically, the use of a medium containing ammonia as sole nitrogen source is expected to induce large quantities of glutamate dehydrogenase, a TPNH-specific enzyme. We have found that cells infected in glycerol-casamino acids medium show the same changes as we earlier reported for glucose-ammonia medium (unpublished results).

Ribonucleotide Reductase

In their initial studies on RNA metabolism in T2-infected cells, Volkin and Astrachan (1956, 1958) had shown that P^{32} incorporated into viral RNA is subsequently incorporated efficiently into DNA. Since T2 RNA is rapidly degraded, and since the substrates for DNA synthesis are deoxyribonucleotides, it was reasonable to propose that the phage-infected cell possess a high capacity for reduction of ribonucleotides to deoxyribonucleotides. This was not immediately apparent in the 1950s, since little was known then about the enzymology of deoxyribose synthesis. However, Cohen, Barner, and Lichtenstein (1961) found that labeled T6-specific RNA is rapidly converted to deoxyribonucleotides in crude cell-free extracts. Sekiguchi and Cohen (1963) showed that a major pathway for degradation of phage-specific RNA involves attack by polynucleotide phosphorylase, an enzyme which generates ribonucleotides at the diphosphate level. At about the same time, Reichard and his colleagues were investigating a complex enzyme from *E. coli* which reduces ribonucleotides to deoxyribonucleotides, with ribonucleoside diphosphates being the preferred substrates (Reichard, Baldesten, and Rutberg, 1961). Thus, breakdown of phage-specific RNA catalyzed by polynucleotide phosphorylase appears to provide a supply of substrates for direct reduction to the deoxy level.

The above studies did not provide a clear indication of whether the efficient reduction of ribonucleotides was catalyzed by the bacterial system or whether a new enzyme was involved. Cohen and Barner (1962) showed that the activity of a crude system for reduction of cytidylic acid increases manyfold after infection by T6. The kinetics of appearance of this new activity were quite different from the timing of synthesis of other phage-induced enzymes. Whereas the formation of a "typical" early enzyme, such as dCMP hydroxymethylase, begins almost immediately after infection, the formation of the new CMP reductive system did not begin until between 5 and 10 minutes after infection. Teleologically this was satisfying, inasmuch as the release of nucleotides from phage mRNA does not begin immediately, and the appearance of the new reductase activity appears to be delayed until it is needed. However, this result did raise the question of whether the increase in activity really did result from synthesis of a viral enzyme, or whether it reflected some other phenomenon, such as activation of the preexisting bacterial enzyme. Biswas, Hardy, and Beck (1965) presented confirmation of the results of Cohen and Barner, using T2-infected cells, but it was not proven until 1969 that the increased reductase activity does result from synthesis of new enzyme protein (Yeh, Dubovi, and Tessman, 1969; Berglund, Karlström, and Reichard, 1969).

As with most of the virus-specific enzymes, both biochemistry and genetics were used to prove that T-even phages direct the synthesis of a new ribonucleotide reductase. Yeh et al. (1969) isolated T4 mutants (called *nrd*) unable to induce the new enzyme, using two methods: the nonspecific "brute force" method earlier used with success to isolate *cd* mutants, and a more sophisticated method involving suppression of the white-halo phenotype in a *wh* mutant. As with *cd* mutants, *nrd* × *wh* double mutants do not give the white halo characteristic of *wh* mutants, presumably because, like *cd* mutants, they fail to accumulate excess dUMP. As expected, *nrd* mutants are defective in ribonucleotide reductase induction, but can form other early enzymes normally. Complementation studies of the mutants isolated revealed the existence of at least two cistrons, suggesting that formation of the new reductase system involves the synthesis of two or more proteins.

Some information on the enzymology of the new reductase system has come from the studies of Berglund et al. (1969). In uninfected *E. coli,* a single enzyme system is responsible for reduction of all four common ribonucleoside diphosphates (cf. Reichard, 1968). The enzyme system is complex, involving the participation of four protein components. Two of these participate in the actual reduction, while a small, heat-stable protein, called thioredoxin, provides the electrons for reduction through reversible oxidation of two cysteinyl residues. The fourth protein, thioredoxin reductase, catalyzes the TPNH-specific reduction of oxidized thioredoxin as shown below. The T4-induced system contains two components, as determined in studies with partially purified enzymes. These are called A and B.

$$\text{Riboside diphosphate} + \text{thioredoxin (SH)}_2 \xrightarrow[\text{(B1 and B2)}]{\text{2 protein fractions}} \text{deoxyriboside diphosphate} + \text{thioredoxin (S-S)} \quad (1)$$

$$\text{Thioredoxin (S-S)} + \text{TPNH} + \text{H}^+ \xrightarrow[\text{reductase}]{\text{thioredoxin}} \text{thioredoxin (SH)}_2 + \text{TPN}^+ \quad (2)$$

The system requires thioredoxin reductase of *E. coli* (or a dithiol, which can chemically reduce thioredoxin (S-S)), but there is no evidence for involvement of *E. coli* thioredoxin. Rather, component B is a phage-specific thioredoxin (Berglund, 1969), while component A is the viral reductase *per se.*

E. coli ribonucleotide reductase is subject to complex activation and inhibition patterns by nucleoside triphosphates. The phage system is also subject to feedback effects, but they are different from those seen with the bacterial enzyme. It should be interesting to learn the biological significance of the altered feedback effects.

The availability of *nrd* mutants allows one to ask rather direct questions about the metabolic role of the virus-specific ribonucleotide reductase system. To date, such investigations have not been reported. However, Warner and Hobbs (1969) have gained some insight through a study of the effects of hydroxyurea on phage growth and DNA replication. This agent has been shown to block DNA synthesis, probably by inhibition of ribonucleotide reductase (Young, Schoechetman, and Karnofsky, 1967; Krakoff, Brown, and Reichard, 1968). DNA synthesis in T4-infected cells is much more sensitive to inhibition than in uninfected *E. coli.* Deoxyribonucleotide biosynthesis is also inhibited in hydroxyurea-treated cells. These findings suggest that the T4 reductase system is more sensitive to inhibition than the *E. coli* system, and that the phage system is utilized preferentially in infected cells.

As seen in Figure 6–2, the *cd, wh, td, nd,* and *nrd* genes all map together in a distinct region of the T4 genome, between gene 32 and the *r*III gene. These genes can be considered functionally as the "thymidylate region," since all of the gene products are involved in thymidylate metabolism. The involvement of T4 genes in pyrimidine metabolism is indicated below. The clustering of at least

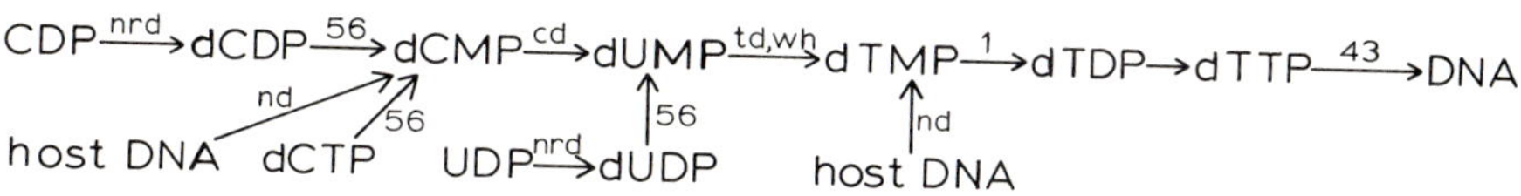

five of these genes is a phenomenon of some interest, even though this region does not contain all of the genes controlling thymidylate metabolism. It has been suggested (Hall, 1967; Yeh et al. 1969) that these five genes might represent a unit for coordinated gene expression. This idea, while superficially attractive, is difficult to defend in view of much evidence that the expression of these genes is not coordinate. Dihydrofolate reductase activity appears more rapidly than that of other phage-induced enzymes (Mathews and Cohen, 1963b; Warner and Lewis, 1966). Moreover, while most early enzymes are synthesized in excessive amounts in infection by DO amber mutants (see p. 158), dihydrofolate reductase is subject to normal control under these conditions (Mathews, 1966a; Warner and Lewis, 1966). Thymidylate synthetase has been reported to be synthesized slightly later than the other early enzymes (Sekiguchi and Cohen, 1964) and we have already mentioned the lag in onset of ribonucleotide reductase appearance, first noted by Cohen and Barner, but more recently observed by Yeh et al. (1969). Some of these data on kinetics of enzyme induction are indicated in Figure 6–4. Whereas some joint control of the enzymes of the "thymidylate region" might be possible, the form of this control is certainly quite different from the mechanism of coordinate gene expression seen in bacterial operons.

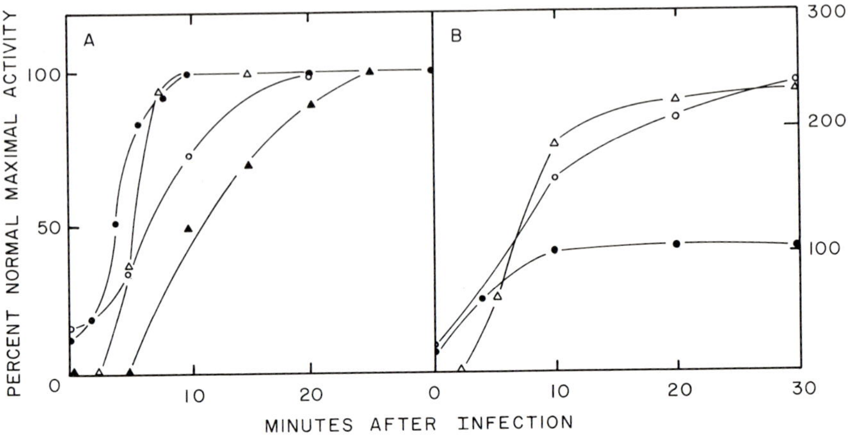

FIGURE 6-4 Kinetics of T-even phage-induced enzyme synthesis. Closed circles, dihydrofolate reductase; open circles, thymidylate synthetase; closed triangles, ribonucleotide reductase; open triangles, deoxycytidylate hydroxymethylase. Panel A. Normal infection. Combined data from Mathews and Cohen (1963b) and Yeh, Dubovi, and Tessman (1969). Panel B. Infection with T4 *am*N82, a DNA-negative mutant. Data from Mathews (1966a).

Thymidine Kinase

Hiraga, Igarashi, and Yura (1967) presented evidence that infection with T2 or T4 leads to induction of a new thymidine kinase, which can be distinguished from the corresponding bacterial enzyme by virtue of its greater heat lability. Induction of the enzyme was also seen in infection of an *E. coli* mutant which lacks thymidine kinase, suggesting that the new enzyme is a viral gene product. However, Okazaki and Kornberg (1964) had earlier reported that there is no change in the activity of this enzyme upon T2 induction. To date these contradictory reports have not been reconciled.

ENZYMES OF POLYNUCLEOTIDE METABOLISM

Deoxyribonucleases

The literature on T-even phage-induced deoxyribonucleases (DNases) is in a somewhat unsatisfactory state. In part, this results from the large number of nucleases induced in infected cells: at least five, and probably more. With this large number of new nucleases, plus the six or more already present in *E. coli* (cf. Shortman and Lehman, 1964; Jorgensen and Koerner, 1965), it is difficult to determine the precise metabolic function of each enzyme. For example, we have already observed that bacterial DNA breaks down after T-even infection. However, we still do not know whether the enzymes responsible are exclusively virus-specific, or whether phage infection somehow unmasks the activity of one or more bacterial nucleases.

The original observation that T2 infection increases the intracellular DNase activity was made by Pardee and Williams (1952). This activation was associated with infection by intact phage (Kunkee and Pardee, 1956). However, the earliest proposal advanced to explain DNase activation was a phage-associated destruction of an RNA fraction which inhibits a bacterial DNase (Kozloff, 1953; Lehman, Roussos, and Pratt, 1962). However, Stone and Burton (1962) presented evidence indicating that the activation of DNase is due to the synthesis of new enzyme protein. Moreover, the level of inhibitory RNA did not change after infection.

Fractionation of phage-infected cell extracts has directly revealed the existence of at least five new DNase activities, specific to the virus-infected cell. These include two exonucleases: one with greatest activity toward partially degraded DNA (Oleson and Koerner, 1964; Short and Koerner, 1965, 1969), and another exonuclease activity which is closely associated with phage-specific DNA polymerase (Short and Koerner, 1965; Goulian, Lucas, and Kornberg, 1968; Nossal, 1969; cf. below). The precise number of distinct endonucleases is somewhat difficult to determine from the literature. Bose and Nossal (1964) described an endonuclease activity from T2-infected cells which preferentially attacks denatured DNA to give 5′ phosphoryl-terminated oligonucleotides. This enzyme is inhibited by a tRNA preparation from *E. coli.* More recently Hurwitz and his colleagues (Feiner et al., 1968; Sadowski et al., 1968; Sadowski and Hurwitz, 1969a, b) have described two T4-specific endonucleases, which they call endonucleases II and IV. Endonuclease II induces single-strand breaks in native DNA, to give 3′ hydroxyl and 5′ phosphoryl termini. Endonuclease IV attacks single-stranded DNA, also giving 3′ hydroxyl and 5′ phosphoryl termini. Neither enzyme has any activity on T4 DNA, whether glucosylated or not. One might expect endonuclease IV to be the same as the endonuclease of Bose and Nossal. However, it is not inhibited by tRNA, as is the enzyme of Bose and Nossal. On the other hand, this might reflect a species difference, since Bose and Nossal were working with T2, while Sadowski et al. were working with T4.

When the report of Sadowski et al. appeared, it seemed reasonable to propose that endonuclease II is the same as one of two T4-specific "nicking enzymes" described by Ando et al. (1969). These enzymes both catalyze reactions leading to 3′ hydroxyl-terminated single-strand breaks in double-stranded DNA. However, in a more recent report, Ando et al. (1970) have reported that both of their enzymes are more active against T4 DNA than *E. coli* DNA. One of the enzymes is specific for non-glucosylated T4 DNA, while the other can attack either glucosylated or nonglucosylated phage DNA. At any rate, the preference of both enzymes

the possibility that the products of these genes form two necessary but non-identical subunits of the same enzyme. Whether this enzyme is a hitherto undiscovered nuclease, or whether it performs some other function, are unknown at this time. Any proposal which is advanced to explain the functions of these genes must take into account the finding of Bernstein (1968) that crosses involving gene 46 and 47 mutants under restrictive conditions show greatly reduced recombination frequencies when compared to those performed under permissive conditions (although Kutter and Wiberg have reported that recombination by gene 46 or 47 mutants is unimpaired).

Quite recently Jorgensen et al. (1970) have isolated another class of T4 mutants which are defective in host cell DNA degradation. These were isolated on the basis of their sensitivity to hydroxyurea, which, it will be recalled, inhibits ribonucleotide reductase (p. 115). Thus, this agent renders infected cells dependent upon bacterial DNA breakdown as the sole source of deoxyribonucleotides (Warner and Hobbs, 1969). Therefore, the breakdown-negative mutants of Jorgensen et al. cannot replicate in the presence of hydroxyurea because of deoxyribonucleotide deficiency. These mutants all map in a single region, the *nd* gene, which is located between genes 32 and 63. Unlike genes 46 and 47, the *nd* function is dispensable for growth, since *nd* mutants can grow in the absence of hydroxyurea. Enzyme assays show that *nd* mutants are defective in the induction of the endonuclease II described by Sadowski and Hurwitz (1969a).

Hercules et al. (1970) have also isolated *nd* mutants, by use of a different technique, and have presented results in substantial agreement with those described above.

DNA Polymerase

In their original paper describing several new enzyme activities in T2-infected cells Kornberg et al. (1959) showed that the level of DNA polymerase, already present in uninfected cells, rises eight- to tenfold in infected cells. Use of hydroxymethyldeoxycytidine triphosphate as a substrate was necessary in these assays, for dCTP was rapidly destroyed by the dCTPase activity present in crude extracts, such that no polymerization was seen. Aposhian and Kornberg (1962) showed that the new polymerase activity could be separated from *E. coli* polymerase and shown to be different by several criteria. The T4-induced DNA polymerase was highly purified by Goulian, Lucas, and Kornberg (1968). It has a molecular weight of 112,000, similar to the molecular weight of the *E. coli* polymerase (109,000). However, it differs from the bacterial enzyme in being unable to promote extensive replication of DNA *in vitro*. The T4 enzyme requires single-stranded DNA in order to be active *in vitro*. The input DNA is replicated once, but the enzyme is unable to initiate new cycles of replication, using the double-stranded DNA which is formed as the product. As we have already noted, the purified T4 DNA polymerase retains an exonuclease activity which seems to be specifically associated with the polymerase molecule. The role of this exonuclease is not known. However, it may be involved in initiation of replication *in vivo,* for this process appears to require the presence of a single-strand break in a double-stranded DNA (cf. Kornberg, 1969). Or, the exonuclease may participate in a repair function, such as excision of faulty material which is inserted by occasional incorrect action of the polymerase.

Amber mutants of T4 defective in gene 43 are unable to induce DNA polymerase (deWaard, Paul, and Lehman, 1965; Warner and Barnes, 1966b). A gene 43 *ts* mutant was shown to induce a DNA polymerase of greater heat lability than the wild-type enzyme, proving that gene 43 is a structural gene for DNA polymerase and not a regulatory gene (deWaard et al., 1965). Most gene 43 mutants are defective in both DNA polymerase and the polymerase-associated nuclease (Nossal, 1969), in confirmation of enzymological data suggesting that the two activities are located on the same protein molecule. However, three gene 43 amber mutants, all of which map at the same site, are able to induce the nuclease, but not the polymerase. These mutations are seen as inactivating the polymerase active site without affecting that of the nuclease.

Ever since the discovery of DNA polymerase in the late 1950s (cf. Kornberg, 1961), some controversy has existed regarding its precise role in replication, indeed, whether it is involved at all. In part, this arose from the fact that DNA polymerase can only catalyze polynucleotide chain extension in one direction. Since the enzyme adds 5′ nucleotides at 3′ hydroxyl termini, but has no activity with 3′ deoxynucleoside triphosphates, it can only replicate in the 5′ to 3′ direction. Yet we know from a variety of systems that both strands of DNA are replicated simultaneously; this must involve chain extension in both directions. Another reason for the lack of universal acceptance of DNA polymerase as the replication enzyme was the fact that the reactions it catalyzed *in vitro* could just as easily be visualized as participating in recombination or in repair to damage caused to DNA, as by ultraviolet irradiation. Speyer (1965) presented data which indicated a direct involvement of T4 DNA polymerase in replication. He reasoned that the polymerase made by a *ts* gene 43 mutant, even though it is active at 30°, might not be as well-designed a catalyst as the wild-type enzyme. Specifically, he proposed that a mutant enzyme might select incorrect nucleotides for insertion into replicating DNA more often than the wild-type polymerase, and that this would be reflected in higher mutation frequencies in other genes. Speyer tested this by preparing a series of double mutants containing an *r*II marker and different *ts* mutations in gene 43. Spontaneous reversion of the *r*II mutation was measured by determining the relative plating efficiencies of the double mutants on *E. coli* K12(λ), a restrictive strain, and on *E. coli* B. The relative plating efficiency for the *r*II mutant alone was 6×10^{-9}, while one of the double mutants showed a reversion frequency, when grown at 27°, of 8×10^{-6}, more than one thousandfold higher. When the temperature was increased to 37°, this value rose to 2×10^{-5}. Not all of the *ts* gene 43 mutations were mutagenic. However, the fact that some of them were is strong evidence that DNA polymerase plays a direct role in base selection. Of course, it is conceivable that faulty base insertion occurs while the mutant polymerase is involved in repair or recombination. However, Speyer and Rosenberg (1968) have recently presented evidence that would appear to rule out the possibility that all of the induced mutations arise during either recombination or repair.

Not only are some gene 43 *ts* mutations mutagenic, but some are antimutagenic, giving lower reversion frequencies than seen when the wild-type polymerase functions (Drake et al., 1969). Why the antimutagenic polymerase has not been selected for by evolution is not clear. Presumably the inherent advantage gained to a virus by lowering the spontaneous mutation rate is offset by some other change which imposes a selective disadvantage; for example, perhaps the rate at which an anti-mutagenic polymerase can function is lowered. To date this question has not been extensively investigated.

Finally, in an extension of the above studies, Drake and Greening (1970) have shown that at least some of the antimutagenic T4 DNA polymerases can suppress the effects of chemical mutagens as well as lowering the spontaneous mutation rate.

Hydroxymethylcytosine-glucosyl Transferases

As we mentioned in Chapter 3, T-even phage DNA is glucosylated, with glucose residues being attached through the hydroxymethyl group of HMC. The glucosylation reaction involves a transfer of the sugar from uridine diphosphate glucose (UDPG) to HMC residues in preformed polynucleotide chains, rather than to free nucleotides (Kornberg et al., 1959). The glucosyl transferases which catalyze this process are induced early in T-even phage infection, and the types of enzymes formed by T2, T4, and T6 conform rather closely to the observed pattern of glucosylation of viral DNA in each phage. For example, all three DNAs contain some glucose residues bound in the α-configuration, while only T4 DNA contains any β-glucosyl residues. All three phages induce an HMC-α-glucosyl transferase, while T4 induces an additional HMC-β-glucosyl transferase (Zimmerman, Kornberg, and Kornberg, 1962; Josse and Kornberg, 1962). T6, it may be recalled, has some 72% of its HMC residues diglucosylated, and, for this purpose, it induces a second enzyme which transfers glucose in the β-configuration to monoglucosylated HMC residues (Kornberg, Zimmerman, and Kornberg, 1961). Although T2 DNA contains 5 percent of its HMC residues in the diglucosylated form, the mechanism of addition of the second glucose has not yet been clarified at the enzymatic level.

The question of the specificity of the glucosyl transferases has received some attention, particularly the questions of why some 25 percent of the HMC residues of T2 and T6 DNA remain unglucosylated and what determines the 70:30 ratio of α- to β-glucosyl linkages in T4 DNA. Each α-glucosyl transferase appears to recognize specific HMC residues. For example, T2 DNA, which has presumably been glucosylated *in vivo* to the limit of the T2 transferase, will not serve as a substrate for the T2 transferase *in vitro*. However, T2 DNA can be further glucosylated by either the T4 or the T6 transferases (Kornberg, Zimmerman, and Kornberg, 1961). Some aspect of DNA secondary structure seems to be a factor in determining whether a particular HMC residue will serve as substrate for a particular transferase, for native DNAs are far better substrates *in vitro* than heat-denatured material (Josse and Kornberg, 1962).

The question of specificity of the T4 glucosyl transferases has been investigated by de Waard, Ubbink, and Beukman (1967). Unglucosylated DNA, prepared by the use of a transferase-negative T6 mutant (see below) was incubated with the α- and β-transferases under different conditions and the products subjected to chemical and enzymatic degradation. It was found that the α-glucosyl transferase cannot glucosylate HMC residues which are linked through the 5′ carbon to another HMC nucleotide. However, the enzyme can react with all other possible sequences, and the β-enzyme seems to have no similar limit on specificity. It had been known for some time that the α-glucosyl transferase can glucosylate *in vitro* positions which are normally β-glucosylated, and *vice versa* (Josse and Kornberg, 1962). de Waard et al. presented evidence that the intracellular magnesium concentration may regulate the ratio of α- to β-glucosyl residues formed *in vivo*. The relative proportion of α- and β-linkages in their *in vitro* product varied widely with the magnesium concentration of the incubation mixture. No such differences were seen *in vivo* when the magnesium concentration of the growth medium was varied widely. However, the intracellular Mg^{++} concentration is probably maintained at a constant level. In this context, it would be of interest to examine the relative extents of glucosylation *in vivo* in cells infected by T4 mutants, such as the DNA delay mutants, which appear to have defects in regulation of internal cation concentrations.

Since the substrate for the glucosyl-transferases is a preformed polynucleotide chain, it follows that nascent DNA synthesized *in vivo* is unglucosylated. Erikson and Szybalski (1964) were able to differentiate between glucosylated and unglucosylated DNAs on the basis of density in a cesium sulfate equilibrium density gradient. Using this criterion, they showed that glucosylation of T6 DNA *in vivo* commences about nine minutes after infection, some two minutes later than the onset of DNA replication.

The question of the biological significance of glucosylation of DNA has received a good deal of attention, particularly insofar as it has presented an explanation for one case of *host-induced modification,* originally discovered by Luria and Human (1952). This phenomenon is somewhat easier to describe by example than to define generally. Luria and Human found that growth of T2 and T6 on certain *E. coli* strains gave rise to progeny phage which were unable to form plaques on *E. coli* B, but which could grow in *Shigella dysenteriae*. Subsequently, it was found that the *E. coli* strains exhibiting this phenomenon were deficient in their ability to synthesize UDPG (Hattman and Fukasawa, 1963; Shedlovsky and Brenner, 1963; Symonds et al., 1963). Thus, phage grown in these strains are unable to glucosylate their DNA. When these phage, bearing unglucosylated DNA, are used to infect *E. coli* B, infection is abortive, in all probability because the DNA is subject to enzymatic degradation, probably by bacterial nucleases. However, *Shigella,* having a different complement of nucleases, can support replication of unglucosylated DNA without breaking it down. Thus, glucosylation appears to provide the function of stabilization of DNA in the environment of a normal host, namely, *E. coli.*

The involvement of bacterial, rather than phage-induced, nucleases in restriction of host-modified phage is indicated by an experiment of Molholt and Fraser (1965). Cells of a restrictive strain of *E. coli* were converted to spheroplasts, a treatment which removes the cell wall and causes release of a number of hydrolytic enzymes, including nucleases, which are located in the "periplasmic space," between the cell wall and the membrane (Neu and Heppel, 1964). Such spheroplasts did support growth of host-modified phage, suggesting that the enzymes involved in degradation of unglucosylated DNA have been removed during formation of the spheroplasts. In line with the suggestion of an involvement of bacterial nucleases in restriction is the finding of Richardson (1966) that two of the known *E. coli* nucleases are far more active *in vitro* with unglucosylated T-even phage DNA. However, the exact bacterial nuclease(s) responsible for restriction in this system has not yet been identified. Of four well-characterized *E. coli* nucleases—endonuclease I and exonucleases I, II, and III—restrictive and permissive strains, including *Shigella* all have comparable levels. Moreover, mutants lacking endonuclease I are fully capable of restricting the growth of host-modified phage (Revel, 1967; Eigner and Block, 1968). Whether restriction involves some other nuclease, or whether it involves a hitherto unrecognized property of one of the known nucleases in its native environment is not yet known.

It was stated above that host-modified phage cannot plate on *E. coli* B. Actually, a small proportion, about 1 in 10^4, of these phage do form plaques. By means of density-gradient separation of whole phages, Boyle,

Ritchie, and Symonds (1965) were able to show that this small fraction of plaque-formers are those which received a large amount of glucosylated DNA, presumably enough to escape restriction, from parental phage during growth of the host-modified phage.

Since the original discovery of Luria and Human, many other cases of host-induced modification have been discovered. For example, phage λ which has been grown in *E. coli* strain C is restricted in strains B or K (Bertani and Weigle, 1953). In this case glucosylation is not involved, since λ DNA is not glucosylated. We shall have occasion to discuss host-controlled modification in λ later in the book. However, the phenomenon is not limited to viral DNAs. Sex factors and colicinogenic factors are also subject to modification and restriction (Glover et al., 1963). Thus, host-induced modification is seen as a "general phenomenon in which DNA may become modified when it is synthesized in one cytoplasm and then undergo restriction upon entering another cytoplasm" (Eskridge, Weinfeld, and Paigen, 1967). Different mechanisms appear to be responsible for different cases of modification. However, the general phenomenon is an important one to understand, inasmuch as it "reveals that the cellular environment can produce alterations in the chemical structure of a genetic material without altering its coded genetic information" (Luria and Darnell, 1967, p. 257).

The existence of a host-controlled modification system concerned with glucosylation has presented an opportunity for isolation of phage mutants defective in induction of the glucosyl transferases. Revel, Hattman, and Luria (1965) isolated such mutants of T2 and T6 by virtue of their ability to form plaques on *Shigella,* coupled with an inability to grow on *E. coli* B. This technique does not work for T4, since a mutant defective in one transferase can apparently be glucosylated to a sufficient extent by the other that it can escape restriction. However, Hosoda (1967) made the accidental discovery that a gene 30 amber mutant of T4 bore a secondary mutation which led to deficiency in α-glucosyl transferase induction. The gene 30 mutation was bred out of this strain to give the single mutant, T4*αgt,* which differs from wild-type only in α-glucosyl transferase induction. This mutant plates with high efficiency both on *Shigella* and on *E. coli* B, implying, as mentioned above, that the remaining β-glucosyl transferase can glucosylate the DNA to a sufficient extent that it is not degraded in *E. coli* B. Starting with this strain, Georgopoulos (1967) isolated a number of mutants that are restricted in *E. coli* B. As expected, these are all double mutants, defective in induction of both transferases, and are designated as T4*αgt βgt.* Mapping of these genes, which was necessarily somewhat involved (Georgopoulos, 1968), shows that the *αgt* gene lies between genes 47 and 55 on the T4 map, while the *βgt* gene lies between genes 41 and 42. Molholt and deGroot (1969) have isolated *am* and *ts* mutants in the *αgt* gene for T2. This gene maps in the same part of the T2 chromosome as does the corresponding gene in T4. The availability of *ts* mutations in this gene enabled Molholt and deGroot to verify that the T2 *αgt* gene is the structural gene for α-glucosyl transferase.

Although it is generally assumed that a major function of DNA glucosylation is to protect phage DNA against enzymatic degradation, there is some evidence that the presence of the hydroxymethyl group in HMC also plays this function, whether or not the DNA is glucosylated. Wiberg (1967) studied DNA synthesis in a UDPG-negative bacterial host. Wild-type T4 synthesizes DNA at a normal rate, although this DNA is unglucosylated. However, it does contain HMC. On the other hand, a dCTPase-negative (gene 56) mutant cannot accumulate DNA to a significant extent, presumably because it forms cytosine-containing DNA, which is rapidly degraded. Thus, the hydroxymethyl group *per se* provides a degree of protection. Although it is difficult to dissect out the precise protective roles of hydroxymethylation and glucosylation, it would appear to this author that hydroxymethylation provides primary protection against phage-induced nucleases, while glucosylation provides resistance to attack by bacterial nucleases. In support of this, Wiberg (1967) found that infection of the UDPG-negative host by a 56 × 46 × 47 triple mutant did allow some phage DNA synthesis. Apparently in the absence of the nuclease(s) controlled by genes 46 and 47, some cytosine-containing DNA can accumulate. We have already discussed the evidence of Molholt and Fraser (1965) which implicated bacterial nucleases in the degradation of unglucosylated DNA, even if it contains HMC. Moreover, this and other work suggests that the restriction by bacterial nucleases occurs during penetration of DNA into the cytoplasm. As we have seen, unglucosylated DNA can be synthesized and packaged, as in infection of UDPG-negative host. However, this DNA is restricted when the glucoseless phage is used to initiate a new cycle of infection, suggesting that injection and/or penetration is the sensitive step. Since many bacterial nucleases are located in the periplasmic space, these considerations lend further credence to the idea of bacterial enzymes as the chief agents in restriction.

DNA Methylase

A wide variety of enzymes which transfer methyl groups from S-adenosylmethionine to nucleotides in preformed nucleic acid chains is now known. Although we do not know the precise role of nucleic acid methylation in any biological system, it is of interest that the activity of a DNA methylase rises manyfold following infection with phages T1, T2, or T4 (Hausmann and Gold, 1966; Sellin, Srinivasan, and Borek, 1966). This increase seems to result from synthesis of new enzyme protein, although this has not yet been proven. Infection with T7 or λ does not change the level of DNA methylase preexisting in the host, while infection with T3, T5, or T6 leads to decreases in this activity. The decline in methylase activity seen after T3 infection is quite precipitous, and was shown to result from induction of an enzyme which cleaves S-adenosylmethionine, to give thiomethyladenosine plus homoserine (Gefter et al., 1966). These results on enzyme induction correlate well with the observed pattern of methylation of phage DNAs. T1, T2, and T4 DNAs do contain methylated bases, as one might expect from the fact that they apparently induce DNA methylase. T7, which has no effect on methylase activity, also has methylated DNA, suggesting that the methylation in this case is carried out by a bacterial enzyme. Finally, the DNAs of phages T3, T5, and T6 are not methylated to detectable levels. The phage-induced DNA methylases are specific for adenine residues in DNA, as expected from the fact that 6-methylaminopurine is the only methylated base so far detected in phage DNA (except, of course, for thymine).

Since methylation, like glucosylation, involves chemical modification of a preformed polydeoxynucleotide, a role has been suggested for methylation in host-induced modification (Gold and Hurwitz, 1963; Arber, 1965). However, no definitive evidence on this suggestion has appeared, at least with regard to T-even phages. Gefter et al. (1966) prepared T2 presumably containing unmethylated DNA, by coinfection of *E. coli* with T2 and with uv-irradiated T3, which induces the S-adenosylmethionine-cleaving enzyme, but which cannot replicate. This T2 preparation was biologically indistinguishable from normal T2, at least by the criteria applied. When this work appeared, it seemed to me that unmethylated T2 DNA might become methylated by bacterial enzymes, shortly after injection, and that this might obscure any changes connected with the absence of methyl groups. In particular, the work of Lark (1968) suggested that *E. coli* DNA must be methylated in order to function as a template for further replication. Therefore, one might expect unmethylated T2 to be deficient in DNA replication in infection of a methyl-deficient host. However, unmethylated T2 does not show a significant reduction in DNA synthesis, relative to normal phage, when used to infect *E. coli* B previously infected with uv-inactivated T3. The same results were seen even with phage which had been subjected to five serial passages through the methyl-deficient host (Mathews, unpublished results), suggesting that T2 DNA does not need to be methylated in order to serve as a template for its own replication, at least in *E. coli* B. In this context, however, it would be of interest to know whether there are any *E. coli* strains which permit growth of normal T2 but which restrict unmethylated T2.

Polynucleotide Kinase

This enzyme, originally discovered in T4-infected cells by Richardson (1965), was also found in T2-infected cells, but not those infected by T1 or T5 (Novogrodsky and Hurwitz, 1966). The enzyme transfers orthophosphate from ATP to 5′ hydroxyl termini of a variety of nucleic acid compounds, including DNA, RNA, various oligonucleotides, and even nucleoside 3′ monophosphates. The specificity for phosphate donors is also very low; in addition to ATP, the enzyme can utilize UTP, CTP, and GTP with approximately equal efficiency (Novogrodsky et al., 1966). Because of the low specificity of the reaction, it is difficult to propose a metabolic role. Isolation of mutants unable to induce the enzyme would be of great value in this regard. However, in the absence of knowledge about a metabolic role, we do not know whether to search for such a mutant among stocks of conditional lethal mutants or whether to assume that such a mutation would not be lethal.

In an earlier review (Mathews, 1969), I suggested that the existence of polynucleotide kinase might account for the discovery of thymidine kinase in T4-infected cells (Hiraga et al., 1967). However, although the enzyme does have a broad specificity for phosphate acceptors, it has no activity with ribonucleosides, though to my knowledge the enzyme has not been tested for its activity with deoxyribonucleosides. Polynucleotide kinase is, however, probably responsible for the increased activity of T4-infected cells to phosphorylate 3′-dTMP to 3′, 5′-dTMP (Harris, Panter, and Symons, 1968).

Polynucleotide Ligase

This enzyme, which was originally discovered in uninfected *E. coli* by Gellert (1967), was found to be increased after T4 phage infection (Weiss and Richardson, 1967a; Becker et al., 1967) due to synthesis of a new enzyme. Discovery of this enzyme excited a good deal of interest, first because it appeared that it might participate in genetic recombination, and secondly, because its existence made it possible to see how DNA polymerase might function in DNA replication, even though it can only catalyze chain extension in one direction (Okazaki et al., 1968). We shall explore these possibilities later in this section and in Chapter 7.

Polynucleotide ligase, as extracted from T4-infected cells, catalyzes the closure of single-strand breaks, or nicks, in a double-stranded polydeoxynucleotide (for the reaction, see Figure 6–1). The enzyme requires the presence of a 3′ hydroxyl group adjacent to a 5′ monophosphoryl group. For the phage-induced ligase ATP is required for the reaction, with the products being AMP and pyrophosphate (it is of interest that the *E. coli* ligase requires DPN instead of ATP, and that it is split during the reaction to give AMP and nicotinamide mononucleotide (Olivera and Lehman, 1967; Zimmerman et al., 1967). The reaction proceeds *via* Formation of a covalently bound enzyme-AMP complex (Weiss and Richardson, 1967b; Weiss, Thompson, and Richardson, 1968). This enzyme, as well as the above-described polynucleotide kinase, have provided useful end-group reagents for analysis of DNA structure (cf. Weiss, Live, and Richardson, 1968).

Polynucleotide ligase is usually assayed with a specially prepared substrate made by nicking T7 DNA through brief treatment with pancreatic DNAse. This DNA is then treated with phosphatase to remove monophosphate groups in the nicks. The 5′ hydroxyl groups are then esterified with P^{32} phosphate, by use of polynucleotide kinase and γ-labeled ATP. The action of the ligase is then measured as incorporation of radioactivity into an acid-insoluble form which is not removed by further phosphatase digestion.

Fareed and Richardson (1967) identified gene 30 in T4 as the structural gene for polynucleotide ligase. This involved primarily the demonstration that *ts* gene 30 mutants induce a ligase which is more temperature-sensitive than the wild-type enzyme. This knowledge has made it possible to ask questions about the biological role of the ligase. Gene 30 amber mutants were originally classified as DNA-negative, or DO (Epstein et al., 1963). However, subsequent work (Warner and Hobbs, 1967) showed that they are able to synthesize a small amount of DNA under restrictive conditions. The nature of this small amount of DNA became a matter of great interest after Okazaki et al. (1968) had proposed their model of discontinuous DNA replication. In this model, DNA polymerase is seen as a catalyst for polynucleotide chain extension. However since it can only function in the 5′ to 3′ direction, DNA chains are synthesized as short fragments, which are grafted together into longer chains through action of polynucleotide ligase. This suggestion led several investigators to examine the size distribution of the small amount of DNA which is formed in nonpermissive infection by gene 30 mutants. As predicted, this DNA is of considerably smaller size than that of mature DNA, suggesting that it consists primarily of "Okazaki fragments" which have not been stitched together by ligase action into full-length strands of DNA (Sugimoto, Okazaki, and Okazaki, 1968; Masamune and Richardson, 1968; Hosoda and E. Mathews, 1968; Newman and Hanawalt, 1968). Although these results are consistent with the predictions of the Okazaki model, one question which this work does not answer is the question of how a 5′ monophosphoryl terminus is generated in the area of the nick which is to be repaired by ligase. Presumably, the 5′ terminus generated by DNA polymerase action would contain a triphosphate group, which would have to be cleaved before the ligase could act. One way around this is a model proposed by Guild (1968) and by Kornberg (1969). In this model (see Figure 6–5), the DNA polymerase functions continuously through the replicating fork, at the point where the two strands of parental DNA are coming apart. The covalently bonded Y-shaped structure which is then generated can then be cleaved by a nuclease, which could generate a 5′ monophosphoryl residue directly. However, any model of this type must take into account the fact that the "Okazaki fragments" formed in T4 DNA replication hybridize with equal efficiency to both strands of T4 DNA (Sugimoto, Okazaki, Imae, and Okazaki, 1969).

Thus, synthesis would appear to be discontinuous along both template strands.

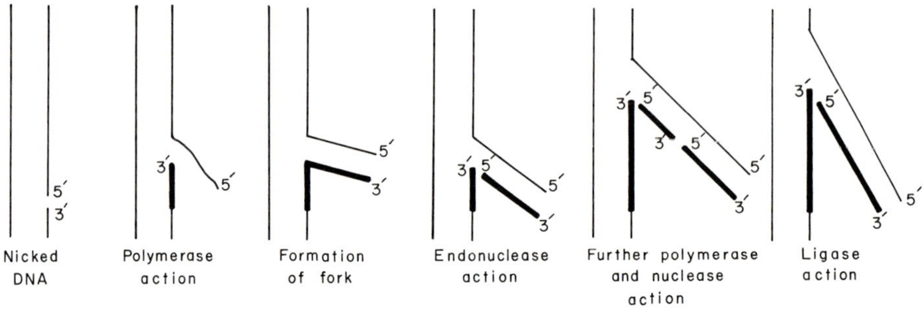

FIGURE 6-5 A model for DNA replication, as proposed by Guild (1968) and Kornberg (1969). Heavy lines represent newly-synthesized DNA.

In support of their model, Okazaki and Okazaki (1969) have shown that short chains complementary to both template T4 DNA strands are synthesized in the 5′ to 3′ direction.

We shall consider the possible role of polynucleotide ligase in recombination in Chapter 7.

Finally, it should be noted that another polynucleotide joining enzyme was discovered in T2-infected cells some three years prior to discovery of polynucleotide ligase (Mead, 1964). From the details presented, this may be the same as the ligase, although Mead failed to provide evidence that his enzyme preparation was strictly associated with phage infection. Since later workers have failed to refer to Mead's paper, it is difficult to discern the relationship between Mead's joining enzyme and polynucleotide ligase.

3′ Deoxynucleotidase and 5′-Polynucleotide Phosphatase

These activities have been found to be associated with T-even phage infection by Becker and Hurwitz (1967), while Harris et al. (1968) found elevated activity of a thymidine 3′-phosphatase in T4-infected cells. The 3′ deoxynucleotidase removes phosphate groups from 3′ phosphoryl termini in DNA, while the 5′ polynucleotide phosphatase removes 5′ phosphate groups from DNA and RNA. The 3′ deoxynucleotidase has no activity with RNA, 5′ nucleotides, or mononucleotides, while the 5′ polynucleotide phosphatase has no activity against any mononucleotides. To date mutants lacking the ability to induce either of these activities have not been isolated, and nothing is known of their physiological roles. It is known that, in order for a DNA molecule to be replicated, it must contain a free 3′ hydroxyl terminus. Therefore, one could propose that the 3′ deoxynucleotidase functions to remove inhibitory 3′ phosphoryl termini. However, as Becker and Hurwitz have pointed out, there is no known mechanism in *E. coli* whereby such termini might be generated.

UV Repair Enzyme

We have already mentioned indirect evidence (Haber, 1966) which suggested that the *v* gene in T4 controls the synthesis of an enzyme capable of repairing uv damage. In an abstract, Setlow and Carrier (1966) reported that the product of the *v* gene is an enzyme capable of excising thymine dimers from uv-damaged DNA. This is supported by more recent work of Friedberg and King (1969), who showed that the *v* gene controls an endonuclease specific for uv-damaged DNA. Sekiguchi and his colleagues (Sekiguchi et al., 1970; Yasuda and Sekiguchi, 1970) have reported extensively upon the properties of this enzyme. It catalyzes the release of thymine dimers from uv-irradiated T4 or *E. coli* DNA and can also bring about single-strand breaks in uv-irradiated DNA, presumably via the dimer excision reaction. Neither T4 *v* mutants nor T2, which has no *v* gene, can induce the dimer-excision enzyme, nor can extracts of cells infected with T4*v* or T2 cause single-strand breaks to appear in uv-irradiated DNA *in vitro*. The product of the related gene *x*, which is also involved in determining uv resistance, has not yet been identified. It does induce the dimer-excision enzyme (Yasuda and Sekiguchi, 1970). Gene *x* maps between genes 42 and 43 (unpublished work by R. Edgar, cited in Speyer and Rosenberg, 1968).

PRESUMED BUT UNDEMONSTRATED ENZYMATIC FUNCTIONS

In addition to the enzymes discussed above, there are a number of processes in T-even phage infection whose existence suggests the appearance of other new enzyme activities. These include the inactivation of at least one *E. coli* transfer RNA, the alteration of *E. coli* valyl tRNA synthetase, and various modifications of *E. coli* RNA polymerase. These functions will be discussed at some length in the next chapter.

7

Expression of T-Even Phage Genes

According to the so-called central dogma of molecular biology, there are three processes in the transfer of genetic information encoded in DNA: replication, transcription, and translation. Since the first two processes require a DNA template, and since, in a singly infected cell the same DNA molecule must be simultaneously used for both processes, at least within the first few minutes, it is clear that there must be some temporal or spatial control mechanisms which keep these processes from interfering with each other. Moreover, we have already seen that different virus-specific proteins are synthesized at different times in the latent period. *A priori* one could expect that the timing of phage protein synthesis could be controlled either at the level of transcription or translation. A further element of genetic control, governing the process of virus assembly, is implied from the fact that the number of genes involved in this process is considerably greater than the number of known components of virus structure, suggesting that some of the gene products play regulatory roles. Finally, we have seen that the final step in the infective cycle, namely lysis, is also subject to genetic control, and that mutation, as in one of the *r* genes, can upset the normal timing mechanism which governs the length of the latent period. In this chapter we shall summarize current knowledge of T-even DNA replication, RNA and protein synthesis, virus assembly, lysis, and control mechanisms governing these processes. In addition, since replication is intimately connected with the process of recombination, we shall also discuss the status of biochemical knowledge of this process. Once again, the reader is reminded that excellent discussions of the genetics of recombination in phage are available elsewhere (cf. Luria and Darnell, 1967; Hayes, 1968).

DNA REPLICATION

We have already commented on some of the unique aspects of T-even DNA replication, at least when compared to the same process as it occurs in cellular organisms. Not the least unusual of these aspects is the failure of daughter DNA molecules to segregate following a round of replication. Instead, as originally shown by Hershey, Dixon, and Chase (1953), newly synthesized DNA accumulates for several minutes before being incorporated into infectious progeny. Toward the end of the eclipse period in T2-infected cells, some 40–80 phage-equivalent units of newly-synthesized DNA are present in what Hershey et al. called the "replicating DNA pool." During virus maturation, DNA is withdrawn at random from this pool for incorporation into infectious units; this occurs at the same rate as replenishment of the pool by continued DNA synthesis. Thus, the "replicating pool" is maintained at constant size, with a continuing excess of 40–80 phage units of noninfectious DNA present per infected cell.

What is the justification for calling the above noninfectious DNA a "replicating DNA pool"? An answer to this question demands that pool DNA be shown to be the precursor for DNA in progeny phages. It was known that DNA in the pool is physically different from mature phage DNA, as shown by its sensitivity to deoxyribonuclease, its nonprecipitability by phage antiserum, and its failure to sediment in centrifugal fields capable of precipitating mature phage. These properties allowed one to independently determine the fate of DNA precursors, with respect both to the pool and mature phage. Kinetic studies on P^{32} assimilation (Stent and Maaløe, 1953; Hershey, 1953) clearly established the precursor-product relationship between pool DNA and progeny DNA. Hershey was able to show that when P^{32} is administered as orthophosphate to a T2-infected culture at some point during the latent period, some 7 to 8 minutes elapse before this material is incorporated into the pool, and an additional period of equal duration occurs before label is detected in infectious progeny. This study also showed that the flow of DNA precursors is essentially irreversible; once phosphorus has passed from pool to progeny, it does not reappear in the pool. Although this may appear to be a somewhat trivial point to belabor, it must be considered in designing any experiment in which isotope incorporation is to be used to measure rates of phage nucleic acid synthesis. Incorporation of labeled precursors into DNA is essentially irreversible, such that kinetics of DNA synthesis can be followed by adding isotope at some point and following its accumulation in DNA. On the other hand, incorporation of labeled precursors into RNA is decidedly reversible, in view of the high rate of turnover of phage messenger RNA. Therefore, in order that kinetic studies of RNA synthesis yield valid results, they must be designed in terms of pulse-labeling experiments, carried out over time periods short enough that the loss of label from RNA synthesized during the pulse is not a significant event. In designing conditions for a pulse-labeling experiment, one should assure himself that incorporation of label into RNA occurs with essentially no lag. Since, as we have seen, there is a significant lag in incorporation of labeled precursors into DNA (Hershey, 1953), it is clear that pulse-labeling is not a suitable technique for determining rates of phage DNA synthesis. For further discussion of these points, see Mathews (1968).

Semiconservative Replication

The above-mentioned differences between modes of viral and cellular DNA replication are primarily centered about the fate of newly replicated DNA, but this information says relatively little about possible differences in the mechanism of replication. Even after the existence of a replicating pool of viral DNA had been well-established, there was no need to postulate that this pool was generated by a different mechanism from that envisaged by Watson and Crick (1953) for DNA replication; namely, separation of complementary strands by unwinding of a duplex structure, coupled with synthesis of two daughter strands complementary to those in the parent molecule. An important prediction of this model, namely that parental DNA is conserved during replication, with half being transferred to each daughter, was proved for cellular systems by the density-shift experiment of Meselson and Stahl (1958). At the time, it seemed a simple matter to extend the same experimental approach to T-even DNA replication. In principle, one would infect cells with phage grown in the presence of bromodeoxyuridine (BUDR) and P^{32}, such that the DNA is denser than normal and radioactive. Infection would be carried out in "light," non-radioactive medium, containing no BUDR. One would then isolate DNA from the infected cells and search for "replicative hybrids" by means of equilibrium density gradient centrifugation. Such hybrids would be characterized by the presence of radioactivity and by a density intermediate between that of heavy parental DNA and light progeny DNA. When this type of experiment was first attempted, however, such hybrids were not detected. This is because parental DNA becomes extensively fragmented during replication, with fragments being dispersed to many progeny phage particles. Presumably, this reflects the high rate of recombination characteristic of T-even phages (see next section). According to one study (Roller, 1964), some 80–85 percent of parental T4 DNA is transferred to progeny, with no individual progeny phage particle receiving more than 8–10 percent of a phage-equivalent unit of parental DNA. Thus, a replicative hybrid, if such existed, would not contain sufficient BUDR for its density to be appreciably altered. However, Kozinski (1961a) reasoned that, if the parental contribution of 8–10 percent is present as a single piece—a proposal which has now received experimental support (Shahn and Kozinski, 1966)—a radioactive hybrid might be detected in DNA which has been mechanically fragmented to give molecules about 8–10 percent or less of the length of mature phage DNA. Such a result was indeed seen when the size of DNA molecules was reduced by sonic oscillation prior to equilibrium density centrifugation. A rationale for this experiment is presented in Figure 7–1. Thus, the replication of phage DNA is semiconservative, as is that of bacterial DNA.

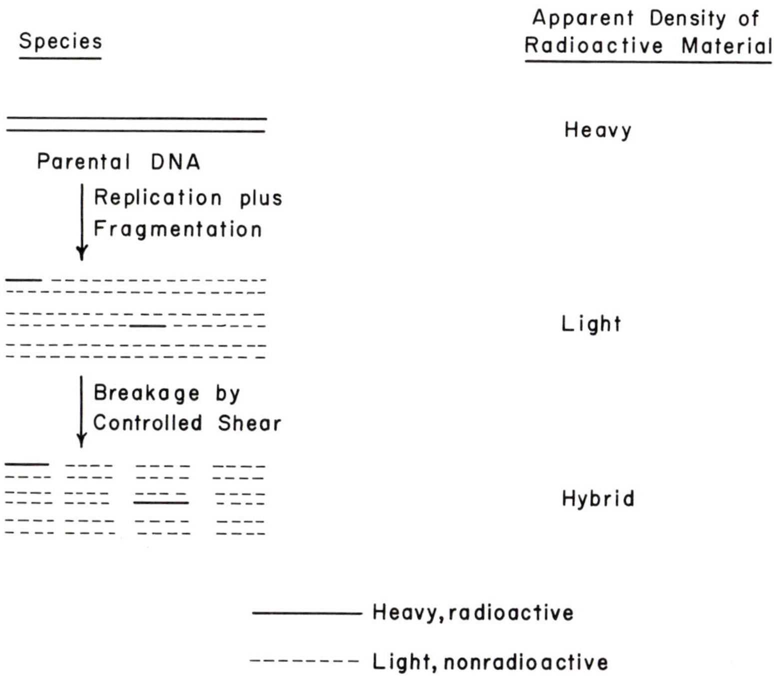

FIGURE 7-1 Demonstration of semiconservative replication for phage T4, as presented by Kozinski (1961a).

Properties of Replicative Intermediates

As with any metabolic sequence, a clarification of phage DNA replication demands that one be able to isolate and characterize intermediates in the process: in this case, DNA species in the replicating pool. This is not an easy task, first, because of the difficulty of extracting DNA from cells without encountering some degradation due to hydrodynamic shear forces and/

or attack by nucleases, and, secondly, because replicating DNA, as it exists in the cell, is complexed with other cellular materials, including protein (Kozinski and Lin, 1965; Miller and Kozinski, 1970; Miller and Buckley, 1970) and membrane material (Earhart et al., 1968). Nevertheless, several investigators have attempted to isolate and characterize replicating T-even phage DNA. Frankel (1966a) examined DNA from infected cells lysed with the aid of EDTA and lysozyme, followed by treatment with sodium dodecyl sulfate (SDS). This DNA differed from that extracted from whole phage, both in sedimentation rate and in chromatographic behavior on methylated albumin columns. This, of course, raises the question of how mature phage DNA is formed, if it does not exist as a free intracellular species.

In a companion study, Frankel (1966b) reported more fully on the sedimentation properties of replicating T4 DNA. The basic experimental approach was to add H^3-thymidine to an infected culture of *E. coli* B3, a thymine auxotroph. At various times after thymidine addition, portions of the culture were lysed by the lysozyme-EDTA-SDS method, under conditions minimizing shear degradation, and centrifuged through sucrose gradients in the presence of P^{32}-labeled reference T4 DNA. The distribution of each isotope through the gradient was determined. Results of a typical experiment are shown in Figure 7–2. Within the first few minutes of infection, a small amount of tritium is incorporated into material which cosediments with the marker DNA. Later, however, almost all of the replicating DNA is present in a species which sediments much more rapidly. At later times, coinciding with the onset of phage maturation, tritium counts reappear in the reference DNA region. This presumably represents material released from newly-formed progeny phages.

The rapidly-sedimenting replicative DNA, or "Frankel structure," was judged to be free of protein, as evidenced by its insensitivity to proteolytic enzymes, and by its failure to be labeled by prior growth in the presence of S^{35}-sulfate. The absence of lipid or other membrane material was assumed, both from the fact that detergent was used in its preparation and from the fact that the density of the Frankel structure is equal to that of mature phage DNA.

The sensitivity of the structure to attack by certain nucleases suggests that it contains some single-stranded regions. Using the equation of Burgi and Hershey (1963), which relates sedimentation rates for polynucleotides in sucrose gradients with molecular weight, Frankel estimated that replicating DNA behaves as though it contains at least 20 phage-equivalent units of DNA. He proposed that much of the DNA in this structure exists as a concatemer, or a covalently linked end-to-end aggregate of phage DNA molecules. The existence of such a structure suggests an attractive possibility for the generation of a population of mature DNA molecules, which, as we have seen in Chapter 3, are circularly permuted and termi-

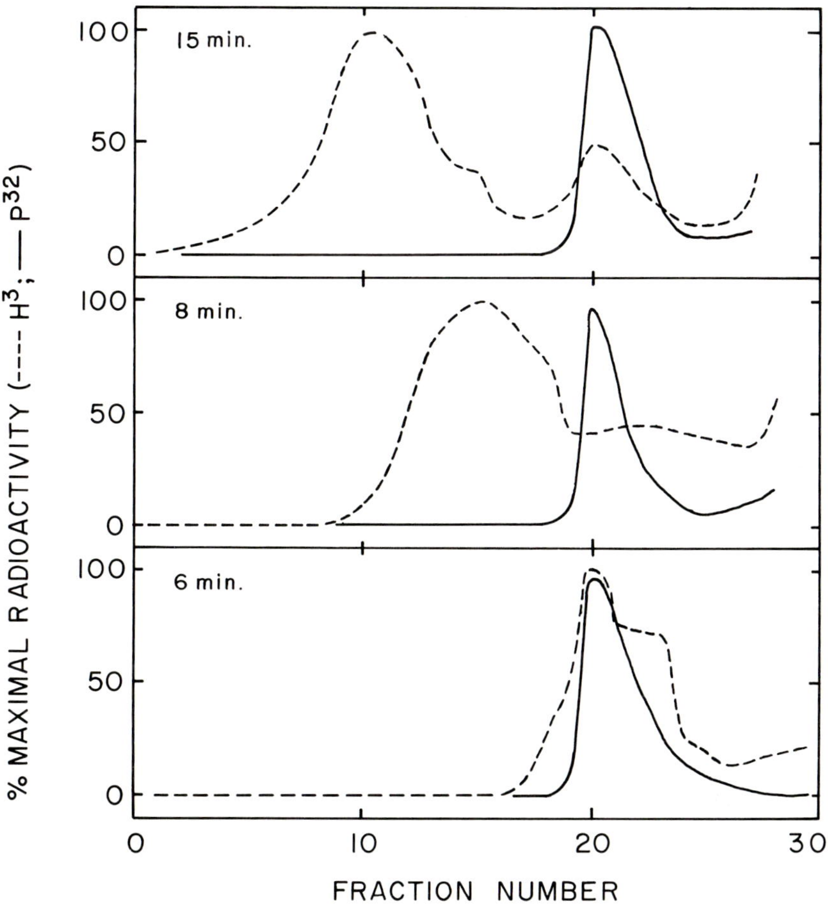

FIGURE 7-2 Sedimentation profiles for H^3-labeled replicating T4 DNA. Cells were harvested at the indicated times, lysed, and centrifuged in the presence of P^{32}-labeled DNA from whole phage particles. Modified from Frankel, 1966a.

nally redundant with respect to base sequence. The model, shown in Figure 7–3, suggests the existence of some mechanism, the "headful cutting device," which recognizes one phage-equivalent length of DNA in the concatemeric structure (which, because of terminal redundancy, constitutes about 102% of a phage genome), and then causes a double-strand break to release a mature-sized DNA model. If we consider the phage head to be the device for recognition of length, with scission occurring when the head becomes full of DNA, this can explain why mature-length DNA molecules are not seen as free intermediates in replication: these molecules are wrapped up in heads at the same time as they are being formed. Further-

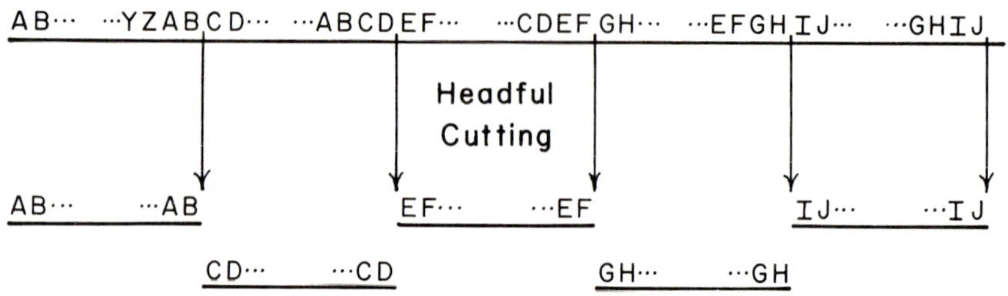

FIGURE 7-3 A scheme for generating a population of circularly permuted, terminally redundant DNA molecules by controlled cleavage of concatemeric replicating DNA.

more, the idea of a headful recognition device is in full accord with the finding of Streisinger, Emrich, and Stahl (1967), that phage bearing long internal deletion mutations show longer regions of terminal redundancy than do wild-type phage: presumably, the headful cutting device must take more DNA from the ends, to compensate for the loss of DNA from the middle of the molecule.

Although the proposal of a concatemeric structure for T4 replicating DNA gave plausible answers to several questions regarding the fate of this DNA, it said relatively little about how such a structure might come into existence, particularly in singly-infected cells. We shall have more to say about this later.

The idea of a concatemeric structure as a replicative intermediate was challenged by Kozinski, Kozinski, and James (1967). These workers used CsCl equilibrium density gradient centrifugation to isolate a replicative hybrid formed by infecting heavy (BUDR-labeled) bacteria in heavy medium with light, H^3-labeled phage. As Frankel had reported, this material sedimented rapidly in sucrose gradients. However, electron microscopic examination revealed a compact, twisted structure not more than twice as long as mature T4 DNA molecules. Kozinski et al. agreed that concatemeric structures might be generated by recombinational events later in infection, but felt that they did not arise via a purely replicative process. On the other hand, since the replicative hybrid had been carried through an extraction procedure plus preparative CsCl centrifugation, there was little assurance that the structure visualized in the electron microscope was completely intact. Moreover, the possible unphysiological effects of BUDR incorporation on subsequent events in replication have not been adequately assessed. This agent is lethal to bacteria, and, as we have seen in Chapter 5, is a potent mutagen. Furthermore, BUDR-substituted phage plate with relatively low efficiency. For these reasons, one must be careful in drawing inferences about normal replication from experiments involving the use of this agent. Fortunately for his concatemer hypothesis, Frankel

(1968a) provided direct confirmatory evidence. When replicating DNA was centrifuged through alkaline sucrose gradients, a treatment which sediments single strands of denatured DNA, some of the strands sedimented as though they were at least four times as long as single strands of T4 DNA. In a later study Frankel (1968b) showed that some strands from vegetative DNA sediment as though they were as much as seven times as long as those of mature T4 DNA.

By means of appropriate pulse-chase experiments, Frankel (1968b) was also able to show that the rapidly sedimenting DNA strands are precursors to phage-length strands, verifying that the Frankel structure is indeed a replicative intermediate. Moreover, when parental DNA is labeled, much of the radioactivity sediments with the Frankel structure.

Direct visualization of replicating T4 DNA has been achieved by Huberman (1968), using both tritium radioautography and electron microscopy. Extremely large, tangled molecules were seen, as shown in Figure 7–4. From the estimated total length of the DNA strands in such structures, Huberman concluded that all, or nearly all, of the phage DNA in a given cell forms part of a single structure (called a complex), at least before phage maturation has commenced.

It can be seen that Frankel's work, which is reproduced in Figure 7–2, showed that T4-infected cells can incorporate H^3-thymidine into material cosedimenting with reference DNA quite early in infection. This incorporation is detectable as early as five minutes, before replication has commenced. The nature of this "early DNA," as we call it, has been investigated in the author's laboratory (Murray and Mathews, 1969a). Early incorporation of thymidine results from addition of nucleotides to parental DNA, to the extent of about 0.06 phage equivalent units of new DNA per parental molecule. The relatively small increase in molecular weight which accompanies this addition does not appreciably affect the sedimentation rate in sucrose gradients, explaining the cosedimentation of early DNA with reference DNA. At present, we do not know whether early DNA is added at the ends of parental molecules or internally, or whether it is added as a single piece to each molecule. Early DNA synthesis commences after infection with no measurable lag, suggesting that its formation is catalyzed by bacterial enzymes, rather than virus-specific enzymes which would have to be synthesized before they could act. Further support for the idea of bacterial enzyme involvement comes from the fact that several DNA-negative amber mutants are fully able to form early DNA in non-permissive infections. Finally, preliminary experiments with temperature-sensitive, DNA-defective bacterial mutants are also in accord with the idea that bacterial enzymes are responsible for early DNA formation (M. Sisson, unpublished results).

Presently, we have no firm ideas on the biological role of early DNA, if, indeed, it has one. Preliminary experiments, in which P^{32} is incorporated

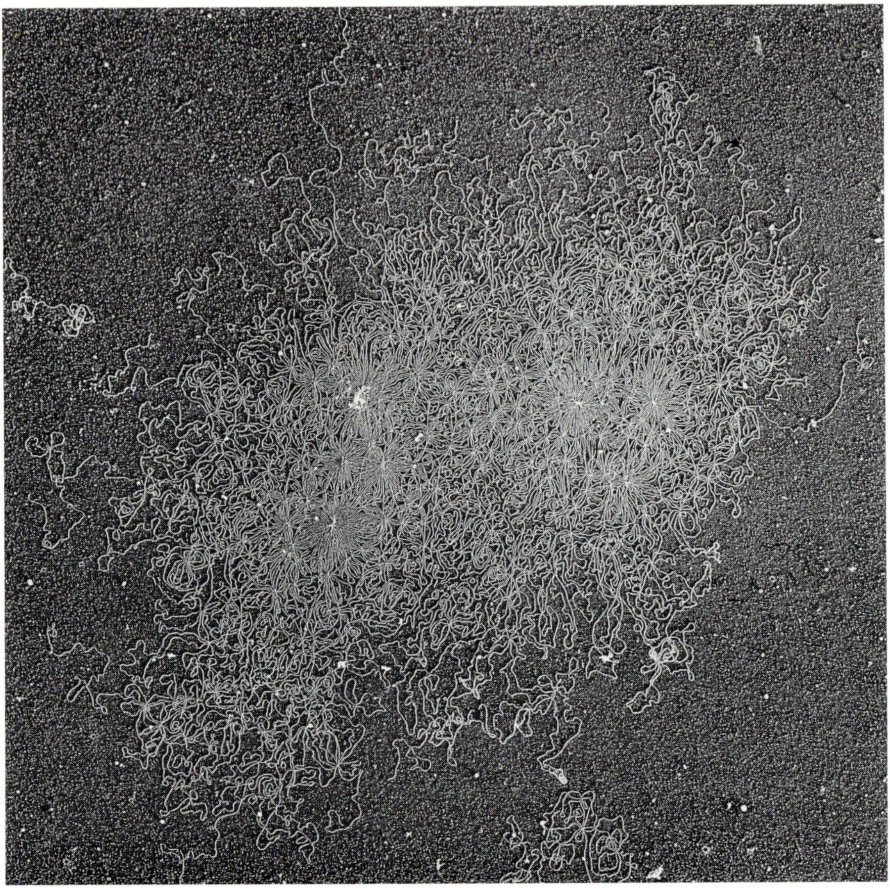

FIGURE 7-4 An electron micrograph of T4 replicating DNA (from Huberman, 1968. I am grateful to Dr. Joel A. Huberman and the Cold Spring Harbor Laboratory of Quantitative Biology for permission to reproduce this photograph).

into early DNA, and viability of infective centers followed as decay of the incorporated isotope occurs, suggests that early DNA formation does play an essential function in the overall replication of T4. A similar conclusion was drawn from experiments in which BUDR-containing early DNA was selectively inactivated by near-ultraviolet irradiation (Mathews, 1970a).

DNA Replication and Cellular Topography

As we have already mentioned, both in Chapter 4 and Chapter 5, there is a strong possibility that T4 DNA replication occurs at some membrane attachment site. Although the nature of such a site remains quite obscure, recent evidence does suggest an association between membranous material and replicating DNA. Tremblay, Daniels, and Schaechter (1969) have described a method for isolation of DNA attached to the cell membrane. Although it has not been firmly established that this material truly represents a replicating complex, all evidence so far obtained appears to be in accord with this idea. Labeled parental DNA from T4D becomes associated with this membrane complex immediately after infection. At later times, however, label is lost from the complex, presumably as the replicating DNA starts to mature. On the other hand, in a nonpermissive infection by T4 *am*N82, a DNA-negative gene 44 mutant, the parental DNA becomes attached to the membranous material, and very little of it is released late in infection. This suggests that the removal of parental DNA from the membrane complex is a specific consequence of replication, although other interpretations can be entertained.

Further evidence for an involvement of membrane in T4 DNA replication comes from work of Frankel et al. (1968), who isolated a T4-specific DNA polymerase from membranes of infected cells. This new enzyme is not the same as the well-characterized DNA polymerase which is the product of gene 43 (see p. 121), for it is formed in nonpermissive infections by T4 *am*X5, an amber mutant defective in genes 41 through 45. Moreover, the membrane polymerase does not require a DNA primer in order to function *in vitro*. Formation of the polymerase does not appear to result from unmasking of a latent membrane-bound bacterial enzyme, for enzyme appearance is sensitive to the action of chloramphenicol. However, formation of the enzyme occurs with essentially no lag, and maximal activity is reached by five minutes, with a steep decline occurring later. In this context, however, the rapid formation of both the early DNA mentioned above and the membrane-bound polymerase suggest some relationship between the two events. Unfortunately, however, no information is yet available on the metabolic role of either early DNA or the membrane polymerase.

The work of Frankel et al. (1968) also presented direct evidence for the synthesis of specific membrane proteins in T4 infection, although it is impossible to tell at this stage whether they are directly involved in DNA replication. The approach was to label proteins before infection with C^{14}-leucine and after infection with H^3-leucine. A partially purified membrane preparation was subjected to disc gel electrophoresis, and the distribution of both isotopes determined. It appeared from these experiments that several components are formed in infected, but not uninfected cells.

In addition to "membrane material," parental phage DNA also becomes complexed with phage-coded proteins, as early as three minutes after infection (Miller and Kozinski, 1970; Miller and Buckley, 1970). Comparison of DNA-complexed proteins with the total complement of early phage-induced proteins, by means of disc gel electrophoresis, shows distinct differences, indicating that specific proteins are utilized in forming the complex. Protein and DNA in the complex are "colinear," i.e., protein molecules are bound all along the length of DNA molecules, rather than being attached at the ends or some other specific sites.

From the number of components known to be involved in DNA replication, it should be clear that the cellular geography in the vicinity of replicating T4 DNA must be very complex indeed. Included in the "replicating complex" must be T4 DNA itself, partially unwound at the replication point, with daughter strands attached; ample supplies of the four needed deoxynucleoside triphosphates; some membrane material, which has not yet been well-defined; and several proteins as indicated in the previous paragraph. We have already mentioned the possible roles of DNA polymerase, DNA ligase, and an endonuclease in replication, plus the indirect evidence that dCMP hydroxymethylase forms part of a replicating complex. Undoubtedly, additional proteins are present as well (cf. Miller and Buckley, 1970). At any rate, the question can be raised whether phage proteins in the replicating complex are site-specific, i.e., is a phage-coded protein molecule which forms part of the replicating complex only able to function in the vicinity of the genome which spawned it, or can it diffuse through the cell and facilitate the replication of other phage genomes? Although this might seem a trivial question in view of the existence of intergenic complementation, it has been asked in a biochemical way (Kozinski and Kozinski, 1967). *E. coli* B was coinfected with H^3-labeled T4D and P^{32}-labeled T4 *am*N122, a gene 42 mutant which is unable to induce dCMP hydroxymethylase. Both labeled genomes replicated, indicating that the hydroxymethylase induced by the wild-type phage was able to facilitate the replication of the hydroxymethylase-negative genome. In retrospect, this result should hardly be surprising in view of the fact that Mathews et al. (1964) had estimated that a T6-infected cell contains 6000 molecules of hydroxymethylase.

A slightly different question was posed by Eckhart (1966), namely, the question of whether a T4 genome could replicate using the complement of early proteins induced by the heterologous phage T2. Cells infected with T2 were superinfected with T4 in the presence of chloramphenicol, to block formation of T4 early proteins. Replication of T4 genetic markers was measured by determining their mutability by BUDR (recall from Chapter 5 that BUDR must be incorporated during DNA replication in order to exert its mutagenic effect). Although individual T4 markers were replicated, Eckhart's evidence indicated that entire T4 genomes were not replicated. Rather, the evidence suggested that those markers which did replicate were able to do so only by virtue of their prior incorporation into a T2 genome by recombination. As one might expect, the efficiency of this process is rather low.

Growing Points in Replicating T4 DNA

Using the technique of pulse-labeling with H^3-labeled bromouracil, a technique originally developed for analyzing bacterial chromosome replication (Bonhoeffer and Gierer, 1963), Werner (1968a,b) has obtained a large amount of information about the number and distribution of DNA replication points in T4-infected cells. Using this procedure, one labels infected cells with H^3-BU of known specific activity, and then continues growth for a time in light, nonradioactive medium. The total radioactivity incorporated during the pulse provides a measure of the total length of DNA synthesized during a labeling period. A portion of the labeled DNA is reduced by controlled shearing to fragments of known size. The density of this sheared material is compared with that of light DNA, using equilibrium density gradient analysis, and the density increment is used to calculate the lengths of individual BU-substituted regions. When one makes appropriate allowance for the incomplete substitution of BU for thymine and for the fact that two daughter strands are being generated per replication point, the total length of DNA synthesized per cell during a pulse, divided by the length of each individual region, gives the number of growing points per cell. The rationale for this type of experiment is illustrated in Figure 7–5.

From this type of experimental approach, Werner could calculate that in cells infected at 25° C, the number of growing points per cell increases linearly, from zero at 10 minutes after infection to about 60, at thirty minutes. Thereafter, the number remains constant. The rate of movement of a growing point is about 0.08 phage-equivalent lengths per minute in the presence of thymine, 0.05 in the presence of BU. Both values are close to the value determined by Bonhoeffer and Gierer (1963) for the rate of *E. coli* DNA synthesis. The work of a number of investigators (cf. Lark, 1966; Cooper and Helmstetter, 1968) shows that the number of replication points per *E. coli* chromosome is one, or a small number, depending upon growth conditions. Thus, the rapid rate of DNA synthesis in infected, as compared with uninfected, cells is due to an increased number of DNA growing points, rather than to an increased rate at which a single growing point can traverse a DNA molecule.

The distance between growing points can be calculated from the length of time required for a given point to be passed by two successive growing points. This in turn can be measured from the time required for H^3 in an H^3-BU labeling experiment to pass from a region of hybrid density (one heavy strand, one light strand) to one of high density (two heavy strands). Werner calculated from this type of experiment that growth points are distributed at intervals of 0.15–0.19 phage-equivalent lengths. Since the number of growing points per phage-equivalent unit of DNA in a cell is between 1.0 and 1.5, this suggests that the growing points are clustered.

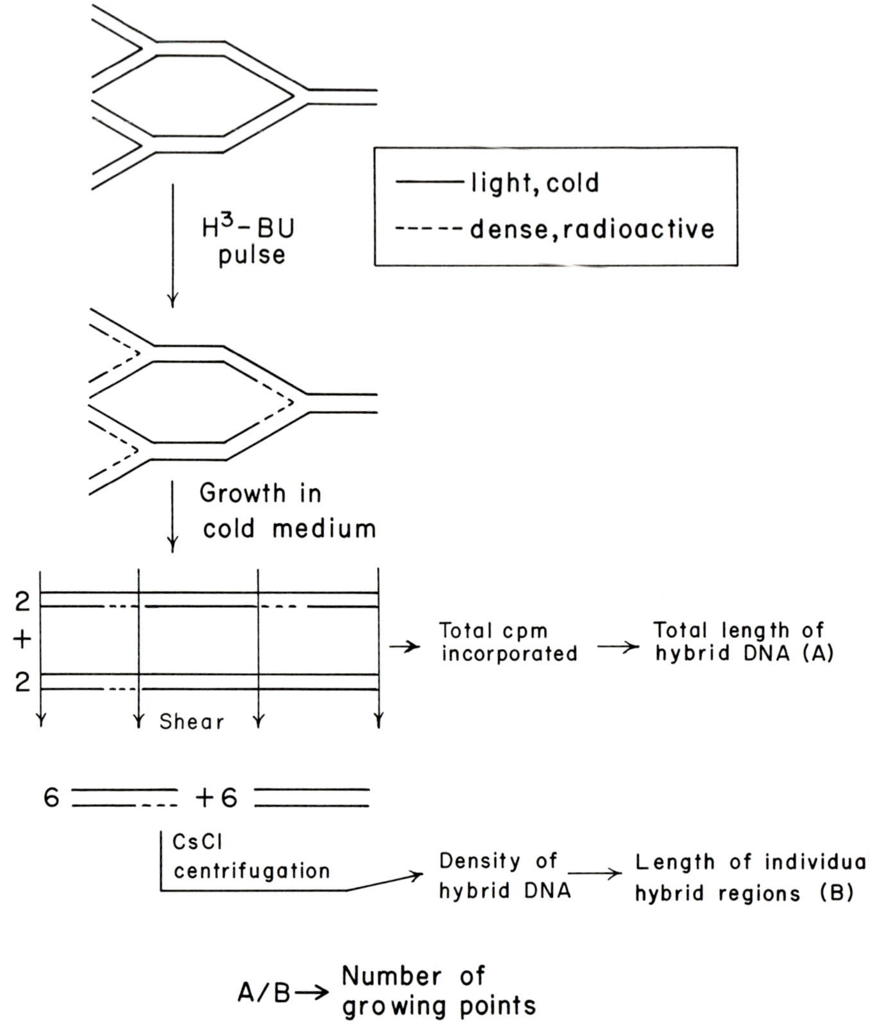

FIGURE 7-5 A rationale for the experiments of Werner (1968a, b), on the distribution of growing points in T4 replicating DNA.

Circular Models for T4 DNA Replication

From his observations, Werner has suggested a circular model for T4 DNA replication, as shown in Figure 7–6. He envisions a primary circularization of paternal DNA, occurring *via* recombination between the terminally redundant ends. Initiation occurs at some specific site, and a growing point begins to traverse the molecule. After the growing point has moved some

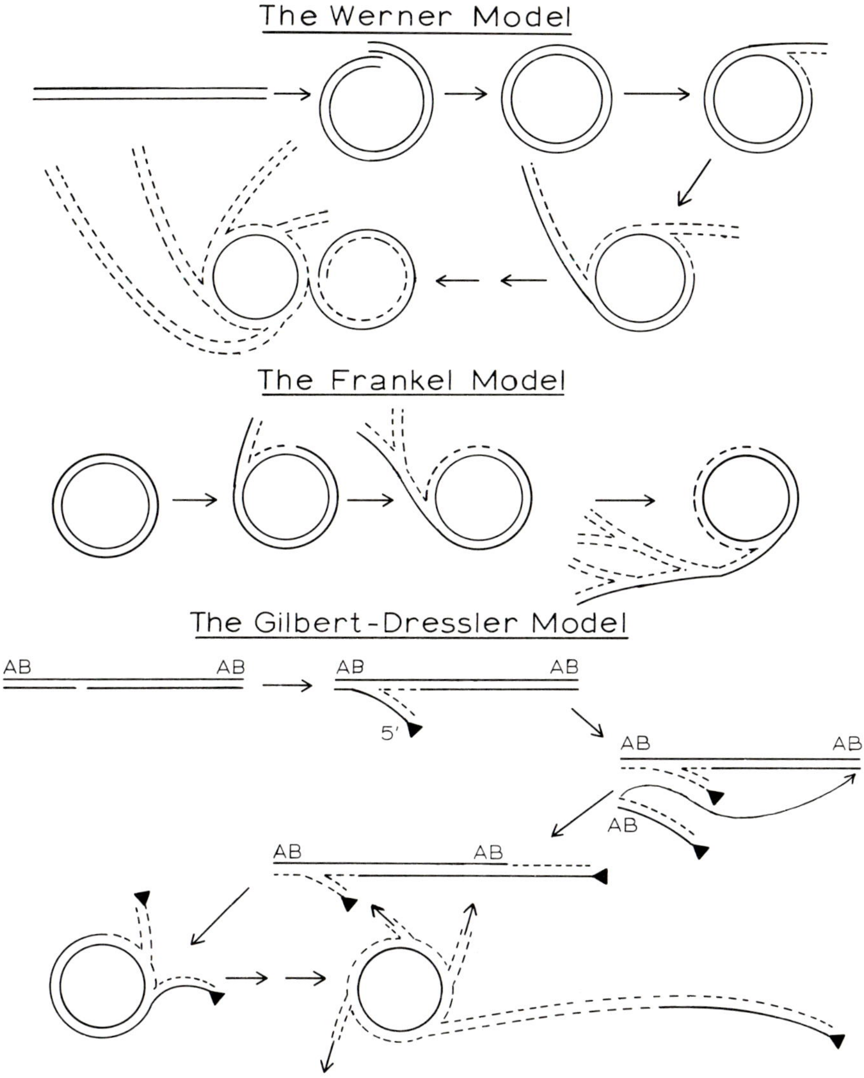

FIGURE 7-6 Circular models for T4 DNA replication. The black triangles represent sites for binding of 5′ termini to membrane.

0.15 to 0.19 phage-equivalent lengths, a new growing point is initiated from the same site at which the first was started. New growing points continue to start, giving a "pinwheel" type of replicative intermediate. When a growing point has traversed the entire circle, it can start around again, and thus generate concatemeric structures. Presumably, these structures can recombine with each other, to give the entire replicating complex a rather involved tertiary structure. This model is in reasonably good agreement with conclusions one might reach from inspection of Huberman's pictures, which show centers somewhat more open in structure than the periphery. In further accord with the circular model, electron microscopic examination of radioautograms of replicating DNA pulse-labeled with H^3-thymine indicate that most of the labeled material is incorporated near the center.

Frankel (1968b) has also advanced a circular model to explain T4 DNA replication. However, as seen in Figure 7–6, this differs from Werner's model in that only a single point traverses the circle, and that subsequent growing points are initiated from the physical ends of the protruding strands. At the same Cold Spring Harbor Symposium at which Frankel and Werner presented their models for T4 DNA replication, Gilbert and Dressler (1968) advanced a general hypothesis to explain DNA replication in a variety of systems, the so-called "rolling circle model." Actually, this model is very similar to that proposed by Werner, with two modifications. First, Gilbert and Dressler propose that the 5′ end of one of the parental strands is attached to a membrane site, thus preventing repair of the nick which must be created in order for replication to begin. This "anchoring" may prevent recombination between the initially opened region and a homologous region on the replicating structure, an event which could complicate the orderly travel of growing points. The other modification is that Gilbert and Dressler do not propose circularization of a T4 genome as the initial event. Rather, they propose that linear DNA molecules begin to replicate, and that circularization occurs a little later, as shown also in Figure 7–6. This modification was made to explain data of Doermann and Boehner (1963) on recovery of genetic markers in single infections. Moreover, this is in accord with data of Kozinski (1961) and of Roller (1964) on the dispersion of parental DNA among progeny phage. The parental DNA is dispersed among the earliest-maturing phage, with relatively short parental fragments being incorporated into individual progeny particles. Finally, this aspect of the model is in agreement with the finding of Frankel (1968a) that a sizable proportion of the strands of replicating T4 DNA are shorter than those of mature DNA. However, it should be observed that such strands could just as easily be generated by nicking of the replicating DNA, a process known to occur (Kozinski and Lorkiewicz, 1967).

Although the circular model for T4 DNA replication does account for several puzzling aspects of this process, including the way in which a concatenated intermediate might be generated from a single linear DNA molecule, it does leave a number of questions unanswered. For example, the model predicts that only one parental strand, namely, the one which becomes opened early in infection, will be incorporated into progeny. However, Kozinski (1969) has found that parental label which is incorporated into progeny hybridizes equally well with both strands of

T4 DNA, separated preparatively by the method of Guha and Szybalski (1968). Thus, both parental strands appear to participate in replication. However, one can propose either that parental material on both strands is dispersed by recombination, or that the initial replication point has an equal probability of moving in either direction. The latter suggestion seems untenable in view of recent work on the origin and direction of T4 DNA replication (Mosig and Werner, 1969; Mosig, 1969). These workers studied DNA replication in cells singly infected with light T4 particles. As will be recalled from Chapter 5 (p. 89), these are phages containing DNA molecules about two thirds the length of mature T4 DNA. The distribution of incomplete nucleotide sequences among a population of these molecules is random. In single infection by these phages, one third of the particles are unable to initiate DNA replication. Of the two thirds which do initiate replication, the average amount of DNA synthesized is 50 percent of the input DNA. These results are consistent with the existence of a unique initiation site on the T4 genome. Among a randomly permuted collection of incomplete molecules, one third would be expected to lack this site, and, in the remainder, only one growing point is formed (although the reasons why secondary growing points are not initiated are obscure). Replication ceases when the growing point reaches the end of the incomplete molecule. Genetic studies, in which the frequency of markers replicated in these incomplete phage infections was measured by mixed infection with light particles and whole phages, suggested that the initiation sites lies between genes 42 and 43, and that the direction of replication is clockwise, as the T4 map is usually drawn.

A potentially serious objection to the circular model is that, with the exception of our demonstration of attachment of a very small amount of "early DNA" to parental DNA (Murray and Mathews, 1969a), no one has been able to demonstrate covalent attachment of newly replicated DNA to parental DNA (discussion by Werner, Kozinski, and Frankel, following paper by Werner, 1968). Moreover, the circular model treats each parental DNA molecule as an independently replicating entity, saying nothing about possible interactions between parental molecules in a multiply infected cell. Yet the electron micrographs of Huberman (1968) suggest that essentially all of the replicating DNA in an individual infected cell forms part of the same complex. Moreover, there is direct evidence for formation of covalently bonded joint structures formed from distinct parental molecules (Anraku and Tomizawa, 1965). These were detected by coinfecting cells with BU-labeled phage and homologous phage labeled with P^{32}, and, by means of appropriate density gradient analysis, demonstrating the existence of covalently bonded structures bearing both labels. Such structures were originally seen as possible intermediates in recombination (see next section), but there is no reason to discount their possible involvement in replication.

Further objections to the circular models for T4 DNA replication have recently been presented by Miller, Kozinski, and Litwin (1970). These workers have confirmed the finding of Frankel (1968a, b) that much of the replicating DNA pool contains polynucleotide strands longer than those of mature T4 DNA. However, such strands cannot be detected after infection until such time as about 20 phage-equivalent units of DNA have been synthesized per cell. Moreover, the addition of chloroamphenicol at 5 minutes, a process which inhibits recombination without affecting replication, prevents the formation of long strands. Therefore, Miller et al. argue that the long strands are formed by recombination between newly synthesized DNA molecules, rather than by the primary act of replication. In addition, these workers reported that parental DNA matures (i.e., is incorporated into progeny phage) at the same rate as newly synthesized DNA. The rolling circle model predicts that 50 percent of the parental DNA (i.e., one strand) would be "sequestered" in a form unavailable for maturation. Unfortunately, in their discussion of this experiment, Miller et al. ignored the earlier work of Hershey (1953), who reported that parental DNA is preferentially matured. At any rate, the evidence against the circular models of Figure 7–6 seems quite substantial. On the other hand, it is difficult to conceive of any way in which T4 DNA, initiating replication from a specific site not located on the end of the molecule, could undergo a complete round of replication in the absence of some circularization step.

Characterization of Early-Function Genes

It is clear that any model which is successfully advanced to explain T4 DNA replication will have to take into account a large amount of data from diverse sources, some of it conflicting, and most of it confusing. Of possible help in clarifying this area should be studies on conditional lethal mutants blocked in various stages of the overall process. By examining the species formed in nonpermissive infections, one can, perhaps, hope to detect new intermediates which accumulate under these conditions. Some work along these lines was reported by Frankel (1966c), who studied the properties of intracellular DNA in infection by amber mutants in gene 32 (DNA-negative; unable to form joint molecules of parental DNA; Tomizawa, Anraku, and Iwama, 1966); gene 46 (DNA arrest; host cell DNA breakdown-negative; see Chapter 6); and gene 33 (maturation-deficient). Neither the gene 32 nor the gene 46 mutant forms a rapidly-sedimenting species in a restrictive host, although the gene 46 mutant forms an ample amount of new DNA which cosediments with mature phage DNA. The maturation-defective gene 33 mutant, on the other hand, forms an intracellular species which sediments about twice as rapidly as that formed in single infection. We have confirmed this with amber mutants in both genes 33 and 55 (also maturation-deficient), and have shown that, even as late as 30 minutes after infection there is no detectable loss of material from this structure of phage DNA-sized pieces, as would be expected if maturation were allowed to proceed (Murray and Mathews, 1969b).

Using similar techniques, Oishi (1968) has reported a preliminary analysis of the function of gene 41 (DNA-negative phenotype). The small amount of DNA made in nonpermissive infections by this mutant is single-stranded and of low molecular weight. This would appear to lend credence to the Okazaki model of DNA replication (Okazaki et al., 1968), which proposes that DNA is formed in short, single-stranded pieces by DNA polymerase, and these pieces are stitched together with the aid of DNA ligase (see p. 129). Oishi's work seems to implicate the action of the gene 41 product as well in this latter process. This is of interest in light of recent data of Scotti (1969), who has examined the time in an infectious cycle during which each of several T4 "early" genes are expressed. This is done by infecting cells at a permissive temperature and testing to see at what time a shiftup to a nonpermissive temperature will have no effect on the final burst size. The shapes of the curves obtained with one gene 30 (ligase) mutant and one gene 41 mutant are similar, suggesting that the two gene products participate in the same function.

The author's laboratory has been interested in the function of gene 44, a gene whose product is essential for any replication, beyond early DNA synthesis, but whose biochemical function is unknown (Murray and Mathews, 1969b). In cells multiply infected by T4 *am*N82, a gene 44 mutant, a polynucleotide structure is formed which sediments rapidly in sucrose gradients, though not as rapidly as the Frankel structure which is formed in productively infected cells. This structure contains both parental DNA and H^3-thymidine which has been incorporated into early DNA. The sedimentation rate of this "N82 structure" is dependent upon multiplicity of infection, suggesting that a substantial portion of the parental DNA within an infected cell comes together to form an aggregate molecule. We are uncertain of the nature of this aggregate, but, since the structure is extremely sensitive to shear degradation, it may be linear. There are, admittedly, formidable problems in seeing how a circularly permuted collection of parental DNA molecules within a single cell can form a concatemeric aggregate which maintains regularly redundant gene order, as demanded by the headful cutting model of Streisinger et al. (1964) and Frankel (1966b). Clearly, a knowledge of the nature of the N82 structure demands more understanding of the nature and biological function of early DNA. Such information seems well worth seeking, however, since we believe that the N82 structure may well be an intermediate in normal replication. This belief comes from studies with T4 *ts*B110, a temperature-sensitive gene 44 mutant (Mathews, unpublished results). When a culture is infected with this strain at 43°, a nonpermissive temperature, and shifted at 10 minutes to 30°, a permissive temperature, DNA

have already discussed (see p. 135). The key finding of this work, for our present purposes, was that parental DNA is dispersed among progeny phages, with each progeny particle which contains parental material having about 8 to 10 percent of a phage-equivalent unit of parental DNA. *A priori* it could be argued that the parental DNA is broken down to nucleotides which are then reutilized from DNA synthesis, and that this process has nothing to do with recombination. However, the fact that Kozinski's shearing experiment generated material of hybrid density indicated that parental DNA is dispersed in large pieces; if parental material were degraded and reutilized, at random, there would not be sufficient material incorporated into any fragment to materially affect its density (see Figure 7–1). Moreover, Shahn and Kozinski (1966) have more recently provided direct evidence that the contribution of parental DNA to a progeny molecule is transferred as a single piece. Dispersion of parental DNA in this fashion can only occur via some process of breakage and joining.

So—breakage-reunion does occur in T4-infected cells. But does this have anything to do with genetic recombination? The copy choice model demands that replication be occurring in order for recombination to be seen. However, recombination is seen, even when DNA replication is severely depressed by adding fluorodexyuridine to infected cells (Séchaud et al., 1965; Simon, 1966). On the other hand, inhibition of DNA synthesis by FUDR is not complete (cf. Sekiguchi and Cohen, 1964; Mathews, 1966a), so that a more direct test was necessary to rule out the copy choice model. Tomizawa and Anraku (1964a) found that recombination between two *r*II markers is increased manyfold when energy metabolism in a cell—and presumably DNA replication as well—is blocked by addition of cyanide to infected cells. Using this system, Tomizawa and Anraku (1964a, b) attempted to demonstrate the existence of physically recombinant molecules, which, from the predictions of the breakage-reunion model, would contain DNA from two or more parental molecules. Cells were mixedly infected in the presence of cyanide with P^{32}-labeled phage (light, radioactive) and BUDR-labeled phage (dense, nonradioactive). DNA from the infected cells was subjected to cesium chloride density-gradient centrifugation. A significant proportion of the DNA contained both labels, as judged by the fact that some of the radioactivity banded in a position denser than that of light DNA. The parental components of these "joint molecules," as they were called, were attached by hydrogen bonds, for they could be separated by application of heat or high pH. From considerations of the temperature needed to disrupt joint molecules, as well as studies on shear fragmentation and enzymatic digestion, it was possible to estimate that the region of overlap involves between 20 and 5000 nucleotide pairs (Tomizawa, 1967).

In considering the structure of a hydrogen-bonded joint molecule, two possibilities appear most likely: either a double-overlap or a single-overlap model, as shown in Figure 7–8. Potentially, the single-overlap molecules can be converted to covalently bonded structures, if a limited amount of DNA synthesis can occur, whereas it is difficult to conceive of a similar event if the joint molecules are of the double-overlap type. Therefore, Anraku and Tomizawa (1965) explored the nature of joint molecules formed in the presence of FUDR. In DNA extracted at 15 to 20 minutes after infection all of the joint structures were of the hydrogen-bonded type, whereas at later times (30 to 45 minutes), most of the joint structures seen were covalently bonded, as shown by the fact that they were not disrupted by heating in the presence of 1% formaldehyde. The most likely route for formation of such structures is illustrated in part (a) of Figure 7–8.

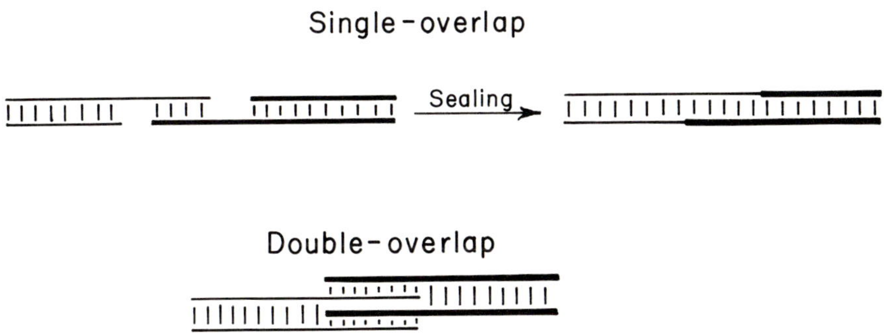

FIGURE 7-8 Single-overlap and double-overlap structures as possible intermediates in recombination.

Although most of the physically recombinant DNA molecules are of the covalently bonded type, it remained possible that genetic recombination generates noncovalently bonded overlap structures, which then mature. If this is so, then one would expect some DNA molecules from mature phage to contain single-stranded interruptions. Moreover, a phage population containing a high frequency of heterozygotes of the nonterminal-redundancy type might be expected to have a relatively high proportion of its DNA molecules containing such nicks, since the heterozygous regions would probably be located in the overlap region. However, phage DNA was found to contain no single-strand interruptions, as judged by band centrifugation in alkaline cesium chloride (Tomizawa and Anraku, 1965). Moreover, the same result was seen even in a phage population in which the proportion of r/r^+ heterozygotes was increased to 5 percent by means of growth of phage in the presence of FUDR, a technique previously shown to increase the frequency of heterozygotes (Séchaud et al., 1965).

Enzymological Aspects

The above evidence would appear to clearly establish breakage-reunion as the general mechanism for recombination in T4. It now becomes appropriate to ask which phage enzymes and other gene products are involved in the process. Tomizawa, Anraku, and Iwama (1969) found that the product of gene 32 is required for the initial formation of hydrogen-bonded joint molecules. When infection was carried out with a gene 32 amber mutant labeled with P^{32} and with the same mutant labeled with BUDR, no joint structures were seen. Although the action of the gene 32 product was not known, it is required for successful DNA replication as well as for formation of joint molecules (Warner and Hobbs, 1967; Kozinski and Felgenhauer, 1967). Moreover, gene 32 mutants are recombination-deficient (Berger, Warren, and Fry, 1969). The gene 32 product has been purified (Alberts et al., 1968), and studies of its mode of action *in vitro* are currently under way.

In considering the intracellular events leading to the formation of joint molecules, and in particular the role of gene 32, it becomes appropriate to ask questions about the fate of parental DNA. The single-overlap model, which seems favored by the evidence described above, predicts that single-strand breaks, or nicks, will be introduced into parental DNA as a prerequisite to formation of a joint molecule. Presumably, part of the DNA between two adjacent nicks would then be digested by a nuclease, so as to expose single-stranded regions, necessary for formation of intermolecular complexes (see Figure 7–9). Exchange of genetic material would then occur, generating hydrogen-bonded joint structures of the type postulated by Tomizawa. (In this context, we should note that genetic investigations by Drake (1967) indicate that the length of the hypothetical pairing region involved in genetic recombination is between 100 and 5000 base pairs, of the same order of magnitude as that involved in the overlap region of the Tomizawa structure); also, the involvement of nuclease digestion after nicking is in accord with the previously mentioned finding that mutants defective in genes 46 and 47 are recombination-deficient (Bernstein, 1968).

As one might expect from the above discussion, parental DNA does become nicked following injection into a host cell (Kozinski and Lorkiewicz, 1967; Kozinski, Kozinski, and James, 1967). This is determined from the fact that intracellular parental DNA sediments more slowly in alkaline sucrose gradients than denatured DNA from mature phage, although no differences are seen when sedimentation is carried out through neutral sucrose gradients. Later in infection, the nicks become covalently closed. Phage-coded enzymes appear to be involved in both nicking and repair, since addition of chloramphenicol (CM) early in infection prevents the nicking reaction, while addition of the drug somewhat later inhibits

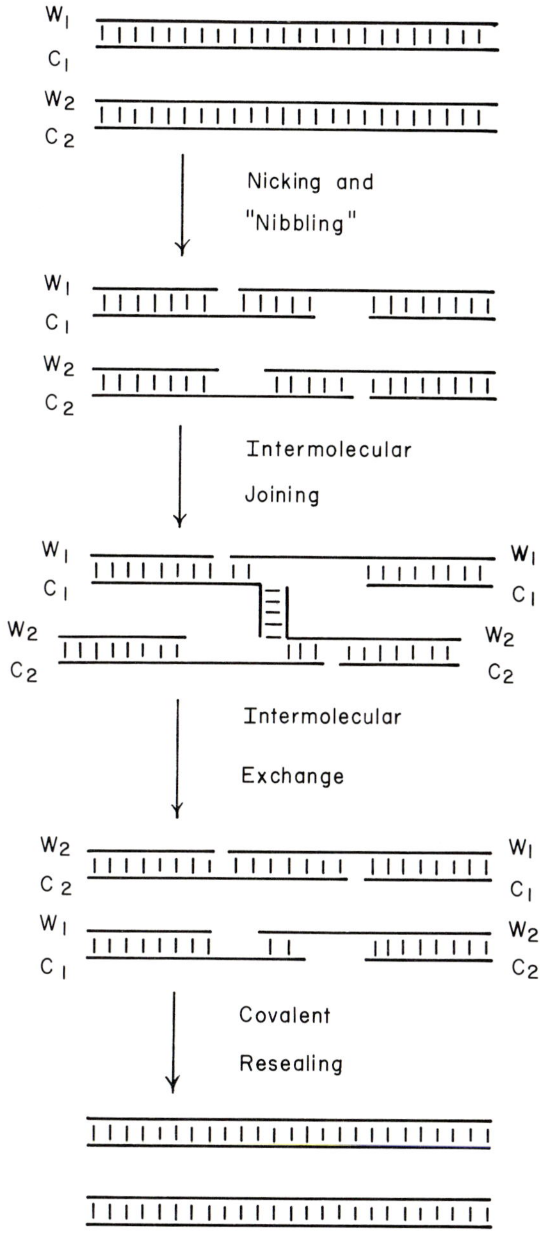

FIGURE 7-9 A schematic model for recombination. The symbols w_1 and w_2, and c_1 and c_2, serve to identify the two strands in each of the parental DNA molecules.

the repair of nicked intracellular DNA. Kozinski's interpretation is that the nicking enzyme (which has not yet been identified among the various nucleases which are known to be T4-induced) is formed quite early in infection, and that CM inhibits its synthesis. The repair enzyme (or enzymes) are formed somewhat later, as shown by the fact that CM can be added after the nicking has commenced, and still prevent repair. In this context it is worth noting that CM can, under certain defined conditions, "rescue" a gene 30 (DNA ligase) mutation (Kozinski and Mitchell, 1969). When CM is added 5 minutes after infection with a gene 30 mutant under restrictive conditions, a pool of DNA fragments, of molecular weight around 20 million, accumulates. Upon removal of the CM, this pool can mature to form infective progeny. In control cultures not treated with CM, intracellular DNA is degraded to small pieces, and very few of the infected cells form infective progeny. Kozinski and Mitchell suggest that at 5 minutes DNA synthetic enzymes have been formed, but not enough of the nicking enzymes to degrade much of the intracellular DNA. In this light, at least one metabolic role of DNA ligase is seen as repair of viral DNA which happens to be attacked by phage-coded nucleases. When the nuclease(s) is absent, the ligase becomes dispensable (or, perhaps, replaced by the host-cell ligase).

Studies on the fate of parental DNA in nonpermissive infection by T4 *am*A453, a gene 32 amber mutant, show that it is cut, rather than nicked (Kozinski and Felgenhauer, 1967). Intracellular DNA from such infected cultures sediments more slowly than mature DNA in neutral, as well as alkaline, sucrose gradients. In addition, part of the parental DNA becomes acid-soluble, indicating significant degradation to low-molecular-weight material. Finally, the cut fragments of A453 intracellular DNA show no single-stranded regions, as judged by several criteria. One reasonable explanation of these data is that in infection by a gene 32 mutant the exonuclease digestion proceeds too far, such that there remain no single-stranded regions, necessary for formation of an end-to-end overlap structure. By this reasoning, the action of the gene 32 product is seen as controlling the exonuclease digestion of strands of nicked DNA, preventing it from going too far. However, any specific theory of gene 32 action should take into account the fact that the gene 32 product is also required for replication. In this context, it should be observed that purified gene 32 product binds strongly to single-stranded DNA (Alberts et al., 1968). One possible explanation for its action in replication is that it stabilizes single-stranded regions which exist transiently near the growing point. As such, it might reduce the susceptibility of these regions to nucleolytic attack.

More recent results by Alberts and Frey (1970) support the above conclusion. The gene 32 product has now been purified to homogeneity. It has a molecular weight of 35,000 and is formed to the extent of about

10,000 molecules per cell. The protein binds strongly to single-stranded, but not double-stranded, DNA, with a stoichiometry of about 1 molecule of protein per 10 nucleotide residues. Addition of the gene 32 protein to a solution of denatured DNA, under annealing conditions, results in a 100-fold stimulation of the rate of renaturation. Alberts and Frey visualize a situation in infected cells wherein the reversible binding of the gene 32 protein to DNA results in fluctuating regions of single-strandedness in DNA, a situation essential in order for replication to occur. Seen in this light, the gene protein appears to play a primary role in replication, but its action in stabilizing single-stranded regions in DNA helps to explain its function in recombination as well.

What enzymes are involved in the final sealing of recombinant molecules? From their known actions *in vitro,* it is reasonable to propose that DNA polymerase and polynucleotide ligase play roles in this process. The role of the ligase is of particular interest, inasmuch as it was originally discovered in searches for enzymes which might be participating in recombination. However, results obtained to date in several laboratories are somewhat equivocal. Broker, Anraku, and Lehman (1969) have used a double amber mutant defective in genes 30 (ligase) and 43 (DNA polymerase) to generate hydrogen-bonded structures of the Tomizawa type. These structures, when examined in the electron microscope, show branch points, as might be expected if they represent incomplete recombinant molecules. Moreover, they can be sealed *in vitro* to covalently bonded structures, but purified ligase and polymerase are both required since the structures contain gaps some 300–400 nucleotides long (Anraku, Anraku, and Lehman, 1969). Although this would suggest that ligase and polymerase are the enzymes involved in this phase of repair, Kozinski (1968) has obtained evidence for molecular recombination occurring *in vivo* in non-permissive infection by a gene 30 amber mutant. Moreover, Tomizawa (1967) has shown that hydrogen-bonded joint molecules can be converted to recombinant molecules *in vitro* with the aid of an extract prepared from cells infected with T4 *am*X5, a multiple mutant containing amber mutations in genes 41 through 45. Indeed, some sealing action was evident with an uninfected cell extract. However, it should be recalled that uninfected *E. coli* contains both polymerase and ligase, and the extent to which the bacterial enzymes can function in intermolecular recombination in the absence of either or both of the corresponding viral enzymes is not known.

EARLY AND LATE PROTEINS

We have already indicated that phage-specific proteins fall into two general categories, from the standpoint of time of synthesis in the infected cell. In general, the enzymes of DNA metabolism are formed during the first ten to fifteen minutes of the latent period, while the other proteins, dealing primarily with virus structure and assembly, begin to appear at about ten minutes and continue to be synthesized throughout the remainder of the latent period. Thus, the phage-infected bacterium presents an example of sequential gene expression, which can, in a sense, be studied as a model system for understanding cellular differentiation. In other words, we can ask what mechanisms turn off the formation of early proteins and activate the synthesis of late ones?

DNA Replication and the Control of Protein Synthesis

An involvement of DNA replication in effecting the switch from early to late gene expression has long been suspected. Dirksen, Wiberg, Koerner, and Buchanan (1960) found that irradiation of T2 phage with ultraviolet light prior to infection abolished control of early enzyme synthesis: the formation of early enzymes, which in their hands normally had ceased by 15 minutes, continued for up to two hours with the uv-irradiated phage, with final specific activities attained being four- to sixfold higher than the normal values. Similar results were obtained by Delihas (1961). Although the precise effect of uv irradiation was not clear, it was known that uv-irradiated phage are defective in DNA replication (Vidaver and Kozloff, 1957). Since DNA replication, as we have seen, commences at about seven minutes, it appeared reasonable to propose that some event connected with initiation of replication allowed late genes to be expressed at the expense of early ones. More direct evidence on this point was presented by Wiberg, Dirksen, Epstein, Luria, and Buchanan (1962), who studied the kinetics of early enzyme synthesis in cells infected with DNA-defective amber mutants of T4. Those mutants which were most severely blocked in DNA synthesis gave results very similar to those seen with uv-irradiated T2, while mutants in which DNA synthesis was not completely blocked showed intermediate behavior.

Concomitant with a failure of turn-off of early enzyme synthesis is a failure of late protein synthesis to begin. Thus, phage irradiated with sufficient doses of uv light do not induce the synthesis of virus structural proteins (Watanabe, 1957), or of lysozyme, which has generally been considered as a late protein (Sekiguchi and Cohen, 1964; Ebisuzaki, 1965). Similarly, nonpermissive *E. coli* strains infected with DNA-negative amber mutants of T4 fail to induce detectable levels of lysozyme (Mathews, 1966a). Within the limits of detectability, late gene expression does not occur in such cultures. Sercarz (1966) devised a fluorescent antibody technique for detection of T4 structural proteins in single cells. Even with this sensitive method, no late proteins were seen in cells infected with DNA-negative mutants, although amber mutants defective in late genes were able to form structural proteins.

Several workers have attempted to describe in detail the precise relationship between DNA replication and the activation of the early-to-late "switch." As stated above, Sekiguchi and Cohen (1964) reported that uv irradiation of T6 prior to infection leads to failure of lysozyme synthesis, as well as prolonged and exaggerated formation of early enzymes. However, early enzyme synthesis was shut off normally, and lysozyme was formed, in cells infected in the presence of 5-fluorodeoxyuridine (or FUDR which, it will be recalled, is converted *in vitro* to a potent inhibitor of thymidylate synthetase). On the other hand, DNA synthesis was not completely inhibited in these cultures, and the precise extent of residual synthesis was not determined. Similar studies were carried out in the author's laboratory (Mathews, 1966a), in which an attempt was made to control DNA synthesis by limiting exogenous thymine in a thymine-requiring strain infected by T4 *td*8, a thymidylate synthetase-negative phage mutant. However, the kinetics of early and late protein synthesis in this system were identical to those seen in normal infection. It turns out that DNA synthesis in this system is not completely blocked, for a small pool of thymine nucleotides (about 30 phage-equivalents per infected cell) is released *via* degradation of bacterial DNA. Volkin and Ruffilli (1962) had reported that FUDR, as a secondary effect, leads to an inhibition of host cell DNA breakdown in T2-infected cells. The inhibition in T4-infected cells, as we found out, is much less effective. However, when FUDR was used in conjunction with the thymine-deprived *td*8 system, we found that the onset of lysozyme appearance is delayed by about 5 minutes, and the maximal rate of synthesis is not attained until 20 minutes after infection, at which time about five phage-equivalents per cell of newly-formed DNA is present. Since the corresponding figure for the nonFUDR-treated culture was nine, we could estimate that between five and nine phage-equivalents of new DNA must be synthesized before a late gene can be expressed normally.

In the same study, we carried out experiments on rates of incorporation of leucine-C^{14} into protein under various conditions of blockage of DNA synthesis. The results were in accord with the idea that sustained DNA replication is essential for continued protein synthesis to occur. However, we also found that in the FUDR-treated, thymine-deprived culture, early enzyme synthesis was shut off as in normal infection. Thus, if DNA synthesis is involved in the arrest of early functions, the quantitative requirement is considerably less than that needed for full expression of late genes.

Following the above rather inconclusive results, we attempted to develop a system wherein DNA synthesis could be controlled by exogenous addition of 5-hydroxymethylcytosine or one of its derivatives. For this we used several amber and *ts* T4 mutants which were defective in gene 42 (the structural gene for dCMP hydroxymethylase). However, none of these mutants responded, either to HMC or to the corresponding nucleoside (Mathews, unpublished results), suggesting that *E. coli* cannot carry out either the appropriate deoxyribosyl exchange reaction or phosphorylation.

The nucleotide, 5′hydroxymethyldeoxycytidylate, was also ineffective, even when the permeability of the cells was increased by pretreating them with EDTA, as described by Leive (1965).

Further evidence suggesting that DNA replication might not be a prerequisite for arrest of early enzyme synthesis was obtained in some experiments which I carried out with Richard Kessin (Mathews and Kessin, 1967). We found that one of the gene 42 *ts* mutants mentioned above, namely L13, was completely blocked in DNA synthesis at 44°, within the limits of sensitivity of our methods. Although no lysozyme was formed under these conditions, we did observe normal shutoff of early enzyme synthesis. We also presented evidence that cells infected at 44° retain the capacity to synthesize early enzymes through at least the first 30 minutes of infection. Thus, the "shutoff" of which we speak is not a complete arrest, but rather a decrease in the rate of synthesis of early enzymes.

Further evidence against a direct involvement of DNA replication in the arrest of early functions was provided by P. S. Cohen (1968). In a bacterial mutant unable to concentrate potassium from the medium, protein synthesis, to a much greater extent than either DNA or RNA synthesis, is dependent upon the potassium concentration of the medium. In limiting potassium, the formation of dCTPase is shut off at about the normal time, and DNA replication commences, although at a slightly reduced rate. However, the cessation of dCTPase synthesis is reversible; formation of this enzyme begins again when potassium is added to a high concentration relatively late in infection. Thus, Cohen argues that a repressor of early functions is itself an early gene product whose formation is dependent upon the presence of potassium. However, Cohen does not appear to have asked whether the initial shutoff of dCTPase synthesis is not actually due to a cessation of all protein synthesis in the presence of limiting potassium. In other words, are his observations physiologically significant, or do they merely reflect the gross pathology of the mutant?

Within the past two years it has become possible to control T4 DNA synthesis experimentally by regulating the concentration of thymine available to the infected cells. As mentioned above, my own studies in this area were inconclusive because of the release of thymine nucleotides from degradation of bacterial DNA, a process which was incompletely inhibited in the presence of FUDR. However, the finding that T4 mutants in genes 46 and 47 are unable to complete the breakdown of host cell DNA (Wiberg, 1966) presented a method of controlling this process genetically. Bolle, Epstein, Salser, and Geiduschek (1968a) infected a thymineless *E. coli* strain with a *td*-46 double mutant and showed that the expression of late genes in this system is completely dependent upon exogenous thymine. More recently, Lembach, Kuninaka, and Buchanan (1969) presented similar results, using FUDR treatment in conjunction with the use of a gene 46 mutant. In addition, these latter authors presented evidence that the continued synthesis of late proteins is dependent upon continued DNA synthesis late in infection.

Timing of Synthesis of Individual Proteins

So far we have classified all phage-directed proteins as "early" or "late," depending upon their time of synthesis. Most of the work we have referred to has measured the expression of early or late functions by assaying for a small number of specific phage-coded enzymes. Levinthal and his collaborators (Levinthal, Hosoda, and Shub, 1967; Hosoda and Levinthal, 1968) have devised a method of simultaneously following the synthesis of a large number of individual proteins. Cells are pulse-labeled with leucine-C^{14} or sulfate-S^{35} for two-minute periods. Extracts are prepared and subjected to disc gel electrophoresis. The gels are then dried and subjected to radioautography. Thus, one can, by labeling for various times covering the entire latent period, determine the times at which individual proteins are synthesized (although, of course, the bands seen on the radioautograms are not immediately identifiable as specific proteins). At any rate, in infection at 37°, all proteins formed do indeed fall into two major classes, those whose synthesis begins earlier than 7 minutes after infection, and those whose synthesis begins later. When similar experiments were done at 25° to extend the latent period, it was seen that there is considerable heterogeneity in the times at which the synthesis of specific proteins begin and end. In general, however, there appeared to be four rough classes of phage proteins, which are named A, B, C, and D. Class A proteins are synthesized between 0 and 10 minutes. Class B proteins also are formed from the onset of infection, but their synthesis continues until about 22 minutes at 25°. Class C protein are formed between 5 and 26 minutes, while Class D, the true late, proteins begin to be synthesized at about 20 minutes and continue for the remainder of the latent period. Clearly, the regulation of phage protein synthesis is more complex than a simple division into the two categories of "early" and "late."

Quite recently Lembach and Buchanan (1970) have presented an alternate approach to the timing of synthesis of individual proteins. This involves assaying for specific T4-induced enzymes in protoplasts of infected cells treated with inhibitors of nucleic acid or protein synthesis. Lembach and Buchanan find at least three classes of early enzymes. Messenger RNA for the first class can accumulate in the absence of protein synthesis, while mRNA's for the other classes are either partially or absolutely dependent upon protein synthesis for their formation. These findings are quite in accord with studies on the timing of mRNA synthesis, to be discussed below.

We stated earlier that late proteins are not formed in infection by DNA-negative amber mutants of T4. Levinthal and his co-workers confirmed this and showed also that late proteins are not formed by maturation-defective mutants either, even though DNA synthesis does occur. Thus, DNA replication is seen as a necessary, but not sufficient, condition which must be met in order for late genes to be expressed. In addition, the synthesis of early enzymes in cells infected by MD mutants is prolonged and extended, as in infection by DO mutants. This agrees with the idea, expressed earlier, that the mechanism for arrest of early enzyme synthesis is not directly related to the onset of DNA replication.

TRANSCRIPTION OF PHAGE GENES

Metabolism of T4 mRNA

General properties of T4 RNA. In Chapter 4 we reviewed the principal evidence which established the existence of a virus-specific messenger RNA in T-even phage-infected cells. In this chapter, I would like to further explore the biological roles of T4 mRNA molecules. This section will be concerned primarily with results deduced from *in vivo* experiments. In the next section, we shall discuss the enzymology of T4 synthesis and attempt to see how results of these experiments, which were done largely *in vitro,* can be related to occurrences in the intact cell.

From our current knowledge of the transmission of genetic information, it is clear that messenger RNA, unlike other classes of RNA, is a heterogenerous collection of molecules. Ribosomal RNAs are relatively homogeneous within a species, while transfer RNA molecules, although heterogeneous with respect to nucleotide sequence, all seem to be similar with respect to size and probable three-dimensional conformation. However, each gene (or operon) encoded in DNA directs the synthesis of a messenger RNA molecule which in turn carries the instructions for synthesis of a specific protein, or small group of coordinately controlled proteins. Thus, a population of mRNA molecules should reflect the heterogeneity in size which is known to exist among the protein molecules of a given organism. It has been difficult to obtain reliable experimental data on this point, because of the difficulty of proving that the mRNA which one isolates has not been partially degraded during the experimental workup. In early studies (cf. Spiegelman, 1961), pulse-labeled RNA in T4-infected cells sedimented in sucrose gradients as a broad peak with sedimentation coefficients in the range of 8 to 14S. Much greater heterogeneity was seen, however, when greater care was taken to minimize degradation of the RNA (Sagik, Green, Hayashi, and Spiegelman, 1962). Radioactivity was dispersed throughout the entire gradient, encompassing the region from 4 to 25S. However, Volkin and Riggsby (1969) have recently provided evidence that the bulk of T4 mRNA molecules sediment at about 11S, which would be expected of a single-stranded polynucleotide containing some 600 nucleotides. Using the known residue weights of amino acids and nucleotides, and the knowledge that the coding ratio is 3, this corresponds to a molecular weight of about 20,000 for the average phage-directed protein. This may seem small in comparison with the range of molecular weights which have been determined for individual proteins (see Chapters 3 and 5). However, the average chain length of phage messengers was confirmed by a method not requiring gentle isolation of the RNA. A pulse-labeled culture was extracted to remove extraneous radioactivity, and the RNA remaining was subjected to alkaline hydrolysis. This releases the 3′ terminus as a nucleoside, the remainder of each molecule as a nucleotide. Ion-exchange chromatography of the hydrolysate revealed a nucleotide-to-nucleoside ratio of about 600 to one. Thus, it would appear that most T4 mRNA molecules specify rather small proteins, and, in the case of proteins with higher molecular weights, it may be appropriate to ask whether some of them are composed of smaller subunits.

The fact that intragenic complementation can be demonstrated in many, if not most, T4 genes (H. Bernstein, personal communication) is in accord with this conclusion. On the other hand, the results of Volkin and Riggsby seem to be contradicted by demonstrations, involving polyacrylamide gel electrophoresis, that a sizable proportion of phage mRNA exists as high-molecular-weight species (up to 2.2 million) in cells infected with T4 (Adesnik and Levinthal, 1970), T5 (Sirbasku and Buchanan, 1970) or T7 (Summers, 1969). Another interesting feature of these latter studies is the finding that much of the phage mRNA present at any point in time exists as discrete, and presumably mature, molecular entities, in spite of the known metabolic instability of mRNA, as discussed below.

As we mentioned in Chapter 4, messenger RNA is metabolically unstable. This probably results, at least in part, from its lack of secondary structure, which renders it more susceptible to nucleolytic attack than the more highly ordered ribosomal or transfer RNAs. Fom an anlysis of the products of degradation of T6 mRNA *in vitro,* Sekiguchi and Cohen (1963) concluded that the enzymes responsible for this process are polynucleotide phosphorylase, which generates nucleoside 5′-diphosphates, and a potassium-activated phosphodiesterase, which generates 5′-monophosphates. With regard to the latter enzyme, it is of interest that mRNA is more stable in potassium-starved cells than in well-nourished cultures (Cohen and Ennis, 1966).

More recently Kuwano et al. (1969) have described a new enzyme in *E. coli,* called RNase V, which must be seriously considered as a candidate for the enzyme which degrades mRNA *in vivo.* This enzyme is specifically active against mRNA preparations *in vitro.* Moreover, the enzyme is inhibited by agents such as chloramphenicol and streptomycin, which have long been known to prolong the lifetime of mRNA *in vivo.* Finally, the enzyme exists in intact cells in close association with the ribosomes.

Several workers have attempted to quantitate the metabolic instability of T4 mRNA. Greene and Korn (1967) measured the rate of loss of radioactivity from pulse-labeled RNA after administration of actinomycin D to infected cells which had been rendered permeable to the drug by treatment with EDTA. Label is lost as a first-order process, with a half-life of 3.5 minutes at 37°. No differences in rates of decay were seen at different times in infection. Similar results have been seen in the author's laboratory, in pulse-chase experiments involving uracil-C^{14} as a labeling agent (Mathews, 1968). And, as we shall see in the next section, the same conclusion has been reached in studies involving quantitative DNA-RNA hybridization.

Temporal control of transcription. We have discussed above the fact that individual phage proteins are sythesized at rather specific times during the latent period. *A priori* one could expect control of this process to be exercised either at the level of transcription or translation. In other words, either (a) different portions of the genome are transcribed at different times after infection, or (b) the entire genome is transcribed throughout the latent period, but only certain messengers are translated at particular times. By the rather gross criteria of sedimentation profiles in sucrose gradients and quantitative nucleotide composition, no significant differences are observed between RNA pulse-labeled early or late in infection. Kano-Sueoka and Spiegelman (1962) obtained the first evidence that different messengers are synthesized in different portions of the latent period. The approach was to pulse-label one culture early in infection with uridine-C^{14} and another culture with uridine-H^3 late in infection. The cells were mixed, and RNA was extracted and fractionated on a methylated albumin-kieselguhr (MAK) column, which resolves nucleic acid species primarily by differences in size and nucleotide composition. Elution profiles for the two isotopes were different, a clear indication that different mRNA species are formed at different times in infection.

It became possible to state the above conclusion in quantitative terms, with the development of rapid methods for measuring the extent to which an RNA sample could form a specific hybrid molecule with denatured DNA (Nygaard and Hall, 1964; Hall, Nygaard, and Green, 1964). As had been previously shown, RNA pulse-labeled early in T2 infection (from 0 to 6.5 minutes) hybridizes specifically with T2 DNA, but not with *E. coli* DNA. When the hybridization is carried out in the presence of excess unlabeled RNA isolated at 6.5 minutes, complex formation with labeled RNA is completely inhibited, as shown in Figure 7–10. This is to be expected from the fact that all of the molecular species present in the labeled RNA are also present in the unlabeled preparation. However, when the unlabeled competitor is prepared from cells harvested late (19 minutes) in infection, complete inhibition of hybridization was also seen. On the other hand, when "late" labeled RNA (i.e., pulse-labeled from 15

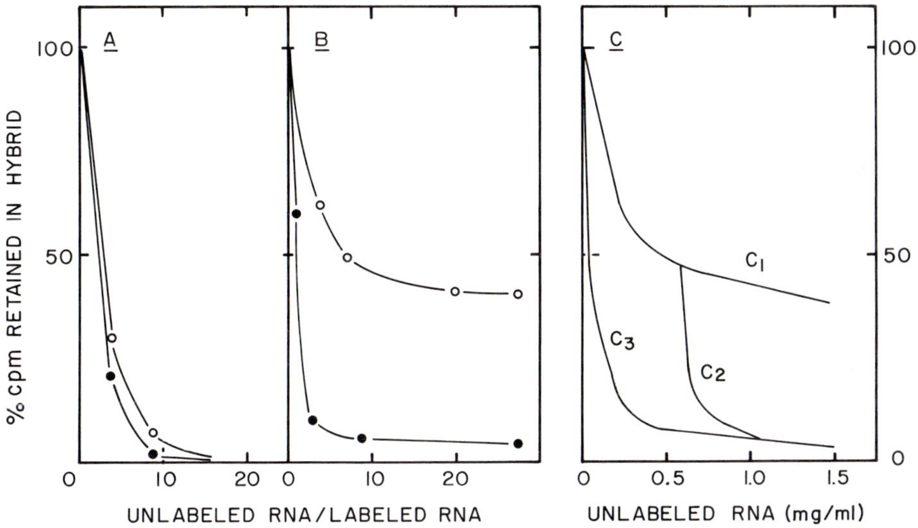

FIGURE 7-10 Use of hybridization-competition to study the control of phage gene transcription. Panel A. Pulse-labeled early RNA was hybridized with phage DNA in the presence of excess unlabeled early RNA (open circles) or late RNA (close circles). Panel B. The same experiment, carried out with pulse-labeled late RNA (Data for these two panels from Hall, Nygaard, and Green, 1964). Panel C. Hybridization with mixed competitors. Late pulse-labeled RNA was competed with unlabeled early RNA (c_1), unlabeled late RNA (c_3), or a mixture containing 0.6 mg/ml of early RNA and varying amounts of late RNA (c_2). The ordinate indicates the total concentration of competing RNA (early and late). Modified from Bolle et al. (1968b).

to 19 minutes) is treated with excess "early" unlabeled RNA, inhibition of hybridization is not complete, and tapers off with about 40 percent of the labeled RNA still complexed. These results show that, within the limits of detection, all of the RNA species being synthesized early are still present in the cell at a late time, while there are some messengers being synthesized late which are not present early. Thus it would appear that the control of gene expression is being exercised both at the levels of transcription and translation.

Using a method called "annealing with mixed competitors," Bolle, Epstein, Salser, and Geiduschek (1968b), further increased the sensitivity of the above method. A series of hybridizations of late labeled T4 RNA was carried out in the presence of a constant amount of unlabeled early RNA, sufficient to exclude most of the early RNA species present from the hybrids. Varying amounts of unlabeled late RNA or additional early RNA were added to different tubes, the objective being to compare their competing ability against the remaining late species whose binding was not inhibited by the early unlabeled RNA present. The slopes of the initial portions of the competition curves were compared to calculate the relative amount of late RNA species in the early unlabeled preparation (see Figure 7–10). By this criterion, it was estimated that the late species are several hundredfold—perhaps several thousandfold—less abundant than the early species at early times in infection.

It is possible, *a priori,* to suggest that the above results reflect at least in part a difference in stability between early and late mRNAs. For example, the presence in infected cells at 19 minutes of a species which could completely inhibit the binding of early RNA might reflect the possibility that early RNA is less rapidly degraded than late RNA, so that it is still present at 19 minutes, even though it is not being synthesized. If this is so, then long pulse-labeling periods might be expected to lead to preferential labeling of the longer-lived species. However, Bolle et al. found that RNA pulse-labeled from 1 to 20 minutes behaved the same as RNA labeled from 15 to 20 minutes, indicating that the great bulk of T4 messenger molecules have equal metabolic stabilities.

In the paper discussed above, Bolle et al. also mentioned the existence of several classes of early RNA, whose synthesis is activated at different times within the first few minutes of infection. These include "pre-early" species, formed between 0 and 1 minutes at 36°, "early" RNA, formed between 1 and 5 minutes, "quasilate" species formed primarily between 5 and 20 minutes, and "late" RNA, whose formation begins at about 15 minutes and continues for the duration of the latent period. Similar evidence for the existence of more than one species of early RNA has also been presented by Grasso and Buchanan (1969). Thus, the pattern of transcription of the T4 genome *in vivo* approximates the complexity of the timing of synthesis of individual phage proteins, which we discussed on p. 161.

In addition to its time of synthesis, pre-early mRNA species are distinguished from the other classes of early RNA by virtue of the fact that these are the only T4 messengers which can be formed in the absence of protein synthesis (*cf.* Salser, Bolle, and Epstein, 1970). We shall discuss the significance of this a bit later.

No matter how many classes of early messenger actually exist, the work of Bolle et al. directly established that the intracellular concentration of late species present early in infection is negligible. This is at variance with conclusions reached earlier by Edlin (1965), on the basis of genetic data. Edlin found that transient exposure of 5-fluorouracil to nonpermissive *E. coli* strains infected with T4 amber mutants led to phenotypic reversion of the amber mutations; productive infection was seen, even though the mutant genotype was not altered. As we discussed in Chapter 4, this probably results from incorporation of FU into mRNA, whence it is occasionally translated as cytosine. Edlin found that mutants in late-function genes could be reverted by pulses of FU administered early in infection, and from this he concluded that late messengers were being synthesized early in infection. However, as shown by Haselkorn, Baldi, and Doskoçil (1967), FU which is incorporated into early RNA during a pulse can subsequently be turned over and reincorporated into late RNA during a chase. Since Edlin did not prove that all of his administered FU was washed out before the infected cells were plated, it is entirely possible that his results were due to turnover of FU-containing early RNA, with incorporation of FU into late RNA subsequent to the pulse.

A potentially more serious conflict was presented by the results of Bautz and his collaborators (Bautz and Reilly, 1966; Bautz, Kasai, Reilly, and Bautz, 1966; Kasai and Bautz, 1967). These workers devised techniques for the measurement of mRNA molecules specific to particular genes. The method requires the existence of a phage mutant containing a deletion of an entire gene. Such mutants are available for the *r*II gene and the *e* gene (which, it will be recalled, is the structural gene for lysozyme). Denatured DNA from one such mutant is packed in a nitrocellulose column, over a layer of denatured wild-type DNA. Pulse-labeled RNA is passed through the column, and gene-specific mRNA is assayed as that material which passes through the top layer, but is bound at the second layer (i.e., which binds to wild-type but not to deletion DNA). By this test, Bautz and his colleagues found that *r*II-specific RNA, as expected for an early-function messenger, is synthesized throughout the latent period. However, these workers found that the lysozyme gene is transcribed within the first three minutes of infection. Synthesis of this RNA then shuts off and resumes at about 7 minutes. Apparently, there exists some translation-level control mechanism early in infection which prevents synthesis of active lysozyme, even though the messenger is present. This control mechanism, whatever its nature, must be quite effective, for a DNA-negative amber mutant also showed the early burst of lysozyme RNA synthesis (but no late synthesis), even though, as we have seen, no lysozyme is formed under these conditions of infection.

Bautz and his co-workers proposed that each late gene is transcribed once quite early in infection and that some feedback mechanism acting at the level of a coupled transcription-translation process, prevents further late mRNA synthesis during the early period. However, the findings of Bolle et al. appear to be at direct variance with this. It is now clear, however, that lysozyme is not a typical late gene product, and that the regulation of lysozyme formation cannot be studied as a true model for control mechanisms governing the expression of all late genes. This conclusion came from studies originating in Szybalski's laboratory, where it was found that the two strands of denatured T4 bind polyuridylic acid differentially, and that this permits a preparative separation of the two strands, the so-called H and L strands, by density-gradient methods (Guha and Szybalski, 1968). By measuring the extent of hybridization of pulse-labeled RNA with each strand, Guha and Szybalski found that all of the RNA made early in infection hybridizes with the L strand, while later in infection more and more of the RNA is transcribed from the H strand. This raises the possibility that all early-function genes are transcribed from the L strand, and late-function genes from the H strand. The activation of late gene expression, then, is seen as the result of some strand-selection mechanism which allows the H strand to be transcribed. In accord with this idea is the fact that the direction of transcription of all known late-function genes is clockwise as the map is nomally drawn (Sarabhai et al., 1964; Stahl et al., 1966), while the early-acting *r*II gene is transcribed counterclockwise. Similarly, the *e* gene is transcribed counterclockwise, like an early gene. Kasai, Bautz, Guha, and Szybalski (1968) found that lysozyme-specific mRNA, as well as *r*II-specific RNA, binds only to the L strand. Thus, although lysozyme *per se* appears as a late gene product, its structural gene is subject to some special type of early-gene control. Thus, it is inappropriate to study lysozyme synthesis as a case of late gene expression, for the control of this gene is probably more or less unique.

With the use of deletion mutants spanning the *e* and *r*II genes, respectively, and extending into adjacent regions, and with technical improvements in their procedures, Kasai and Bautz (1969) have considerably extended their findings. During the early burst of *e* gene transcription, synthesis of the lysozyme messenger is controlled coordinately with the transcription of neighboring genes. At later times, the lysozyme region is transcribed more rapidly than these adjacent regions. The reason why the *e*-specific messenger made early is not translated early is still unexplained. However, as one possible explanation, Kasai and Bautz suggested that the early-formed lysozyme messenger might comprise part of a polycistronic RNA in which the reading frame for lysozyme synthesis is out of phase, or the initiating codon in lysozyme on RNA is not recognized for some other reason. Later, perhaps, a modification in RNA polymerase (cf. below) permits recognition of an initiation site near the region corresponding to the N-terminus of the lysozyme molecule. In this light, the early transcription of the *e* gene is seen as a "harmless accident" (Schmidt et al., 1970). It is also possible that the entire *e* gene is not transcribed during this period (Salser, Gesteland, and Ricard, 1969).

Kasai and Bautz (1969) also determined the number of *e* and *r*II mRNA molecules per cell at different times in infection. The maximum number of *r*II messengers synthesized (from 4 to 6 minutes) was 63 copies per cell, whereas the maximum number of *e* messengers made was 20 per cell, between 12 and 15 minutes. Since the multiplicity of infection was 3, and since Bremer and Yuan (1968b) had shown the rate of RNA chain growth in T4-infected cells is 28 nucleotides per second, this implies that a given cistron is being simultaneously transcribed by several different RNA polymerase molecules. This conclusion is entirely in accord with studies on the transcription of T4 DNA *in vitro* (Bremer, Konrad, and Bruner, 1966).

The above-mentioned work of Kasai and Bautz (1969), and that of Bolle et al. (1968b) appears to rule out conclusively the possibility that all late genes are transcribed once early in infection. What, then, is the mechanism for activation of late gene transcription at the appropriate time? We have already discussed the involvement of DNA replication in the activation of late protein synthesis. From this it is reasonable to expect that the synthesis of the late messengers also requires DNA replication. Several groups have reported that DNA-negative amber mutants of T4 are unable to form late mRNA, using as a test the fact that RNA pulse-labeled late in infection by these mutants can be completely excluded from DNA-RNA hybrids by early unlabeled RNA (Zograf, Nikiforov, and Shemyakin, 1967; Bolle *et al,* 1968a; Mathews, 1968). In addition, Bolle et al. demonstrated two other systems in which late gene transcription is replication-dependent: (1) cells infected by ultraviolet-irradiated T4 (wild type); and (2) *E. coli* B3, infected in the absence of thymine by a *td*-gene 46 double mutant of T4. As mentioned on page 160, use of this system circumvents the problems encountered in the author's laboratory (Mathews, 1966a), of inability to make DNA synthesis completely thymine-dependent because of the release of thymine nucleotides from breakdown of bacterial DNA.

More recently, Riva and Geiduschek (1969) have presented evidence that continued DNA replication must occur for continued formation of late messengers. When DNA-negative *ts* mutants were shifted from 30° to 43° at 18 minutes after infection, both DNA synthesis and the formation of late mRNA came to a halt. Curiously, an exception to this rule was a *ts* gene 44 mutant, which could continue late messenger production after a temperature shiftup. In view of the work of Scotti (1969), however, it is reasonable to expect that a shiftup at 18 minutes would not abolish DNA synthesis in a gene 44 mutant.

Not only must DNA be synthesized, but it must be biochemically "competent," in order to serve as a template for late gene transcription. We mentioned in Chapter 6 that gene 46–56 or 47–56 double mutants can accumulate a reasonable amount of cytosine-containing DNA, since dCTPase is absent, as is the function responsible for degradation of cytosine-containing DNA. However, these mutants cannot form late mRNA (Bolle et al., 1968a), suggesting that cytosine-containing DNA is not a suitable template for transcription of these regions. Recent work of Kutter and Wiberg (1969), in which the relative cytosine-HMC content was controlled by infecting with a *ts* gene 56 mutant at different temperatures, also indicates that DNA must contain HMC in order to serve as a suitable template for late genes.

Clearly, DNA replication is a necessary condition which must be fulfilled if late genes are to be transcribed under normal conditions of infection. However, it is not a sufficient condition. In Chapter 5 we mentioned the existence of genes 33 and 55, mutations in which are maturation-deficient (MD); DNA replication occurs, but no detectable late proteins are formed. The exhaustive study of Bolle et al. (1968a) also showed that gene 33 and 55 amber mutants are unable to form late mRNAs. Although the nature of these gene products is not known, this work indicates that they act as some type of positive control element which must act upon replicating DNA in order to convert it to a competent template for transcription of late genes. We shall discuss further the functions of genes 33 and 55 in the next section, in connection with studies on T4 transcription *in vitro*.

The gene 33 and 55 mutants present one example of "uncoupling" of DNA replication from late gene transcription; DNA replication occurs, but no late gene transcription is detectable. The converse situation has been noted in gene 30–46 and 30–47 double mutants (Riva, Cascino, and Geiduschek, 1970); late gene transcription occurs, even though DNA replication is undetectable. Riva et al. suggest that the "nuclease" reaction controlled by genes 46 and 47 is normally involved in maintaining "competence" of DNA for late transcription, even in the absence of replication, but that DNA ligase, controlled by gene 30, counteracts the nuclease effect. Thus, competence can only be maintained, in the absence of gene 46 or 47 action, when the ligase also fails to function. What the concept of competence actually means in biochemical terms awaits further clarification of the normal functions of genes 46 and 47.

In contrast to most of the above work, which indicates that late transcription and DNA replication are normally coupled, Sauerbier and Braütigam (1970) have reported that late gene expression continues after tempeature shiftup of a *ts* gene 42 mutant (DO; controls dCMP hydroxymethylase). From this they conclude that replication and late transcription are not coupled. I cannot accept this conclusion, however, for it is evident that Sauerbier and Braütigam have not achieved a complete inhibition of DNA synthesis in their system following the temperature shiftup. As we have already seen, complete expression of late genes occurs in the presence of but a small amount of DNA replication (cf. Mathews, 1966a).

As one possible mechanism involved in the early-to-late switch, Shalitin (1967) has proposed that leakage of polyamines from infected cells, a process which has been reported to occur (FerroLuzzi Ames and Ames, 1965), allows the transcription of regions which are normally blocked in the presence of these substances. Presumably this occurs because the leakage allows DNA replication to begin. In support of this, Shalitin presented hybridization-competition data which suggest that late RNA is not formed in cells infected in the presence of putrescine. However, Cohen (1966) has reported that intracellular levels of polyamines actually rise after infection. Moreover, the studies of Bolle et al. (1968a) on the requirements which must be met before a T4 DNA template can serve for transcription of late genes, particularly the need for at least two positive control elements, would suggest that the basic regulatory mechanism is somewhat more complex than that proposed by Shalitin, although changes in the intracellular polyamine concentrations might well be an important step in regulation.

Relaxed vs. stringent control. It has been known for several years that RNA synthesis in most strains of *E. coli* is dependent upon the availability of amino acids. When an auxotroph is starved for an amino acid which it cannot synthesize, or when an inhibitory amino acid analog is present, the rate of RNA synthesis is reduced to a few percent of the normal level. Strains showing this phenomenon are said to be under stringent control of RNA synthesis. Stringent control does not result directly from a blockage of protein synthesis, since the addition of chloramphenicol to amino acid-starved cells relieves stringency, and allows RNA synthesis to proceed at near-normal rates. Instead, some aspect stemming directly from the presence or absence of a complete set of amino acids seems to be involved. The property of stringency is determined genetically at a single locus, the RC gene (Stent and Brenner, 1961). Mutations in this gene give rise to strains which are called relaxed; such strains show no inhibition of RNA synthesis in the absence of amino acids. The biochemical basis for relaxed and stringent control have not yet been elucidated. For a relatively recent review of this subject, the reader is referred to Edlin and Broda (1968).

Several workers have investigated the extent to which T-even phage nucleic acid synthesis is dependent upon the presence of amino acids. Kellenberger, Lark, and Bolle (1962) showed that T-even DNA synthesis is subject to stringent control, but no data for RNA synthesis were presented. Sekiguchi and Cohen (1964) presented evidence that, in a polyauxotrophic strain of *E. coli* 15, phage RNA can accumulate to higher-than-normal levels in the absence of a required amino acid. They concluded, therefore, that phage mRNA synthesis was not subject to stringent control. (In this context, it should be observed that recent studies on uninfected bacteria seem to indicate that stringent control is directed specifically toward ribosomal and transfer RNAs and does not directly affect bacterial mRNA formation). Using the fluorouracil rescue technique for measuring T4 mRNA synthesis, however, Edlin (1965b) concluded that this process is subject to stringent control, but that the degree of stringency depends upon the bacterial strain used. Similar variability with respect to strain used was seen by several others workers (Sharp and Green, 1968; Olsnes and Hauge, 1968; Friesen, 1969). An additional complicating factor arises from the point, discussed in Chapter 4, that the synthesis of one or more phage-specific proteins must occur in order for the formation of bacterial macromolecules to be completely inhibited. Sharp and Green (1968) found that T2 infection of an amino acid-deprived strain in stringent control led to a stimulation of RNA synthesis. This would suggest that different mechanisms exist for maintaining stringent control in infected and uninfected cells. At any rate, it is clear that the apparently contradictory results obtained in different laboratories cannot be resolved until the biochemical basis for stringent and relaxed control has been better clarified.

Enzymology of Phage mRNA Synthesis

The work described in the previous section established two central aspects of mRNA metabolism in T-even phage-infected cells: (1) Regulation of gene expression is mediated in large measure by temporal control of the transcription of the phage genome; and (2) in order for most late-functioning genes to be transcribed, both DNA replication and the positive action of at least two gene products must occur. While these ideas were being put forth, workers in other laboratories have attempted to clarify the enzymological bases for these complex regulatory phenomena. Specifically, (1) Does transcription of the viral genome utilize bacterial or virus-specific enzymes? (2) Are different RNA-synthesizing enzymes used for synthesis of different classes of phage-specific mRNA? and (3) Can any of the regulatory mechanisms operative *in vivo* be reproduced *in vitro,* and, if so, what can this tell us about intracellular control mechanisms?

As an initial approach to these questions, workers in several laboratories (Khesin et al., 1962; Green, 1964; Luria, 1965; Geiduschek et al., 1966) examined the RNA product formed when DNA from mature phage particles was offered as template to purified preparations of the DNA-dependent RNA polymerase of *E. coli* (Stevens, 1960; Chamberlin and Berg, 1962; Furth, Hurwitz, and Anders, 1962). Although results from these four laboratories did not agree in every particular, it was generally found that hybridization of the radioactive *in vitro* product to T4 DNA could be completely inhibited by excessive amounts of unlabeled *in vivo* RNA, isolated five minutes after infection. In other words, only early RNA species are formed in this system. (Interestingly, a recent report by Bishop and Robertson (1969) shows that purified RNA polymerase of *Micrococcus lysodeikticus* can transcribe both early and late regions of the T4 DNA molecule *in vitro.*) Furthermore, it was inferred, although not proven at the time, that this RNA synthesis *in vitro* is asymmetric, i.e., that only one strand of a given region of double-stranded DNA is transcribed. Extensive evidence for asymmetry has now been obtained (cf. Brody, Diggelmann, and Geiduschek, 1970). These studies, then, laid the groundwork for asking what changes in the DNA template *in vivo,* or in the RNA polymerase, or in the intracellular environment, permit the transcription of "late" regions of the T4 genome.

As one approach to this latter question, Hall and Crouch (1967) and Snyder and Geiduschek (1968) examined RNA synthesis *in vitro,* in crude extracts prepared by gentle lysis of infected cells undergoing late RNA synthesis at the time of harvesting. Such systems do synthesize late RNA, identified as such by its failure to be completely excluded from a hybrid with T4 DNA in the presence of competing RNA prepared from cells harvested at 5 minutes after infection. Such systems, of course, are relatively complete and do not show dependence upon added template. However, Snyder and Geiduschek were able to directly demonstrate a requirement for the action of the product of gene 55. Both groups had found that lysates of DNA-negative amber mutants fail to make late RNA *in vitro,* as expected from their inability to form these species *in vivo.* A centrifugal pellet fraction containing RNA polymerase was prepared from cells infected with a gene 55 amber mutant. As expected, this system makes no late RNA. However, when this pellet was briefly incubated with a supernatant fraction from cells infected with a gene 42 mutant (DO; this fraction contained no RNA polymerase activity), late RNA was formed. A series of control experiments confirmed that the agent in the supernatant fraction which allows this to occur is, as expected, the gene 55 product. This confirms *in vitro* what Bolle et al. (1968a) had demonstrated *in vivo,* namely, that the gene 55 product is a positive control element whose action is a prerequisite to the expression of late genetic functions (specifically, the synthesis of late messengers).

Although the mechanism of action of the gene 55 product is still not known, several recent studies have illuminated the question (Pulitzer, 1970; Pulitzer and Geiduschek, 1970; Diggelmann, Pulitzer, and Geiduschek, 1970). The gene product is continuously required to exert its positive effect, but blockage of its effect, as in temperature shiftup experiments involving some gene 55 *ts* mutants, is reversible. Although the gene 55 product apparently acts directly upon phage DNA in exerting its positive control effect, it has no direct action in the maturation of replicating DNA.

By way of asking whether a virus-coded RNA polymerase is at all responsible for transcription of the T4 genome *in vivo,* Skold and Buchanan (1964) and Furth and Pizer (1966) assayed extracts of infected and uninfected cells for RNA polymerase activity, using T4 DNA as a template. Both groups found that the activity of this enzyme declines markedly after infection, and they concluded that the phage directs the formation of an inhibitor of RNA polymerase. Walter, Seifert, and Zillig (1968) explored this phenomenon further, and reported that it results from a structural modification in the preexisting bacterial RNA polymerase, which renders it less able to respond to T4 DNA *in vitro,* although there are no pronounced changes in the activity of the enzyme with calf DNA as template. RNA polymerase was purified from infected and uninfected *E. coli.* Examination of the subunit structure of the preparations by disc gel electrophoresis in 6 M urea showed that one subunit of the "T4 enzyme" had a slightly higher mobility than the corresponding subunit of the *E. coli* enzyme. However, no immunological differences between the two preparations were seen. This would suggest that the structural modification in the enzyme is not a major one; perhaps it results from addition of new material to the subunit whose mobility is altered, rather than complete replacement of that subunit.

The above results suggest that T-even phages do not synthesize virus-coded RNA polymerases *de novo,* but rather modify the preexisting cellular enzymes to meet their own needs. With the availability of the related antibiotics rifamycin, rifampicin, and streptovaricin, it became possible to obtain direct evidence on this theory. These agents all have a common mode of action, although they differ in their effectiveness with different bacterial species. All three antibiotics inhibit RNA synthesis by binding to some portion of the RNA polymerase molecule (Wehrli et al., 1968; Sippel and Hartmann, 1968; Mizuno et al., 1968; Tocchini-Valentini et al., 1968). Haselkorn, Vogel, and Brown (1969) isolated a rifampicin-resistant mutant of *E. coli* B. RNA synthesis in this organism is insensitive to rifampicin, both *in vivo* and *in vitro.* Furthermore, the pattern of sensitivity or resistance to rifampicin is maintained after T4 infection; phage RNA synthesis is completely inhibited by rifampicin in infected sensitive cells, and completely unaffected by the drug in cultures of the resistant mutant. This indicates that the *E. coli* RNA polymerase molecule, or at least that part of it which is responsible for rifampicin sensitivity, is used for the synthesis of all classes of T4-specific RNA. Essentially the same findings were obtained by Mizuno and Nitta (1969), using streptovaricin. Moreover, Geiduschek and Sklar (1969) obtained identical results in parallel studies on the *Bacillus subtilis* phage SPO1.

What is the nature of molecular modifications of the *E. coli* RNA polymerase, which allow it to transcribe different templates at different times? The beginning of an answer to this question appeared early in 1969, with a report by Burgess, Travers, Dunn, and Bautz (1969). In brief, these authors found that RNA polymerase normally contains a factor, which is relatively easily removed from the enzyme, and which confers upon the enzyme a specificity for recognition of particular DNA templates. This in turn suggested that control of transcription might be exerted through the action of different factors, which bind to the same RNA polymerase molecule and impart differing template specificities. The implications of this idea, both for regulation in T4 and in higher organisms, are relatively enormous, and it should surprise no one to read that this is currently one of the most active areas in biochemistry. What follows is merely a progress report, written in early 1970, which is certain to be out of date by the time this book is published.

What Burgess et al. (1969) observed was that *E. coli* RNA polymerase, when purified by chromatography on a phosphocellulose column, loses most of its activity for transcribing T4 DNA *in vitro* but retains its activity with calf thymus DNA. However, addition to this "PC-enzyme" of a particular fraction from the column restores its ability to function with T4 DNA. The active material in this latter fraction, which was termed sigma factor, is a protein which complexes with RNA polymerase. It has no detectable RNA polymerase activity when tested separately. Preliminary evidence suggested that sigma acts at the level of initiation of transcription. In other words, the sigma-PC-enzyme complex is able to transcribe T4 DNA *in vitro* because it can recognize initiation sites on this template to which the PC-enzyme cannot respond. This in turn raised the intriguing possibility that the activity of RNA polymerase *in vitro* is regulated by a series of sigma factors. In other words, the changes in transcription patterns known to accompany T4 infection *in vivo* could be explained by formation of new sigma factors, or modifications of the existing ones, which allow the same basic RNA polymerase molecule to recognize and initiate transcription of different groups of genes at different times. As we shall see, most of what has been published since the report of Burgess et al. is in general agreement with this formulation. However, very little is known at this writing about the structure of sigma factors or their precise mode of action. One possible action for sigma is to cause the local unwinding, or opening, of the DNA double helix at a promoter region, or initiation site for transcription (Hinkle and Chamberlin, 1970). Whether or not this is the effect of sigma, there is now good evidence that sigma causes RNA polymerase to bind only at promoter sites (Bautz and Bautz, 1970). Complexes thus formed are resistant to rifampicin, allowing titration of DNA samples with RNA polymerase, to determine the number of promoters per DNA molecule. Results of such titrations are as follows: for T4 DNA, 25 promoters (pre-early sites only); for T5 DNA, 35 promoters; for T7 DNA, 2 promoters; and for λ DNA, 3 promoters.

It has probably not escaped the reader's attention that the modification of *E. coli* RNA polymerase *in vivo* after T4 infection (Walter et al., 1968) results in an enzyme with the same template specificity as that produced by removal of the sigma factor. Bautz and Dunn (1969) showed that the T4-modified polymerase lacks sigma factor, and that addition of sigma to the T4-polymerase *in vitro* causes it to regain activity with T4 DNA as template. Moreover, the same authors (Dunn and Bautz, 1969) showed, by means of a pyrophosphate exchange reaction catalyzed by *E. coli* RNA polymerase, that sigma factor acts at some stage prior to formation of the first phosphodiester bond in transcription: in other words, a confirmation that sigma exerts its effect at some stage of initiation, possibly the step at which the first nucleoside triphosphate is bound to the polymerase-DNA complex.

The idea that sigma factor is involved in RNA chain initiation was questioned by di Mauro et al. (1969). These workers found that the sensitivity or resistance of RNA polymerase to inhibition by rifampicin is determined by some aspect of the structure of the PC-enzyme; specifically, PC-enzyme from rifampicin-resistant cells is insensitive to the drug, while PC-enzyme from sensitive cells is inhibited. These workers also showed that the effect of rifampicin on the PC-enzyme from sensitive cells is at the level of RNA chain initiation. From this they suggested that the rifampicin-binding component of PC-enzyme is the agent controlling chain initiation. However, chain initiation involves several steps, and at present there is no basis for concluding that sigma factor and the rifampicin-binding component cannot both be participating in initiation, at different phases of the process.

By now, the astute reader will be asking, "Why, if the function of the T4-modified polymerase is to transcribe T4 DNA *in vivo,* does it lose most of its activity with this template when assayed *in vitro*?" The answer to this question probably involves in part the production of at least one modified sigma factor, which allows initiation at new sites on T4 DNA, but which is removed during the conventional isolation procedure. However, there appears to be some aspect of the timing of transcription which is inherent in the structures of the RNA polymerase and DNA molecules themselves. When purified *E. coli* RNA polymerase is incubated with substrates and T4 DNA for a brief period, the product can be excluded from a DNA-RNA hybrid either by excess RNA isolated from cells harvested at one minute (one-minute RNA, containing pre-early species), or by five-minute RNA (containing early species) (Milanesi, Brody, and Geiduschek, 1969; Bautz, Bautz, and Dunn, 1969; Brody and Geiduschek, 1970). However, in a longer-term synthesis, the product can be completely excluded by 5-minute, but not one-minute, RNA. Thus, the enzyme selectively transcribes pre-early regions first, both *in vivo* and *in vitro,* before beginning to transcribe early species. In addition, all of the RNA formed by the action of *E. coli* RNA polymerase on T4 DNA *in vitro* hybridizes with the L strand of T4 DNA, just as do pre-early and early RNAs formed *in vivo* (Grau, Guha, Geiduschek, and Szybalski, 1969). The nature of these innate control mechanisms is not known. However, it may involve a limited number of initiation sites for transcription, located next to pre-early genes on the L strands, but not next to early genes. Thus, RNA polymerase containing *E. coli* sigma factor would have to traverse the pre-early regions, transcribing as it went, before it was granted access to the early regions. Consistent with this interpretation are two recent findings of Milanesi, Brody, Grau, and Geiduschek (1970): (1) Sonically degraded DNA serves as a template *in vitro* for the synthesis of pre-early (or "immediate early") species but loses its ability to transcribe early (or "delayed early") species, as expected if the promoter sites are physically separated from the early genes by the disruption; and (2) even when the period of initiation *in vitro* is limited to a very short time, pre-early species are formed immediately, while early RNA appears only after a brief lag. Presumably this lag represents the time needed for RNA polymerase to traverse the pre-early regions.

When the T4-modified RNA polymerase was used to form RNA *in vitro,* as described above, the product contained pre-early, early, and late species, as determined by hybridization-competition experiments (Bautz, Bautz, and Dunn, 1969). While it is true that this resembles the transcription pattern seen *in vivo* late in infection, the situation observed in this *in vitro* experiment probably bears little resemblance to that seen in the intact cell. First, we have already discussed the evidence indicating that the true template for transcription of late genes *in vivo* is replicating DNA, which is not present in these experiments. Secondly, by means of appropriate conditions permitting the detection of RNA-RNA hybrids, Bautz et al. (1969) were able to demonstrate that much of the RNA made *in vitro* by the T4-modified enzyme consists of antimessegers, i.e., RNA molecules complementary to DNA strands which are not transcribed *in vivo.* The addition of *E. coli* sigma factor to the T4 polymerase restored the specific transcription of early and pre-early regions, along with the appropriate strand selection mechanisms seen *in vivo.* Therefore, the work of Milanesi et al., and of Bautz et al., clearly established that the *E. coli* sigma factor plays a positive role in regulating gene transcription *in vivo,* by allowing pre-early genes to be transcribed before early genes, and by preventing the transcription of late genes. It now becomes appropriate to ask how the enzyme becomes modified during infection so that it can transcribe the late regions at the proper time.

Several approaches to the above question have been carried out. The laboratories of Hall (Crouch, Hall, and Hager, 1969) and Zillig (Seifert et al., 1969) have continued their studies on structural changes in RNA polymerase, which might be related to its assumption of the ability to transcribe late genes. Crouch et al. were able to purify the T4-modified RNA polymerase which can transcribe late regions *in vitro.* This enzyme has different chromatographic behavior on DEAE-cellulose than does the enzyme from uninfected *E. coli.* Protein synthesis is required in order for this modification to occur. Moreover, amber mutants in genes 44 (DO) or 33 or 55 (MD) cannot effect the modification under restrictive conditions, as might be expected if the structural modification is indeed related to the change in template specificity which occurs late in infection. Seifert et al. were able to distinguish at least two processes in the modification of *E. coli* RNA polymerase. The loss in ability to transcribe T4 *in vitro* is quite rapid, being nearly complete within one minute after infection, while formation of the altered subunit seen on polyacrylamide gels seems to occur between two and four minutes. The regulatory significance of this alteration is not clear, since it is observed under restrictive conditions with amber mutants in genes 33, 44, and 55, none of which can form late mRNA species. This latter fact would also seem to establish that the subunit alteration is not responsible for the change in chromatographic behavior described by Crouch et al.

It should be noted that Oleson, Pispa, and Buchanan (1969) have observed a transient activation of RNA polymerase within the first minute after T4 infection. While this would seem to be at variance with the results of Seifert et al., Oleson et al. used a gentle lysis procedure for preparation of their extracts. Presumably the activation which they see reflects formation of a T4 DNA-RNA polymerase complex *in vivo,* a complex which is destroyed by the extraction procedure of Seifert et al.

An important paper by Travers (1969) has gone far toward establishing the idea that control of T4 transcription *in vivo* might result from the availability at different times of sigma factors leading to recognition of different initiation sites. Travers reasoned that RNA polymerase as it exists in the intact infected cell would have enhanced activity with a T4 DNA template, rather than decreased activity as observed by other workers. By means of a gentle isolation procedure, he was able to isolate just such an enzyme. Chromatography on DEAE-cellulose yielded polymerase with decreased activity on T4 DNA, similar to the T4-modified RNA polymerase studied in several other laboratories. However, material which was not adsorbed to the column, when added back to this enzyme, led to restoration of activity with T4 DNA, as had been earlier noted in comparable studies with the *E. coli* polymerase (Burgess et al., 1969). However, this latter material, which Travers termed the T4 sigma factor, is different from the *E. coli* sigma factor. Competitive DNA-RNA hybridization tests showed that the T4 and *E. coli* sigma factors stimulate the core RNA polymerase to transcribe different regions of the T4 DNA molecule. It is also of interest that the T4 sigma factor suppresses the ability of *E. coli* polymerase to utilize any non T-even DNA template *in vitro,* although purified enzyme containing the *E. coli* sigma factor also exhibits a pronounced preference for T4 DNA as a template (Su, Weinberg, So, and Davie, 1970). Indeed, this might be partly responsible for the shutoff of bacterial RNA synthesis after infection.

If, as seems reasonable, the T4 sigma factor plays a role in the turn-on of synthesis of a class of messengers, which class is involved? In hybridization studies with separated strands of T4 DNA, Travers showed that both the *E. coli* and the T4 sigma factors promote transcription from the L strand of T4 DNA. It will be recalled that this is the strand from which RNA is transcribed *in vivo* early in infection (Guha and Szbalski, 1968). Since late genes are transcribed from the H strand, the work of Travers would suggest that his T4 sigma factor is involved in the initiation of synthesis of quasi-late messengers. The ability of RNA polymerase to transcribe T4 DNA in the Travers assay increases between 5 and 15 minutes after infection at 30°. If this process reflects the kinetics of synthesis of a single sigma factor, one could reasonably expect it to be acting at the level of quasi-late mRNA synthesis. Another gene product, such as that of gene 55, then, might play a similar role in further modifying the polymerase so that it could start to transcribe late genes from the H strand.

In a more recent study, Travers (1970) has shown that the T4-specific sigma factor stimulates the synthesis of early species of T4-specific RNA *in vitro.* This occurs whether the core polymerase used is the uninfected cell enzyme or the phage-modified enzyme. At any rate, Travers proposed that the T4-specific sigma factor is involved in direction of all T4 transcription after the host sigma factor has ceased to function. In the same study, Travers investigated the effect upon *in vitro* T4 DNA transcription of an *E. coli* protein called the rho factor, which is responsible for termination of transcription (see p. 246). In the presence of the *E. coli* sigma factor, the *E. coli* core polymerase transcribes both early and pre-early genes, as we have seen. However, when rho factor is also present, only pre-early genes are transcribed. This indicates that the termination sites for pre-early genes are close to the initiation sites for early genes, and that, in the absence of instructions to terminate, the RNA polymerase molecules can pass directly from pre-early regions. Schmidt et al. (1970) have presented evidence for the existence of a phage-coded antiterminator protein, which negates the effect of rho. This may be involved in the switch-on of early gene transcription, as well as the early burst of lysozyme mRNA synthesis.

Based upon the above discussion, one might formulate the following tentative picture of early transcriptive events in T4 infection. Phage DNA is injected and rapidly binds the host sigma-RNA polymerase complex. Pre-early genes are transcribed, and the T4-specific sigma factor and any proteins involved in modification of the host core enzyme are formed. Perhaps because the modified polymerase binds the T4 sigma factor more tightly than the host factor, the phage sigma displaces the *E. coli* sigma, and early gene transcription commences. The formation of the antiterminator protein may also be involved here. At any rate, this replacement of *E. coli* sigma by T4 results in an arrest of pre-early transcription. Later changes in transcription patterns are controlled by the other positive control factors, including the products of genes 33 and 55, as well as by the onset of DNA replication.

Although it must be again emphasized that little is presently known about the molecular nature or action of sigma factors, the recent work on these agents presents an attractive, if incomplete, explanation for the timing of T4 transcriptions. However, in a sense, the picture which emerges merely pushes back the limits of knowledge a little bit. For example, we must now explain what factors regulate the time of appearance of different sigma factors and other control elements.

Expression of Phage Genes In Vitro

In Chapter 4 we discussed the early evidence from T-even phage systems which helped to establish the existence of messenger RNA: nucleotide composition akin to that of phage DNA, distinctive sedimentation properties, and ability to form specific hybrids with phage DNA. At the same time as this work was being done, work on *in vitro* amino acid-incorporating systems was establishing that information encoded in base sequences of synthetic polynucleotides could provide the instructions specifying the amino acid sequences of polypeptides synthesized in these systems. This latter work culminated in the demonstration that a natural RNA, namely, that derived from RNA phage particles, could direct the synthesis, *in vitro,* of recognizable phage proteins (see Chapter 13). In order to establish the validity of the messenger RNA concept for DNA-based life forms, it was necessary to show that an RNA fraction with the properties of mRNA could similarly prime the synthesis of a recognizable protein *in vitro,* and that such synthesis was dependent upon the added RNA. Such a feat was first accomplished by Salser, Gesteland, and Bolle (1967), who demonstrated the formation of T4 lysozyme in an amino acid-incorporating system programmed with RNA extracted from *E. coli* late in infection by T4. RNA isolated early in infection was ineffective in this system, suggesting that the translation-level control mechanism presumed to be operating *in vivo* (Kasai and Bautz, 1967) is also effective in the cell-free system. Specific dependence upon a lysozyme messenger in the added RNA fraction was shown by the fact that RNA from suppressor-negative bacteria inflected with an *e* gene amber mutant failed to stimulate active lysozyme synthesis in a cell-free system derived from a suppressor-negative bacterial strain. This permitted the development of a sensitive and simple assay for suppressor activity *in vitro* (Gesteland, Salser, and Bolle, 1967). RNA from cells infected with an *e* gene amber mutant can be translated in a cell-free system derived from bacteria bearing a suppressor mutation. Moreover, such RNA is effective in a suppressor-negative cell-free system, but only in the presence of purified transfer RNA from a suppressor-positive bacterial strain. Thus, these important investigations not only provided the first direct confirmation of the messenger role of phage-specific RNA, but they also confirmed that suppression in several cases is due to synthesis of an altered transfer RNA which can recognize the amber codon.

Coolsma and Haselkorn (1969) have also demonstrated T4 lysozyme synthesis *in vitro.* Moreover, analysis by disc gel electrophoresis of all of the proteins synthesized *in vitro* has revealed a basic similarity between the pattern of synthesis *in vitro* and *in vivo.* More recently Wilhelm and Haselkorn (1970) have shown that the activity of this *in vitro* system approaches the rate of protein synthesis *in vivo.* Another *in vitro* system, which synthesizes the major T4 head protein, has been developed by Klagsbrun and Rich (1970). By use of this system, which includes polyribosomes extracted from infected cells, these workers have shown that the head protein, which comprises some 60% of all of the protein synthesized late in infection, is formed upon a special class of polyribosomes, larger than those existing earlier in infection or in uninfected cells.

Several workers have tried to develop phage DNA-dependent protein-synthesizing systems (cf. Doerfler, Zillig, Fuchs, and Albers, 1962). As one might expect, such systems present more technical difficulties, since they require that both transcription and translation occur *in vitro.* Celis and Conway (1968) showed that the products of a T2 DNA-dependent system resembled the proteins formed *in vivo,* by the criteria of migration pattern in disc gel electrophoresis and peptide patterns in fingerprinting experiments. However, none of the proteins formed was shown to have biological activity. More recently, Gold and Schweiger (1969a) have demonstrated the synthesis in a T4 DNA-dependent system of two early enzymes, the α- and β-glucosyl transferases. The requirement for genetically competent DNA was shown by the failure of DNA from glucosyl transferase-negative phage mutants to prime enzyme synthesis in this *in vitro* system. With slight modifications their system was also shown to synthesize lysozyme, a late enzyme, *in vitro* (Schweiger and Gold, 1969). Remarkably, this system retains some control properties akin to those seen *in vivo;* the β-glucosyl transferase activity appears 4 to 5 minutes earlier than that of lysozyme (Gold and Schweiger, 1970).

PHAGE-RELATED CHANGES IN TRANSLATIONAL MACHINERY

Transfer RNA

As we have seen, much of the regulation of gene expression in T-even phage-infected cells occurs at the level of transcription, i.e., different populations of mRNA molecules are synthesized at different times during the latent period. We have also discussed findings which suggest the existence of translation-level control mechanisms as well. These include the early burst of lysozyme mRNA synthesis noted by Bautz and his co-workers (see p. 167), and the fact that early mRNA species continue to be formed throughout the latent period, even after synthesis of early proteins has essentially ceased. Moreover, it is reasonable to expect that some translation-level regulatory mechanism might be operative in the cessation of bacterial protein synthesis immediately after infection. Accordingly, several investigators have examined components of the cellular machinery for protein synthesis, trying to detect post-infection changes which might have physiological significance.

The first such change was observed by Sueoka and Kano-Sueoka (1964). These workers compared the elution patterns from a methylated albumin-kieselguhr (MAK) column of tRNA isolated from infected and uninfected cells. In each column run, the uninfected cell tRNA was charged with a tritiated amino acid, and the infected cell tRNA with the same amino acid labeled with C^{14}. Thus, the elution profiles for infected and uninfected cell tRNA could be compared in a single column run. Of 17 amino-acyl tRNA elution patterns which were checked, a significant difference in the post-infection pattern was seen only with leucyl-tRNA. The metabolic change leading to the alteration seemed to occur between 3 and 8 minutes after infection. Injection of DNA and protein synthesis were both required, since cells infected by phage ghosts, or in the presence of chloramphenicol, were both incapable of evoking the change. Later, Kano-Sueoka and Sueoka (1966) found that this alteration was observed only in T-even phages, and not with any other phages tested. Moreover, stricter attention to the conditions of infection revealed that the alteration in leucyl-tRNA profile is one of the earliest metabolic events in T-even phage infection, occurring between 1 and 2 minutes after infection. The change involves the appearance of a new, more rapidly-eluting, peak of leucyl tRNA, with concomitant reduction in the amount of material in what was the first peak in the elution pattern from uninfected cells. The early elution of the new material suggested that it is of smaller size than normal cellular tRNA.

The regulatory significance of an altered tRNA elution pattern lies in the possibility that the presence of different tRNA molecules under varying conditions allows for the translation of different messages. For example, the genetic code as elucidated in *E. coli* is highly degenerate for leucine, containing six triplets which can respond to this amino acid. In conformance with this high degree of degeneracy, *E. coli* cells contain at least five different tRNAs specific for leucine (Weisblum, Benzer, and Holley, 1962). Suppose that, of the six possible leucine codons as worked out *in vitro,* one appears extensively in bacterial messengers but is used not at all in T-even phage messengers (the other five codons would provide more than enough ways of spelling "leucine"). Now, if infection causes some metabolic event which results in destruction of the tRNA corresponding to that one codon, then the protein-synthesizing machinery of the cell would be able to assemble phage proteins, but not bacterial proteins. The codon for which the corresponding tRNA had been destroyed would become, for all practical purposes, a nonsense codon. This in turn could account for the shutoff of bacterial protein synthesis. One could also envisage a complementary control mechanism, namely, synthesis *de novo* of a virus-specific tRNA molecule, needed to translate certain viral messengers. Such a mechanism could, conceivably, be involved in the intracellular switch from early to late protein synthesis. As we shall see, work performed in several laboratories since 1966 tends to support both formulations, although in no case is there yet a complete demonstration of the regulatory significance of changes in tRNA patterns.

To investigate the regulatory significance of their findings, Kano-Sueoka, Nirenberg, and Sueoka (1968) investigated the codon responses *in vitro* of leucyl-tRNA fractions from infected and uninfected cells. The new species, which is formed early in infection and then disappears, responds to poly UG, while the species which was partly lost responds to the triplet CUG. Biochemical studies (Kan, Kano-Sueoka, and Sueoka, 1968) supported ıge occurs prior to eight minutes after infection, (2) that it requires pro- synthesis in order to occur, and (3) that the new material seen after ction represents *de novo* synthesis, rather than thiolation of a preexist- cellular tRNA species. This work was supported by guanine-labeling lack and Smith, 1968) and P^{32}-labeling studies (Baguley, Bergquist, Ralph, 1968; Daniel, Sarid, and Littauer, 1968). In the latter study, ıdouridine, another specific tRNA component, was detected in the erial labeled after infection.

A convincing demonstration of phage-coded tRNA synthesis was pro- ed by Weiss et al. (1968), who showed that the S^{35}-labeled material ch is formed after infection hybridizes specifically with T4 DNA. More- r, these authors devised a technique for measuring the hybridization of ino acyl tRNAs without deacylation of the bound amino acids. By use this method, and tRNAs charged with tritiated amino acids, it was pos- le to show that at least one leucyl-tRNA and at least one prolyl-tRNA bridize with phage DNA. It was, of course, impossible to tell whether se are the same species which become labeled with S^{35}, so that the total mber of phage-coded tRNA molecules formed is not known, except for fact that it is at least two. This number has been extended with the ent report of Daniel, Sarid, and Littauer (1970) of the additional ex- ence of phage-specific tRNAs for arginine, isoleucine, and possibly tyro- e, as well as leucine.

McClain (1970) has briefly described the discovery of a T4-specific ppressor gene. A strain bearing this gene was isolated as an apparent vertant of a double mutant bearing amber mutations in two widely parated genes, 14 and 48. This gene suppresses amber, but not ochre or GA mutations. Presumably, the suppressor is an altered phage-specific RNA, but, since the amino acid inserted at the site of the amber codon as not yet been identified, we don't know whether or not it corresponds any of the tRNAs listed above.

What is the biological role of these new tRNA species? It is possible that ey provide the element of translation-level control involved in the shift om early to late protein synthetic patterns, by mechanisms akin to those iscussed earlier. As of this writing, however, no work has appeared on the odon responses of these new tRNAs. Certainly, the synthesis *de novo* of RNA molecules might account for the changes in tRNA elution profiles which had been seen earlier by Sueoka and his colleagues, and by Waters and Novelli. However, there is as yet no compelling evidence on this point. ndeed, Salser, Gesteland, and Ricard (1969) have reported that T4 nRNA-directed lysozyme synthesis *in vitro* can occur in uninfected cell xtracts, in the absence of phage-specific tRNAs. Thus, these viral tRNA species would appear to be unnecessary, at least for translation of this particular mRNA.

Since tRNA molecules are gene products, one should, in principle, be

the idea that the new species is formed from the pre
is partially lost after infection. Moreover, two of th
species present before infection show modified chron
after infection. This finding agreed with work of
(1967) who had presented evidence for an early m
existing leucyl-tRNA, plus additional changes in the c
terns, which occur relatively late in infection.

Kano-Sueoka and Sueoka (1968) presented definit
nature of the early modification reaction. The new
called leucyl-$tRNA_F^{leu}$, cannot be deacylated enzymat
sediments in sucrose gradients only half as fast as o
suggesting that it is of less than normal size. This evide
hyperchromicity, suggest that the modification results fr
one or more nicks into the tRNA structure. The molecu
diately fall apart, but its secondary structure is weakene
it does dissociate when the tRNA is bound to a leucine-
in the process of acylation. Still unanswered, of course,
why the modified leucyl-tRNA disappears later in infe
since the leucine cannot be enzymatically removed from
cule *in vitro*. This formulation also predicts the induct
infection, of a very specific nuclease, one which recogni
few internucleotide bonds on a single species of tRNA, a
dozen different species which exist in uninfected *E. coli*
tainly probable that such an event occurs, it should be
known nucleases possess such a high degree of specificity.

The physiological significance of the above modification
by comparing chromatographic patterns of polysome-bou
from uninfected and T2-infected cells (Kano-Sueoka and
Leucyl-$tRNA_1$, the species which undergoes inactivation,
very small extent in polysomes from infected cells, suggesti
used in synthesis of phage proteins. Moreover, studies with
acid-incorporating systems suggested that the codon CUG,
$tRNA_1$ responds, occurs very rarely, if at all, in T2 mess
early modification may well be involved in the shutoff of b
synthesis.

What about the other changes in leucyl-tRNA which had
Work from several laboratories was suggesting that some tR
synthesized *de novo* after T-even phage infection, in spite
dence that the formation of all cellular RNA species ceases
Hsu, Foft, and Weiss (1967) used S^{35}-labeling to specifically
synthesis, inasmuch as tRNA had recently been shown to co
bases. Hsu et al. found that the MAK elution pattern of tRN
labeled after T4 infection differed markedly from the correspo
seen from uninfected cells. Control experiments indicated

able to obtain phage mutants with altered or inactive tRNAs. Such mutants could be used to probe the biological roles of the tRNAs. However, it is very difficult to devise a selection procedure for such mutants, partly because the small size of the genes would tend to reduce mutation frequencies to very low levels, and also because one doesn't know whether these tRNAs perform an essential function or not. If they do, then one would have to look for some type of conditional lethal mutant. One cannot look for amber or ochre mutants, since the generation of a nonsense codon does not block transcription, which would be the only process in synthesis of an RNA gene product. However, one might expect a mutation in a tRNA gene to be temperature-sensitive, if it involves an alteration in one of the mutually complementary regions of the molecule. This could, conceivably, weaken the secondary structure to the extent that the molecule would partially "melt" at a slightly raised temperature. In this latter context, it is of interest that, of the 66 genes for which conditional lethal mutations are known in T4, there is only one, namely gene 40, for which *ts,* but no *am* mutants are presently in existence (Edgar, 1968; King, 1968). It would be of interest to see whether gene 40 is a structural gene for a tRNA molecule.

Amino Acid-Activating Enzymes

Neidhardt and Earhart (1966) reported that when an *E. coli* mutant with a temperature-sensitive valyl-tRNA synthetase activity is infected with T-even phages, a heat-stable enzyme can be isolated from the infected cells. This suggests that a new or altered valyl-tRNA synthetase might be formed as a consequence of infection. Later work (Earhart and Neidhardt, 1967), indicated that several other amino acid-activating enzymes are not altered as a result of infection. Chrispeels et al. (1968) examined the change in valine-activating enzyme in some detail. Chloramphenicol inhibits the transformation. However, a density-prelabeling experiment using deuterated water showed conclusively that the valyl-tRNA synthetase in infected cells is formed almost entirely from material present in the cell before infection. Molecular properties of the altered enzyme suggested that it might have resulted from dimerization of the host cell enzyme. More recently, Neidhardt et al. (1969) have isolated an amber mutant of T4 which cannot effect the transformation. This would suggest that the modification involves addition of a protein to the preexisting valyl-tRNA synthetase. Direct evidence for this has now been obtained (Marchin and Neidhardt, 1970). Under conditions so far tested, the modification function appears to be dispensable for normal phage growth.

Polypeptide Chain Initiation

A priori one could propose that the regulation of phage-directed protein synthesis might be mediated in part through alteration of the cellular machinery for initiating the translation of polypeptide chains. Eisenstadt and Lengyel (1966) had shown that trimethoprim, an inhibitor of bacterial dihydrofolate reductases, inhibits protein synthesis in whole cells of *E. coli.* This occurs because the resultant block in tetrahydrofolate formation leads to a deficiency of 10-formyl tetrahydrofolate, needed in the synthesis of N-formylmethionyl-tRNA, the agent which functions in initiation of all *E. coli* proteins (Capecchi, 1966). It was of interest to perform such experiments with phage-infected cells, to get an indication of whether phage protein synthesis is also initiated *via* N-formylmethionine. However, phage dihydrofolate reductases are far more resistant to inhibition by trimethoprim than is the *E. coli* enzyme (Mathews, 1967a), such that one could not block tetrahydrofolate formation *in vivo.* On the other hand, the availability of dihydrofolate reductaseless mutants of T4 presented a means for obviating this difficulty. Klein, Mathews, Eisenstadt, and Lengyel (1967) reported that trimethoprim blocks protein synthesis in cells infected by a *wh* (reductaseless) mutant of T4. However, we subsequently realized that trimethoprim also blocks DNA synthesis, presumably because tetrahydrofolate is required for the deoxycytidylate hydroxymethylase reaction. Conceivably, the effect we had noted might result merely from an inhibition of late protein synthesis in the absence of DNA synthesis. Accordingly, we constructed a double mutant containing both the *wh* mutant and a *ts* mutation in gene 43 (Klein et al., 1968). Thus, one could examine protein synthesis at a restrictive temperature, where DNA synthesis was not occurring. We still observed inhibition of protein synthesis, suggesting that the N-formylmethionine mechanism is operative in the initiation of phage protein synthesis. From analysis of N-terminal residues of proteins synthesized in T4-infected cells, Miovic and Pizer (1968) reached the same conclusion. Direct support for a formylmethionine-dependent initiation mechanism has recently been presented by Gold and Schweiger (1969b), in their studies on the synthesis of T4 DNA glucosyl transferase *in vitro.*

Ribosomes

Hsu and Weiss (1969) have recently shown that ribosomes extracted from T4-infected cells are far less active *in vitro* in translating RNA from *E. coli* or from the RNA phage MS2 than are ribosomes from uninfected cells. Schedl, Singer, and Conway (1970) have presented similar results. The ability to translate T4 mRNA *in vitro* is unimpaired. The agent responsible for this alteration in ribosomal template specificity was isolated and shown to have some properties of a protein. Addition of this factor to ribosomes from uninfected *E. coli in vitro* caused them to assume the template specificity characteristic of ribosomes extracted from infected cells. The nature of the factor is presently unknown, as is the biological significance of the ribosomal alteration. However, it is reasonable to assume that it is involved in the shutoff of bacterial protein synthesis, and it may also be responsible for the fact that T4 infection interferes with the intracellular replication of RNA phages (Hattman and Hofschneider, 1967). Moreover, since Smith and Haselkorn (1969) have shown that several new proteins are associated with ribosomes extracted from T4-infected cells, it is reasonable to propose that the factor of Hsu and Weiss is one of these proteins.

MATURATION AND VIRUS ASSEMBLY

As we have already seen, the structure of a T-even phage particle is exceedingly complex, particularly when compared to other viruses. Table 3–2 listed twelve protein components present in a T4 particle. When one takes into account the fact that some of these components actually contain two or more distinct protein moieties, we see that the total number of recognized proteins which are structural components of T4 is probably close to 20. Since the virion also contains DNA and several low-molecular-weight compounds, such as dihydropteroylhexaglutamate and polyamines, the question of how these components are assembled into infective virus assumes immense magnitude. Certainly, it is difficult to invoke a spontaneous process, akin to crystallization, which is responsible for the formation of simpler viruses, such as tobacco mosaic virus. Rather, it seemed necessary a priori to postulate some ordered mechanism for construction of the various subassemblies—heads, tail tubes, plates, and so forth—followed by their attachment to one another, again in some ordered process. Direct support for this idea comes from the physiological genetics of T4. Of the 66 genes for which conditional lethal mutants have been described, some 46 play roles in the formation and assembly of viral components. While some products of these genes, it is true, may be minor structural constituents not yet detected biochemically, it is reasonable to expect that some are enzymes or other factors which act to promote specific reactions in virus assembly. As we shall see, there is now good evidence that this is so.

Because of the complexity of the T4 particle, the relative ease of isolating some of the viral substructures in pure form, and the excellent genetic system available, a number of investigators have studied T4 assembly as a model for the understanding of how other large supramolecular structures are put together. It is possible, of course, that the mechanism of assembly of other such structures, such as mitochondria or ribosomes, might turn out to bear no resemblance to that of T4 phage. However, because of the large number of known genes involved in the latter process, one can begin to approach the question of morphogenesis in terms of the sequence of interactions between gene products, an approach which cannot presently be exploited as successfully in other systems. Hopefully, the work on T4 will provide general insights which might further our understanding of how large cellular organelles are assembled.

The earliest definitive knowledge of the stages of T-even phage morphogenesis came from direct observation of ultra-thin sections of infected bacteria in the electron microscope (Kellenberger and Séchaud, 1957; Kellenberger and Kellenberger, 1957; Kellenberger, Séchaud, and Ryter, 1959). These studies confirmed the idea that the principal viral substructures are formed more-or-less independently of one another before being assembled into mature phage. Such a conclusion had already been reached, although less directly, in studies on the phenomenon of phenotypic mixing (Novick and Szilard, 1951). This is exhibited when two dissimilar phages, such as T2 and T4, mixedly infect the same cell. Some of the progeny bear the adsorption properties of T2, but the genetic constitution of T4, while others bear the reverse characteristics. This is most easily explained if one assumes that a fully-formed T2 tail, which determines the host range of the phage, can become attached to a head which encases a T4 genome, and vice versa.

One of the most striking aspects of phage morphogenesis which was pinpointed by the electron-optical work, and one which is still but poorly understood, is the condensation of newly replicated DNA and its packing into the head. Early in infection DNA is seen as very long fibrils which subsequently begin to condense into "packets," about the size and shape of a phage head. However, these packets are not originally wrapped in the head membrane; rather, packets and empty head membranes both seem to exist as free intermediates (see p. 189). The factors inducing newly replicated DNA to condense into a form amenable to packing inside a head which is but 0.2% of its own length are quite obscure, although some have suggested an involvement of the internal protein in condensation. Particularly intriguing is the fact, previously mentioned in connection with DNA replication, that packaging of DNA into heads involves some process which recognizes a rather specific length of DNA, the "headful," and cleaves it from a multimeric replicating structure.

The development of conditional lethal mutants of T4 gave enormous impetus to studies on viral morphogenesis. At first this took the form of electron microscopic examination of the incomplete structures accumulating in nonpermissive infection by conditional lethal mutants (Epstein et al., 1963). For example, mutants in genes 34, 35, 36, 37, or 38 accumulated phage particles which were morphologically complete except for the absence of tail fibers. Free fibers are also seen in lysates of mutants in genes 34, 35, or 36, but not gene 37 (Eiserling, Bolle, and Epstein, 1967). From these observations, one could conclude that genes 34–38 participate in the synthesis of tail fibers, with gene 37 being the structural gene for a major fiber component, and that the addition of fibers to the otherwise complete virion is probably the final step in phage assembly. In addition, the discovery of 46 genes whose action is necessary for synthesis and assembly of the components of the virion put into sharp focus the complexity of the problem, definitely ruling out the idea that T4 morphogenesis might be a relatively simple self-assembly process.

Edgar and Wood (1966) made the important discovery that the morphogenesis of T4 could be made to occur *in vitro.* In an initial experiment, purified fiberless phage (see preceding paragraph) were incubated with an extract of cells infected with a mutant defective in gene 23 (concerned with head formation). Infectivity rapidly increased some one thousandfold over the initial level, suggesting that the gene 23 extract was providing all the material needed for formation of fibers, and for their attachment to the incomplete virus particles: in other words, genetic complementation *in vitro.* This was verified by electron microscopic examination of the particles formed during this incubation. This type of experiment was subsequently repeated with many different pairs of late-function mutants, providing a basis for classification of the genes involved. Although not all of the steps of virus assembly have yet been demonstrated *in vitro,* sufficient information has been obtained from those steps seen that all of the 46 known morphopoietic genes have been classified into fourteen different complementation groups. Mutants in genes which fail to complement *in vitro* are presumed to be defective in the same general step of phage assembly, and are assigned to the same group. Complementing mutants are believed to be individually defective in the assembly of different substructures, or in different steps of the synthesis of the same structure Direct support for this idea comes from electron-microscopic examination of the incomplete phage parts formed when the genes in question cannot act or be complemented.

A combination of evidence from *in vitro* complementation tests and examination of incomplete phage structures by electron microscopy, sucrose gradient centrifugation, and serological tests has been used to order the 46 morphopoietic genes (King, 1968; Edgar and Lielausis, 1968; Wood et al., 1968; King and Wood, 1969; Wood and Henninger, 1969). One can now draw a flow diagram, or "metabolic pathway," for the assembly of T4. Such a diagram, based upon that presented in the view of Wood et al. (1968), is presented in Figure 7–11. Several aspects of this diagram are worthy of comment. First, as was inferred from earlier evidence, we see that heads, tails, and fibers are assembled independently of one another, with the attachment of fibers being the final step in virus assembly. Secondly, the formation of heads seems to occur in at least four stages. The first recognizable intermediate in head formation is the empty head membrane. The product of gene 31 participates in shaping the head. In the absence of this gene product *in vitro,* subunits of head protein aggregate in random fashion (Laemmli, Beguin, and Gujer-Kellerberger, 1970). A second step, which is blocked in the failure of genes 16, 17, or 49 to act, is the filling of this head with DNA. (In fact, empty heads formed at a restrictive temperature in infection by a *ts* gene 49 mutant can be filled with DNA after a temperature shiftdown (Luftig and Wood, 1969)). A third feature of this morphogenetic pathway is the fact that the tail is

of gene 9. Little is known about the biochemical nature of this action; certainly no discernible morphological changes result from this process, but the particles are converted to active substrates for fiber addition. The process of fiber attachment is of interest, inasmuch as it appears to involve the action of an enzyme. As stated above, the morphopoietic gene products may be either major or minor structural components, or enzymes. Some virus-induced enzymic pathway, for example, must be responsible for the synthesis of the phage-specific folate compound, dihydropteroylhexaglutamate. However, except for the fiber-attachment reaction, none of the other steps in phage assembly has been shown by biochemical criteria to be enzyme-catalyzed. In the case of fiber attachment, however, the product of gene 63 is required, and all available evidence suggests that this gene product is an enzyme (Wood and Henninger, 1969). Kinetics of gene 63 action in *in vitro* complementation tests, particularly the dependence of rate of complementation on temperature, are those expected for an enzyme-catalyzed process. Moreover, the factor in extracts which promotes the complementation (presumably the gene 63 product) is heat-labile and of high molecular weight. Interestingly, although the action of the gene 63 product is a late function, the protein itself begins to be formed early in infection, and its formation continues throughout the latent period. Also noteworthy is the fact that the gene 63 product can be replaced *in vitro* by indole (Nashimoto and Uchida, 1969), which presumably induces an allosteric transition in the fiberless phage, such that they are converted to suitable receptors for tail fibers.

One can ask whether the sequential nature of T4 assembly reflects sequential expression of the genes concerned with the process, or with simultaneous gene expression followed by sequential gene action. Most evidence favors the latter idea, particularly the finding of Hosoda and Levinthal (1968) that all late proteins begin to be synthesized at the same time and continue to be formed throughout the latent period. Thus, the sequential nature of phage assembly results from the fact that a given structural intermediate must be acted upon by a specific gene product in order to be converted to an intermediate which can serve as substrate for the action of the following gene product. Of course, all of the information gained thus far about the sequence of morphogenetic steps tells us little about the biochemical nature of these steps. Undoubtedly, a rich lode of information waits to be mined. However, because of the very small amount of material involved in some steps (e.g., the estimate that each T4 particle contains but 6 molecules of bound folate per virion), a biochemical elucidation of these steps will probably not be an easy task.

LYSIS, LYSIS INHIBITION, AND THE rII FUNCTION

The final step in the growth cycle of T-even phage is lysis, or rupture of the cell, with release of the matured progeny phage to the medium. The immediate causative event is the action of phage lysozyme (the *e* gene product) on the bacterial cell wall, such that enough of the cell surface is dissolved to destroy the structural integrity of the cell. However, it is becoming increasingly clear that the timing of lysis is subject to a complex control which involves the actions of several phage genes. For example, it has long been known that the presence of phage lysozyme in the cell is not alone a sufficient condition for lysis to occur. We have already mentioned the phenomenon of lysis inhibition (Doermann, 1948b), in which secondary infection of an infected cell leads to prolongation of the latent period for up to several hours, with a corresponding increase in the phage yield per infected cell. Yet ample amounts of lysozyme are present in these cells, exceeding the amounts present in non-lysis-inhibited cultures at the time of lysis (Rutberg and Rutberg, 1965; Bode, 1967). So, in order to understand how lysis is controlled, we must ask what other gene products participate, aside from lysozyme, and what is their action?

A reasonable current explanation of lysis is that it results from a disturbance of an equilibrium which exists between continued formation of cell surface components and cell wall destruction by lysozyme. One factor which must come into play, and which may play a key controlling role, is an alteration of the cytoplasmic membrane which could permit lysozyme to penetrate to the exterior of the cell and begin to function. Such an alteration is presently most difficult to define in biochemical terms, but a large body of experimental evidence suggests that proper control of membrane function is crucial to many phage-directed processes, including lysis (see Chapter 4).

The "equilibrium theory" of lysis was most clearly enunciated by Champe (1963) and by Mukai, Streisinger, and Miller (1967). These workers stated that some events unrelated to lysozyme synthesis lead to the "metabolic death" of the cell at a particular time. In the presence of lysozyme, and in the absence of lysis inhibition, the cessation of metabolism which results leads to failure of the infected cell to repair continued damage to the cell wall, and the bacterium ruptures. In line with this idea, Mukai et al. found that oxygen uptake in cells infected with lysozyme-negative T4 ceases at about the same time as lysis would occur normally, even though lysis does not take place in these cells.

However, superinfection of such a culture, a treatment which normally leads to lysis inhibition, prolongs the period of oxygen uptake and permits extended phage growth. It will be recalled that *r,* or rapid-lysis, mutants are unable to establish lysis inhibition. As expected from this, Mukai et al. found that prolongation of the period of oxygen uptake and phage growth is not seen when the primary phage is an *e*-*r*II double mutant. Thus, this study established that some event, unrelated to lysozyme synthesis, causes the cessation of metabolism at a preset time. The action of the *r* genes must be involved in this latter process, since an *r* mutant showed no extension of metabolism, even when superinfected by an *r* + phage.

In a companion study Séchaud, Kellenberger, and Streisinger (1967) carried out electron microscopic observations on cells infected by lysozyme-negative mutants. Cells infected by an *e-rII* double mutant show extensive degradation of the membrane by 60 minutes after infection and apparent loss of much of the cytoplasmic contents, even though lysis has not occurred, nor progeny phage been released. By contrast, cells infected under conditions leading to lysis inhibition show no significant changes in membrane structure at 60 minutes. Studies on leakage of P^{32}-labeled material from both cultures confirmed that the apparent change in membrane structure which occurs in the absence of lysis inhibition is also reflected in a far greater loss of cellular components. Thus, the metabolic death of the cell is closely associated with a breakdown of membrane permeability. Presumably, the latter event permits intracellular phage lysozyme to reach its sites of attack in the cell wall. However, these studies say nothing about cause-and-effect relationships; does the change in membrane trigger metabolic death, or vice versa? Or does some other event trigger both of these? A posible answer to these questions has been put forward in a recent study by Cronan and Wulff (1969). These workers have found that the intracellular levels of free fatty acids (FFA) increase about fiftyfold, beginning at 10–13 minutes after T4 infection. This material is derived from cellular phospholipid, which is localized almost exclusively in the membrane (Lennarz, 1967). Protein synthesis is required for this change to occur, suggesting that a phage-induced enzyme, such as a phospholipase, might be responsible. Interestingly, superinfection which leads to lysis inhibition can delay the production of excess FFA. This would suggest that the breakdown of cellular phospholipid is closely related to the dissolution of the membrane, which had earlier been observed by Séchaud et al. (1967). Moreover, this phenomenon could be directly responsible for the metabolic death of the cell, since high levels of FFA inhibit respiration (Borst et al., 1962) and uncouple oxidative phosphorylation (Lehninger, 1965). Moreover, phospholipids are required for electron transport in preparations of *E. coli* membranes (Cunningham and Hager, 1968).

Josslin (1970) has recently described another class of T4 mutants affecting lysis. These mutants map at a single site, called the *t* gene, which lies between genes 38 and 52. The t mutants are unable to lyse their hosts spontaneously, even though ample lysozyme is formed. It seems reasonable to propose that the *t* gene codes for an enzyme such as a phage-specific phospholipase, which is responsible for the breakdown of the bacterial membrane. Preliminary evidence for the existence of such an enzyme has recently been obtained (M. A. Wells and C. K. Mathews, unpublished observations).

As stated above, cells infected with *e* mutants fail to lyse spontaneously. However, stocks of *e* mutants occasionally give rise to plaque-forming particles. Many of these, of course, are either true or pseudowild revertants, able to induce an active lysozyme. However, Emrich (1968) found that some of these plaque-formers induced no detectable lysozyme activity. Further investigation showed that these phage, which retained the original *e* mutation, also bore a secondary mutation in what she termed the spackle, or *s,* gene. Indirect evidence suggested that *s* mutants are defective in some stage of virus-directed synthesis of cell wall material. Apparently, *s* mutants, being unable to repair cell wall damage, lyse even in the absence of the *e* gene product, as might be expected from the equilibrium theory of lysis. The reason for implicating the cell wall, rather than the membrane, in the action of the *s* gene, is the fact that Séchaud et al. (1967) had already shown that lysis does not necessarily result from destruction of the membrane alone. However, this is rather indirect reasoning, and, as Emrich pointed out, one still doesn't know why *e s* mutants lyse spontaneously; if the *s* gene product is needed to repair cell wall damage, what is causing the damage if not lysozyme?

Since *e s* double mutants can lyse their host cells spontaneously, one can prepare phage stocks without having to lyse the cells artificially, as must be done with *e* s^+ strains. The course of infection by these double mutants is very similar to that seen with wild-type phage (Emrich and Streisinger, 1968), indicating that phage lysozyme, the product of the *e* gene, need not be present in order for infection to be initiated successfully. Therefore, if adsorption and penetration involve some enzymatic action on the bacterial cell wall near the attached tail tip ("boring a hole;" see p. 45), the causative agent is not the *e* gene product.

The equilibrium theory of lysis has been challenged by Couse (1968), primarily on the basis of two observations: (1) Cells infected with an *e* gene amber mutant become resistant to lysis by penicillin at about 20 minutes after infection; since penicillin interferes with cell wall formation by inhibiting mucopeptide synthesis (Wise and Park, 1965), this would suggest that cell wall formation is not occurring past 20 minutes, for, if it were, one would expect penicillin lysis to occur. (2) Addition of cyanide at 20 minutes or earlier prevents lysis from occurring, whereas the equilibrium theory would predict that lysis should occur shortly after cyanide addition, since the cell wall could no longer be repaired. Couse favors a "clock" mechanism, in which the time of lysis is determined by some undefined event occurring early in infection. However, it would appear that her observations can be explained as follows: (1) The onset of resistance to penicillin lysis precedes the metabolic death of the infected cells (Mukai et al., 1967) by just a few minutes. Therefore, even taking into account the fact that Couse's experiments were done at 30°, one would expect cell wall synthesis to cease at just about the same time as the penicillin experiments suggest that it does. Moreover, the probable existence of virus-specific changes in cell wall biosynthesis, as inferred from the existence of *spackle* mutants (Emrich, 1968), would, it seems to me, seriously cloud the interpretation of any experiments involving the use of penicillin. Suppose, for example, that phage-induced cell wall biosynthesis is normally resistant to this agent. (2) The effects of cyanide can be most easily explained by assuming that the infected cells before 20 minutes contain insufficient lysozyme to induce lysis. Couse's own data on the effects of chloramphenicol on lysis would appear to support this point.

Of course, the clock of which Couse speaks could be the mechanism, such as phospholipid hydrolysis, which causes the cell to die metabolically at 25–30 minutes after infection at 37°. In general accord with this idea is the finding of Yamagami and Endo (1969) that the capacity for lysis inhibition is lost in filamentous *E. coli* cells. It has long been known that certain agents, such as ultraviolet light and mitomycin C, cause *E. coli* to become filamentous. Apparently, this results from continued growth of the cells, coupled with a failure of the cells to divide. Yamagami and Endo found that a number of agents which cause filament formation also lead to loss of lysis inhibition. The filamentous cells appeared to be equal to normal cells in their resistance to lysis by externally added agents, such as EDTA and/or lysozyme. Thus, a loss in cell wall rigidity does not appear to be the cause of the loss of lysis inhibition. Rather, Yamagami and Endo believe that the filaments are more sensitive to some internal change, probably at the membrane, which allows access of intracellular lysozyme to sites of cell wall digestion.

All of the above results, it must be admitted, leave us in a somewhat uncertain state. Certainly lysis is triggered by the hydrolysis of the mucopeptide layer of the cell wall by lysozyme. However, the nature of the membrane alteration which permits lysozyme to reach the cell wall, and the events triggering this alteration, have not been completely clarified. As mentioned both here and in Chapter 4, there is increasing evidence that T4 infection triggers the synthesis of new membrane and cell wall components, and that these events might be involved in controlling the length of the latent period. However, with the exception of the preliminary work on phospholipid metabolism which was cited in Chapter 4 and the study of Cronan and Wulff (1969), none of these virus-induced changes has been defined biochemically, and it is unlikely that we will really understand the genetic control of lysis until they are.

As another approach to the above questions, a number of workers have attempted to clarify the biochemical function of the *r*II gene product. In Chapter 5 we discussed the paramount role played by the *r*II system in furthering our understanding of gene structure and function. However, when one considers the other systems which have proven most valuable to molecular geneticists—such as the alkaline phosphatase system of *E. coli,* the lactose operon, the T4 lysozyme gene, the panoply of human hemoglobin variants, and the tryptophan synthetase system of *E. coli*—one sees in each case that the gene products are well-characterized proteins about which considerable structural information is available—in most cases the entire amino acid sequence. However, in the case of the *r*II gene, no one has even been able to isolate the product of the gene, much less determine its structure. Of course, the existence of nonsense mutants in the *r*II gene, and the phenomenon of phenotypic reversion by 5-fluorouracil (Benzer and Champe, 1962) strongly implied that the product of the *r*II gene was a protein. However, direct evidence on this point was not obtained until recently, when McClain and Champe (1967) were able to isolate a peptide fragment of the *r*IIB cistron product, by means of appropriate amber mutants and a double-labeling technique. As of this writing, the complete *r*II protein has not yet been isolated. This may reflect the possibility that the *r*II protein is bound to the cell wall or membrane. If this is so, then standard protein fractionation techniques, which depend upon the solubilization of the protein under study, would be expected to prove ineffctive.

Most of the work on the biochemistry of the *r*II function has been directed toward the question of why *r*II mutants are unable to replicate in *E. coli* strains lysogenic for phage λ. Garen (1961), in an exhaustive study of this problem, found that the reactions of phage development occur normally for the first 10 minutes. Following this, however, there is a generalized metabolic arrest, as reflected by decreased oxygen uptake and decreased isotope incorporation into DNA and protein. The presence of high concentrations of magnesium (0.05 M) in the medium, however, reverses the metabolic block. Garen suggested that the metabolic defect is due to leakage of cations from the cell, a process which can be reversed by providing massive cation concentrations in the medium. Later it was found that other cationic substances, notably polyamines, can also reverse the *r*II block, with spermidine being effective at much lower concentrations than magnesium (Brock, 1965; FerroLuzzi-Ames and Ames, 1965). In order for maximal effectiveness, spermidine must be added before 8 minutes. FerroLuzzi-Ames and Ames also showed that putrescine leaks from a λ lysogen infected either with *r*II or r^+ phage. However, leakage stops at about 7 minutes in r^+-infected cells, but continues in *r*II-infected cells. This would suggest that *r*II-infected bacteria are unable to carry out the "sealing reaction" which normally decreases the permeability of the bacterial membrane after the first few minutes of infection (Puck and Lee, 1955).

By testing phage growth in media of widely varying cationic compositions, Sekiguchi (1966) found that the restriction of *r*II growth is caused not so much by a depletion of divalent cations as by the presence in most media of high concentrations of monovalent cations, which compete with divalent cations at sites for cellular uptake. Thus, *r*II mutants can multiply in low-sodium medium in the absence of added magnesium. In agreement with other workers, Sekiguchi also found that permeability, as measured by leakage of diphosphopyridine nucleotide from infected cells, is much greater in *r*II infection of a λ lysogen than in infection by wild-type phage. Sekiguchi also found that *r*II-infected cells show a large drop in incorporation of P^{32} into ATP, and that this effect is seen almost immediately after infection. Since Garen had already shown that oxygen uptake proceeds normally for the first ten to twelve minutes, Sekiguchi concluded that oxidative phosphorylation is uncoupled, rather than completely blocked. This is of interest in light of the apparent connection seen in several systems, particularly mitochrondria, between ion transport and oxidative phosphorylation. It does help to explain the permeability changes seen later in infection, since a gradual ATP depletion would interfere with the action of active transport machinery. Moreover, this could explain the generalized metabolic arrest seen at ten to twelve minutes. However, one is still faced with the question of what is the early uncoupling event which causes ATP depletion.

Buller and Astrachan (1968) presented further evidence implicating the bacterial membrane as the site of the *r*II defect. The *r*II block can be partly reversed by sucrose alone, in the absence of ionic supplements, which would suggest that osmotic stabilization of the *r*II-infected cells can phenotypically reverse the genetic defect. Moreover, *r*II-infected bacteria were seen to swell and lose refractility under conditions where phage replication could not occur. Finally, spermidine was shown to inhibit cell lysis which is caused by phage ghosts. Since the spermidine is presumably not pumped into the cell, Buller and Astrachan concluded that the effect of added polyamines is to stabilize some cell surface site, probably the membrane. In a search for differences in cell surface metabolism which might help to pinpoint the *r*II defect, Buller and Astrachan found that *r*II-infected cells incorporate far more P^{32} into the cardiolipin fraction of phospholipids than do r^{+}-infected cells. However, one doesn't know whether this is a cause or result of the primary *r*II defect.

A somewhat different conclusion about the nature of the *r*II defect has been reached in studies on late gene function in nonpermissively infected cells. Rutberg and Rutberg (1966) had shown that no detectable late gene expression occurs, as determined either by lysozyme synthesis or the formation of tail-fiber antigen. More recently, Sauerbier et al. (1969) found that no detectable late mRNA species are made in *r*II-infected λ lysogens. However, the overall rate of isotype incorporation into RNA does not decrease greatly late in infection. The shutoff of bacterial RNA synthesis in these cells is not complete, and much of the RNA formed late in a restrictive infection by an *r*II mutant is bacterial. Sauerbier et al. concluded that the major defect in *r*II replication is a failure of a virus-induced modification of RNA polymerase to occur, such that the activation of late mRNA synthesis does not occur and the partial shutoff of host RNA synthesis is reversed. This, however, does not explain why phage DNA replication ceases under these conditions. Certainly, DNA replication does not require any detectable late gene expression, as shown by the fact that mutants in genes 33 and 55 (MD) can form ample quantities of DNA in nonpermissive infections (see p. 94). Moreover, the continued abundant incorporation of isotype precursors into RNA is difficult to reconcile with the depletion of cellular ATP starting at time zero (Sekiguchi, 1966), and the generalized metabolic arrest which occurs at 10 to 12 minutes (Garen, 1961).

In another recent study on *r*II gene function, Karam (1969) found that gene 30 mutants (DNA ligase-negative) can replicate if an *r*II mutation is bred into the strain. In other words, T4 ligase is apparently a dispensable function in the absence of an *r*II gene product. Similar results have been presented by Berger and Kozinski (1969) and by Ebisuzaki and Campbell (1970). In addition, Chan, Shugar, and Ebisuzaki (1970) have reported the isolation of non-*r*II T4 mutants which suppress *am* ligase mutations but not *am* *r*II mutations. The effects of mutations in the *r*I and *r*III loci have not been tested. Although Karam suggested that the *r*II gene might control a nuclease whose action is normally counteracted by the ligase, it is really impossible at present to assess the significance of this finding in the light of our other information on *r*II gene function.

8

T-Odd Phages

Whereas recent years have witnessed a massive assault on the biochemistry and genetics of the T-even phages, as discussed in the previous chapters, the T-odd phages, T1, T3, T5, and T7, have been neglected by comparison. However, a growing number of workers are beginning to realize the special advantages presented by these viruses as objects of study for many biological problems. Among these features, of course, is the fact that the T-odd phages contain DNA of the same base composition as that of the host cell—a situation more akin to that seen with animal viruses. Thus, one need not worry about problems such as the hydroxymethylation of deoxycytidylate and the glucosylation of DNA, which one encounters in trying to unravel the replication of T-even phages and which are characteristic only of these viruses. A second advantage, undoubtedly related to the above, is that these phages are smaller and contain less DNA than the T-even phages. In addition, these phages, like the T-evens, are virulent, so that one need not be concerned with problems related to the establishment and control of lysogeny. Thus, the goal of attaining a complete correlation of genes with gene functions assumes somewhat more manageable proportions with the T-odd phages than with several other phages which are currently popular among biochemical virologists.

In addition to the above mentioned advantages, the T-odd phages present several unique biochemical features of considerable intrinsic interest. For example, phage T5 contains specifically located single-strand interruptions in its DNA, and it transfers parental DNA to host cells in a discontinuous fashion, quite different from that seen with the T-even phages. Another example is the induction by T3 of an unusual enzyme which cleaves S-adenosyl methionine.

Of the T-odd phages, T5 is the one which most closely resembles the T-even phages in structure and mode of replication. It is the largest of the T-odd phages, containing a DNA molecule of between 78 and 83 million daltons (Rubenstein, 1968; Bujard, 1969), about 65 percent of the size of a T-even genome. T5 is the only one of the T-odd phages which has been shown to induce several early enzymes of DNA metabolism. As one might expect from this, T5 is the only T-odd phage which brings about a net increase in the rate of DNA synthesis in host cells after infection. T1, T3, and T7, on the other hand, appear to derive most of their raw materials for DNA synthesis from the bacterial chromosome (Labaw, 1953).

Because very little work has been done on the replication of phage T1, I shall confine this chapter to T5 and to the closely related T3 and T7.

T5

As stated above, T5 controls several functions which are similar to those seen in T-even phage-infected cells, including breakdown of bacterial DNA after infection, an increased rate of DNA synthesis, and the induction of enzymes of DNA metabolism. The latter include thymidylate synthetase (Flaks and Cohen 1957; Barner and Cohen, 1959); dihydrofolate reductase (Mathews and Cohen 1963b, Mathews, 1967a); DNA polymerase (Orr, Herriott, and Bessman, 1965; Steuart, Anand, and Bessman; 1968); a tetrafunctional deoxynucleotide kinase, capable of phosphorylating dAMP, dCMP, dGMP, and dTMP (Bessman, Herriott, and Orr, 1965); and at least two deoxyribonucleases (Stone and Burton, 1962; Paul and Lehman, 1966; Fielding and Lunt, 1969) Some other enzymes which one might expect, *a priori,* to be induced, are not; these include polynucleotide kinase (Novogrodsky and Hurwitz, 1966), 3′-deoxynucleotidase (Becker and Hurwitz, 1967), DNA methylase (Hausmann and Gold, 1966), polynucleotide ligase (Becker et al., 1967), and deoxycytidylate deaminase (Mathews, unpublished results). Other possible enzymes of nucleotide metabolism, such as ribonucleotide reductase, dUTPase, or thymidine kinase, have not, to my knowledge, yet been investigated.

Discontinuous DNA Transfer

One of the most unusual aspects of T5 infection is the fact that injection of DNA from parental phage into host cells is a discontinuous process, occurring in two distinct stages. As outlined by the Lannis and their co-workers (see review by Y. T. Lanni, 1968), a fragment containing between 8 and 8.5 percent of the total parental DNA is transferred first, followed by a period in which protein synthesis must occur and, finally, transfer of the rest of the DNA. Let us briefly discuss the experimental work upon which these conclusions are based.

In attempting a "Hershey-Chase" experiment with T5 (see p. 10), in which the transfer of P^{32}-labeled material from parental phage to host cells was followed by means of the Waring blendor technique, Lanni (1960) found that only about 10 percent of the parental label was transferred during the first ten minutes after infection. This did not mean that 10 percent of the phages were injecting all of their DNA, for the proportion of infected cells able to produce progeny phage after blending was only 1 percent. Rather, it appeared more likely that each phage was transferring a small portion of its entire DNA complement during the first few minutes. This so-called first-step-transfer (FST) material was subsequently isolated from blended phage-host complexes and shown to consist of pieces of phage DNA of uniform size, about 8.3 percent of the molecular weight of mature T5 DNA (McCorquodale and Lanni, 1964; Lanni, McCorquodale, and Wilson, 1964). One would then like to know whether these are specific fragments of T5 DNA, and, if so, what genetic functions they control.

Both genetic and biochemical evidence indicates that FST-DNA is indeed a specific fragment of parental DNA, and that it controls two functions associated with the early phase of infection: breakdown of bacterial DNA, and injection of the rest of the molecule. Lanni and McCorquodale (1963) found that blended complexes could bring about degradation of host-cell DNA, but could not initiate phage DNA synthesis. Phage adsorption alone, without FST-DNA injection, did not lead to breakdown, nor was breakdown seen in blended complexes treated with chloramphenicol. Thus, it appeared that the FST-DNA directed the synthesis of a protein whose function was to degrade bacterial DNA. Subsequently, Lanni (1965) found that protein synthesis is also required in order for the transfer of the remainder of the parental DNA to occur. Thus, it would appear that the FST DNA codes for another protein which participates in some as yet unknown way in the transfer of the rest of the molecule. However, several other T5 functions are not controlled by FST-DNA. Blended complexes are unable, for instance, to form the phage-specific enzymes thymidylate synthetase, deoxynucleotide kinase, dihydrofolate reductase, or DNA polymerase (McCorquodale and Buchanan, 1968). Moreover, DNase (measured with heat-denatured DNA as substrate) was not formed by blended complexes (Lanni, 1968). Presumably, the FST-DNA-coded protein which initiates bacterial DNA breakdown is either (a) a nuclease which is not detected under these conditions, or (b) a protein without nuclease activity per se, but which activates a latent bacterial DNase. It is certainly not the DNase of Paul and Lehman (1966), for this enzyme is not formed until after bacterial DNA breakdown is essentially complete. At any rate, it is apparent that the FST-DNA is a specific fragment of T5 DNA, inasmuch as it controls some phage-associated functions, but not others.

The above conclusion is strengthened both by genetic evidence and studies on the synthesis of individual proteins by the disc gel-radioautographic technique of Levinthal et al. (1967). McCorquodale and Buchanan (1968) found that there are three classes of T5-specific proteins, as judged by the time during which each is synthesized. The so-called class I proteins are formed between 1 and 6 minutes after infection; class II between 5 and 20 minutes; and class III from 14 minutes until the time of lysis. Blended complexes containing only FST DNA form class I proteins only, suggesting that the control of class II and III proteins resides in the remainder of the T5 genome. Analysis of transcription reveals a similar pattern of control (Moyer and Buchanan, 1969). There are three temporal classes of T5-specific RNA, the earliest of which corresponds to FST DNA. Supporting genetic analysis has come from the demonstration that certain genes are localized in FST DNA. A number of *am* and *ts* mutants had been isolated and mapped (Fattig and Lanni, 1965; Lanni and Lanni, 1966; Lanni, Lanni, and Tevethia, 1966). The technique for determining whether a given gene is on FST DNA is a "blendor rescue" experiment. Cells are infected with wild-type T5 and then blended. The complexes are superinfected with an *am* mutant and the progeny plated on su^+ and su^- strains. Wild-type recombinants can be formed only if the *am* mutation is in the FST material and can recombine with the wild-type FST DNA already present in the blended complexes. By this criterion one out of six genes tested was localized in FST DNA (Lanni, Lanni, and Tevethia, 1966). Subsequently Lanni (1969) has demonstrated that two genes, called genes I and II, are present on FST DNA. Gene II controls the completion of DNA transfer, since mutants in this gene are able to degrade bacterial DNA but are unable to transfer any more DNA. Gene I probably controls DNA degradation, since mutants in this gene cannot carry out this process. McCorquodale and Lanni (1970) have identified the gene I protein on disc gel radioautograms and have presented evidence for the involvement of this protein both in bacterial DNA degradation and the arrest of host-cell functions. However, gene I mutants are also defective in the completion of DNA transfer. The nature of this apparent pleiotrophic effect is unknown, but it is not due to simple polarity (the inhibition of gene expression distal to a nonsense codon; see p. 304), since the same effect is seen in *ts* mutants of gene I.

It would seem reasonable to expect that FST DNA is physically located at an end of the complete T5 DNA molecule. If so, genes I and II should map near an extremity of the T5 linkage map, which is linear (Fattig and Lanni, 1965). However, it has so far been impossible to map the FST genes with respect to genes on the rest of the map, since all crosses so far attempted between FST and non-FST genes have given maximal recombination values (Lanni, Lanni, and Tevethia, 1966). In other words, there is at present little or no evidence for linkage between FST and non-FST genes, even though T5 unquestionably contains but a single DNA molecule. Of course, this may be expected if the FST markers become physically separated from the rest of the genes at the beginning of infection. Presumably an FST fragment could join with any non-FST fragment following the completion of DNA transfer, regardless of whether the two pieces came originally from the same parental phage particle. However, there is no evidence that the parental DNA is actually cleaved during discontinuous transfer (Y. T. Lanni, personal communication). Rather, the basis for this apparent nonlinkage might partially reside in the unusual structure of T5 DNA, a topic which we shall explore next.

Anatomy of the T5 DNA Molecule

As we have already mentioned, the T5 DNA molecule is unusual in that it contains several single-strand interruptions located at specific points (Abelson and Thomas, 1966). An early clue to this discovery was the finding that, unlike other phage DNAs, T5 molecules break asymmetrically when subjected to hydrodynamic shear (Burgi, Hershey, and Ingraham, 1966), suggesting that the molecule contains weak points located slightly off center. Abelson and Thomas examined the sedimentation patterns of T5 DNA in alkaline sucrose gradients. Whereas most phage DNAs tested showed single peaks under these conditions, indicating an absence of single-strand breaks, T5 DNA showed at least four, and possibly five, components. The same pattern was seen regardless of the host strain used to cultivate the phage, suggesting that the location of the nicks is a specific property of the phage and independent of the bacterium used. From the apparent molecular weight and amount of each component, Abelson and Thomas proposed several possible models for the T5 DNA molecule. All of the models showed four nicks per molecule. Of the entire 38-μ length of the molecule, each model showed one strand containing a 34-μ piece and a 4-μ piece, while the opposing strand contained two segments of 14.4μ each, one 8μ segment, and one 4μ piece. Apparently, there are no nucleotides missing between the ends of two adjacent segments, for T5 DNA can be repaired *in vitro* to give a fully double-stranded molecule, by the action of purified polynucleotide ligase (Jacquemin-Sablon and Richardson, 1968). (This also indicates that the internal breaks are terminated by 3′ hydroxyl and 5′ phosphoryl groups). Such repair apparently occurs *in vivo* as well, for Abelson and Thomas found that T5 DNA extracted from infected cells relatively early in infection also has a complete duplex structure.

Bujard (1969) has investigated the structure of T5 DNA by examining partially and fully denatured DNA preparations in the electron microscope. Regions of partial unwinding under denaturing conditions are seen in all molecules. Presumably these represent AT-rich regions, where the interstrand forces are relatively weak. The locations of these regions are the same in all molecules observed, confirming that T5 DNA molecules contain nonpermuted nucleotide sequences. From the locations of these zones of partial unwinding, Bujard was able to construct a single model of the T5 DNA molecule. As can be seen from Figure 8–1, this model differs from that of Abelson and Thomas in that (a) one strand contains no nicks at all; and (b) the total number of nicks is three, not four. However, in this context, it should be observed that Abelson and Thomas were working with T5(*st*)0, a heat-stable mutant of T5, while Bujard was working with the wild strain. As shown by Rubenstein (1968), heat-stable mutants contain between two and seven per cent less DNA than wild-type T5. The mutant used by Abelson and Thomas has six per cent less DNA than wild T5. Rubenstein's data on the molecular weights of the products of breakage by controlled shear would suggest that a deletion causing an *st* mutation involves loss of DNA from both halves of the molecule. Since the DNA is apparently not deleted in one piece, therefore, and since the location(s) of the deletion(s) is not known, it is very difficult to compare the models of Abelson and Thomas with that of Bujard.

By means of sedimentation analysis of ligase-repaired T5 DNA, and by analysis of the separated strands of this repaired DNA, Jacquemin-Sablon and Richardson (1970) have confirmed Bujard's findings.

Bujard proposed that the left-hand nick in the molecule as seen in Figure 8–1 delineates the FST material from the rest of the DNA. Presumably, then, the opposing strand must also be broken in some unknown fashion during DNA injection, in order for the FST DNA to become

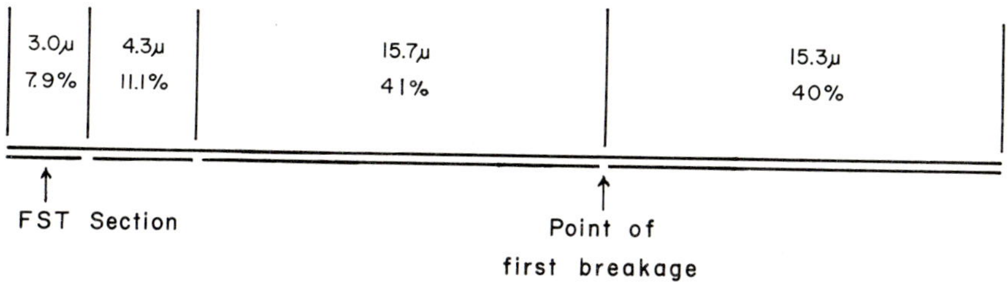

FIGURE 8-1 A model for the T5 DNA molecule, showing the locations of single-strand breaks, as presented by Bujard (1969).

physically separated from the remainder of the DNA (unless, of course, breakage does not occur, but the partially injected parental DNA sits in the incompletely transferred configuration during the time of synthesis of the class I proteins). Another implication of the Bujard model is that the nick located slightly to the right of center is the site of first breakage when T5 DNA is subjected to hydrodynamic shear. Furthermore, the model suggests that DNA is injected left-end first into host cells. It does not, of course, rule out the possibility that the nucleotide sequence of part or all of the FST DNA is repeated at the right-hand end of the molecule. However, the presence of a nick only near the left end would seem to indicate a site of preferential detachment which would, in turn, indicate that FST DNA is functionally asymmetric, if not structurally, since only the left-hand extremity would be injected first and expressed genetically. This site of preferential detachment could also help to explain why genetic markers on FST DNA recombine with such high frequency with nonFST markers. As stated above, the separation of FST DNA from the rest of the DNA simply destroys any physical basis for linkage of markers. Of course, it remains to be proven directly that the left-hand segment shown in Figure 8–1 does indeed represent the FST DNA. However, this sems likely in view of recent genetic analysis by Hendrickson, McCorquodale, and Anderson (1970), who have mapped some 35 complementation groups in T5. These groups are distributed in four regions, all of which gave maximal recombination values among each other. These would appear to correspond to the four regions of the T5 DNA molecule, which are delineated from one another by single-strand breaks.

Metabolism of T5-infected Cells

One is also faced with the question of how the nicks are introduced into DNA during T5 growth and what is their function. This is related to the general problem of the mode of T5 DNA replication, a question particularly pertinent in view of the fact that T5 DNA is not circularly permuted and, quite possibly, not terminally redundant. Thus, its mode of replication may be quite different from that of the T-even phages. Very little work has been done in this area to date. Smith and Skalka (1966) reported that replicating T5 DNA sediments more rapidly in a neutral sucrose gradient than does mature T5 DNA, but that the sedimentation profiles in alkaline gradients are approximately the same. The latter finding would suggest that replicating T5 DNA does not contain concatemeric structures. Tanyashin (1968) reported that replicative T5 DNA, prepared by a different method, is no larger than mature T5 DNA. However, Tanyashin did not discuss the discrepancy between his work and that of Smith and Skalka. Moreover, neither author presented a model of T5 DNA replication on the basis of available data.

Some preliminary work has been done on patterns of transcription during T5 development (Lanni and Szybalski, 1969). T5 DNA was denatured in the presence of poly(U, G) and the fragments separated by CsCl gradient centrifugation and fixed to filters. RNA pulse-labeled at different times in infection was hybridized to the various fragments of T5 DNA. Just as seen with T4 and with λ (p. 242), different segments of the T5 genome are transcribed at different times in infection. Also, somewhat unexpectedly, complementary regions of T5 DNA are apparently transcribed, since fragments accounting for over 70 percent of the DNA hybridize with T5 mRNA. This would suggest that some "antimessengers," as well as messengers, are made *in vivo* in normal infection. It seems highly unlikely, in view of the present status of knowledge of the genetic code, that both complementary regions of a given gene code for functional proteins.

Sirbasku and Buchanan (1970) have published a combined disc gel electrophoretic-radioautographic study of T5 transcription.

Two other recent studies of T5 biochemistry are worthy of mention. It has long been known that calcium ion at relatively high concentration (0.001 M) is required for multiplication of T9. Moyer and Buchanan (1970) have found that this cation is necessary for the normal sequential control of T5-specific protein synthesis. In low calcium (0.0001 M), DNA is injected normally, but only class I proteins are made, plus a trace of class II. Increasing the calcium concentration to 0.001 M leads to shutoff of class I synthesis and formation of normal amounts of class II and III proteins. Hendrickson and McCorquodale (1970) have started to identify the products of genes in the nonFST region. deWaard, Paul and Lehman (1965) had already identified the T5 gene controlling DNA polymerase. Hendrickson and McCorquodale have now examined *am* mutants in other genes for their ability to induce dihydrofolate reductase and deoxyribonuclease. Mutants deficient in DNase induction have been found; these map very close to the DNA polymerase gene, while mutants defective in dihydrofolate reductase synthesis map at a distant site. Both classes of mutants are defective in DNA synthesis. This is of interest, particularly inasmuch as dihydrofolate reductase-negative mutants of T4 are not lethal, nor are they severely affected with regard to DNA synthesis (Mathews, 1967b).

T3 AND T7

As mentioned earlier, these phages are quite similar, both morphologically (Fraser and Williams, 1953) and serologically (Adams and Wade, 1954). They are genetically related as well, for their DNAs cross-hybridize (Shildkraut et al., 1962) and genetic recombination has been reported between markers on T3 and T7 (Hausmann and Gomez, 1967). However, as we shall see, there are some quite pronounced differences between these two viruses.

Of these two phages, T7 has been the object of more systematic study, from both genetic and biochemical standpoints. T3 and T7 are the smallest of the T-phages, with very short tails (Luftig and Haselkorn, 1968). The molecular weight of T7 DNA has been reported by various workers to have values between 23.4 and 27 million (cf. Misra, Sinha, and Das Gupta, 1969), and T3 DNA is of similar size. This is sufficient genetic information, in a double-stranded DNA molecule, to code for 25 to 30 proteins of average size (i.e., about 50,000 MW). In addition, the DNA of T7 is very easily isolated in undegraded form, and it has no single strand breaks (Studier, 1965). Because it has been so well characterized, T7 DNA has been used as a molecular weight standard in many physical studies of other DNA molecules, and it has been used as a "typical" substrate in a number of investigations of enzymes which act upon DNA.

T7 Genetics

Two research groups have independently isolated and mapped large number of T7 amber mutants (Hausmann and Gomez, 1967; Studier, 1969). Fortunately for readers of the T7 literature, these investigators have pooled their data and presented a unified nomenclature (Studier and Hausmann, 1969). All of the several hundred T7 amber mutants thus far isolated fall into 19 complementation groups. The map is linear, and the genes have been numbered serially, with gene 1 at the left-hand extremity of the map and gene 19 at the right-hand end (see Figure 8–2). The map at present is some 200 recombination units long. Since the size of the T7 DNA molecule would suggest that the complete T7 chromosome contains some 5 to 10 undiscovered genes, the total map may turn out to be longer than that.

Preliminary studies of gene function (Studier, 1969; Hausmann and Gomez, 1967) indicate a high degree of ordering of the T7 map. Genes 1 through 6 control DNA synthesis, while genes 7 through 19 control phage assembly and other late functions. We shall have more to say presently about the functions of individual genes.

Hausmann and Gomez (1967) isolated DNA-defective mutants of T3, as well as of T7. Interestingly, no cross-complementation was observed between T3 and T7 amber mutants. This emphasizes the distinctions between these phages, which exist in spite of their genetic relatedness. T3 and T7 have the power to exclude each other in mixed infections, for most cells mixedly infected with both phages will support the growth of only one of them. The mechanisms responsible for this mutual exclusion are not known at present.

The difference between T3 and T7 is seen even more directly when one examines the ability of these phages to induce virus-specific enzymes in infected cells. Both phages bring about a rise in deoxyribonuclease activity (or activities; Hausmann and Gomez, 1968). Extracts of cells in-gel-radioautographic technique, bear this out (Studier and Maizel, 1969). As with T5, there are three classes of proteins, as judged from their time of synthesis: "earliest," "early," and "late." Gene 1 is the only gene affecting the synthesis of proteins other than its own product. Amber mutants in gene 1 form only the "earliest" proteins, without allowing turn-on of the formation of the early and late classes. This raises the possibility that gene 1 controls the synthesis or modification of a sigma factor for RNA polymerase, which would, in turn, allow the initiation of transcription of early and late genes. Recent work of Summers and Siegel (1969) is entirely in accord with this idea. First, experiments with rifampin, a rifamycin-like antibiotic, indicate, as seen with T4, that at least part of the *E. coli* RNA polymerase molecule transcribes all T7 messages *in vivo.* Secondly, purified *E. coli* RNA polymerase transcribes only early RNA species *in vitro,* as does RNA polymerase purified from cells infected with a gene 1 amber mutant, while RNA polymerase purified from wild-type T7-infected cells transcribes both early and late regions. Finally, RNA polymerase from cells infected with a *ts* gene 1 mutant transcribes T7 DNA asymmetrically *in vitro* at 30°, but this asymmetry is lost at 43°. Thus, gene 1 codes for a protein which is not necessary for catalytic activity of RNA polymerase but is necessary for control of template specificity: in other words, a phage-directed sigma factor (or, possibly, an entire RNA polymerase molecule).

Extension of the above work has shown a direct involvement of gene 1 in the synthesis of other T7 messengers *in vivo* (Siegel and Summers, 1970). Of the 13 characterized T7 mRNA species, only 3 are formed in unsuppressed gene 1 amber mutants. Moreover, the isolated gene 1 product binds to *E. coli* RNA polymerase, but only in the absence of the *E. coli* sigma factor (Summers and Siegel, 1970).

The disc gel-radioautographic studies of Studier and Maizel (1969) also showed that the total number of proteins formed in T7-infected cells is between 25 and 30. Of these, 9 are seen in purified phage preparations. The sum of the molecular weights of these 9 proteins is 559,000. Determination of the size of incomplete proteins formed by various amber mutants indicates that these are individual proteins, rather than aggregates containing more than one subunit per molecule. The amount of double-stranded DNA needed to code for this much protein is 11.2 million daltons, or about 43 percent of the coding capacity of the T7 DNA molecule. Studier and Maizel concluded that the 25–30 phage-directed proteins seen in extracts of infected cells represent virtually all of the coding capacity of the T7 genome. From gel patterns seen with extracts of cells infected with amber mutants, it has been possible to correlate specific protein bands with 12 out of the 19 known genes. Moreover, from the molecular weights of peptides formed in infection by amber mutants, and from the mapping data of Studier (1969), it has been possible to orient the T7 DNA molecule and the genetic map with respect to each other, as shown in Figure 8–2.

understanding of these factors, and the factors leading to the breakage of the lysogenic state.

THE CONCEPT OF LYSOGENY

Historical Background

Although the above general statements appear quite straightforward in terms of our present state of knowledge, the development of the concept of lysogeny followed a tortuous path. Although it had been known since the early 1920s that certain bacterial strains produce plaques when plated under proper conditions, most workers felt that this represented contamination of the bacteria with phage. The idea of Wollman (1928), namely that lysogeny represented the replication of certain stable bacterial genes at the expense of the others, was scoffed at. However, McKinley (1925) demonstrated that lysogeny persists even when cells are grown in the presence of specific phage antiserum, ruling out the existence of phage as an extracellular contaminant. Later, Burnet and McKie (1929) showed that artificial lysis of lysogenic bacteria failed to release infective phage, thus ruling out the possibility of phage as an intracellular contaminant. Nevertheless, the idea that viral genes are associated with cellular genes in lysogenic cultures was not generally accepted until after the definitive experiments of Lwoff and his collaborators carried out in the late 1940s and early 1950s (for a more complete discussion of this era, see Lwoff, 1966). Of special interest in the development of this field is the negative influence exerted by the American phage workers during this period. To quote Lwoff (1966): "In 1949, the new school of American virologists . . . condemned lysogeny. . . . Like those wisps of cloud that a breath of wind dispels, the problem is blown away from the temple of science and a smell of sulfur is left floating in the air. Lysogeny has become a heresy."

The key experiments of Lwoff involved microscopic examination of single lysogenic cells of *Bacillus megaterium* through many generations of growth (Lwoff and Gutmann, 1950). A single cell was cultured in a drop of medium on a microscope slide. After each cell division one of the daughters was plated on a nonlysogenic indicator strain, to see whether it was still capable of producing phage. This capacity was maintained through at least 19 generations. Thus, if phage were a contaminant of the lysogenic culture, there would have to have been at least 2^{19} phage present in the initial cell, in order for both daughter cells at the 19th generation to still be capable of producing virus. Since this was clearly impossible, one had to conclude that the capacity for producing virus (which was termed *prophage*) is maintained in the lysogenic cell, replicating with the chromosome and being transferred to both daughter cells.

In this study, it was also observed that a cell would occasionally disappear from the culture fluid. When this happened, about 100 phage particles would appear. This suggested that the lysogenic cell occasionally becomes *induced* to undergo a lytic cycle of phage replication. Lwoff felt that the frequency with which cells in a lysogenic culture become induced is a function of some environmental factor. Accordingly, Lwoff, Siminovitch, and Kjeldgaard (1951) searched for environmental stimuli which could trigger a lytic response in a large fraction of a population of lysogenic cells. After much effort, they found that small doses of ultraviolet light, when applied to a lysogenic culture of *B. megaterium,* led to induction of virus production in at least 99.9 percent of the cells. We now know that many agents can act as inducers, but that the efficiency of induction depends to a large extent upon the phage-host system used. Most of the effective inducing treatments, such as uv or x-irradiation, mitomycin C, or thymine starvation, appear to be effective by virtue of their interference with nucleic acid metabolism. As we shall see in Chapter 10, induction seems to result from an upsetting of opposing forces at work in the cell, one maintaining continuance of the lysogenic state, and the other trying to initiate virus production.

General Aspects of Lysogeny

We shall review the establishment and maintenance of lysogeny in some detail in Chapter 10. It is my purpose here to merely mention some of the most pertinent aspects of this process, as a backdrop to our discussion of the replication of temperate phages.

The genome of a temperate phage contains some genes whose action promotes viral replication and other genes whose action promotes lysogenization. As we have already seen, infection can lead either to a round of viral growth or to lysogenization. The "decision" is based upon the relative activities of these two sets of genes immediately after infection. This in turn depends to some extent upon environmental conditions; lowering the temperature, for example, usually increases the frequency of lysogenization.

One of the genes involved in lysogenization codes for a protein immunity substance, or repressor, which inactivates at least two gene clusters (operons) whose action is necessary for subsequent steps in replication. If sufficient repressor is formed within a cell to block replication, the phage genome becomes associated with the bacterial chromosome, almost certainly by a process of linear insertion into the chromosome. Recognition of the site for insertion is probably accomplished by means of base-pairing interactions between specific sites on the phage and bacterial chromosomes and by the binding of a specific enzyme involved in integration. Once inserted, the phage chromosome, or prophage, is maintained as a stable genetic entity, replicating in synchrony with the host chromosome to which it is attached.

Maintenance of the lysogenic state requires continuous formation of the repressor in quantities sufficient to block viral replication. For this reason, a lysogenized cell, or lysogen, is immune to superinfection by a homologous phage. Immunity, the failure of an injected chromosome to replicate, is distinct from resistance, which is the failure of a cell to absorb to a bacterium. When the level of active repressor drops below the level needed to maintain lysogeny, a cycle of lytic growth ensues, starting with excision of the prophage from its insertion site, probably by a mechanism similar to reversal of the insertion pathway. This occurs spontaneously at low frequency (between 10^{-2} and 10^{-5} per cell per generation time). For this reason, one can enumerate either temperate phage particles or lysogenic cells by plating on a lawn of susceptible bacteria. However, in both cases the plaques are usually turbid, because some of the cells in the vicinity of the plaque become lysogenized, and, being thereby immune, continue to grow.

As stated above, the rate of spontaneous loss of immunity is usually low. However, as Lwoff et al. (1951) found, this frequency can be increased, in some cases to essentially 100 percent, by the use of various artificial inducing agents. The way in which these agents lead to loss of activity of the repressor is still not known with certainty.

By far the most widely studied temperate phage is the coliphage λ, first isolated by Lederberg (1951; also see Dove, 1969a, for a discussion of λ geneology). Most of what we say about lysogeny and about temperate phage replication will be derived from studies on this virus, although we shall talk about related phages, notably ϕ80 and 434, and also phage P22, which infects *Salmonella typhimurium.* Lambda is a tailed phage of much simpler structure than the T-even phages. The particle weight of a λ virion is 57×10^6 daltons, of which 32×10^6 daltons is DNA (Dyson and Van Holde, 1967; Studier, 1965; Burgi and Hershey, 1963).

To perhaps an even greater extent than with T4, our understanding of the metabolism of λ-infected cells has followed the isolation and characterization of λ mutants defective in functions related to replication or lysogeny. For some reason, perhaps related to the early notions of "lysogeny as heresy" (p. 216), the flow of biochemists into T-even phage research preceded the flow into temperate phage work by at least five years. Comprehension of much that follows, therefore, demands at least passing familiarity with microbial genetics. Once again the reader is referred to the excellent treatise of Hayes (1968), and also to the more specialized reviews of Echols and Joyner (1968), Signer (1968), and Dove (1968).

TEMPERATE PHAGE CHROMOSOMES

As mentioned above, the λ DNA molecule has a molecular weight of about 32×10^6. This is a double-stranded DNA, of "normal" base composition, some 17.2μ in length (MacHattie and Thomas, 1964). A DNA molecule of this size can be estimated to code for about 40 proteins of average size. As we shall see, about three quarters of this information has been accounted for in terms of known genes.

A Biological Test for λ DNA

A great deal of information about the λ DNA molecule, and its relations to the genetic map and the control of viral gene expression, has come from the availability of an assay for the biological activity of λ DNA (Kaiser and Hogness, 1960). DNA is isolated from a λ strain bearing a suitable genetic marker. This DNA is added to a culture of cells which has been infected with another λ strain. The test measures the ability of the added DNA to "transform" the infected cells. A certain amount of the added DNA is incorporated into replicating phage genomes by recombination, and the biological activity of the added DNA is measured by the frequency with which its markers are found in the viral progeny.

Structural Features of λ DNA

The above test has been used to show that λ DNA molecules are not circularly permuted with respect to base sequence (Kaiser, 1962). DNA from a strain bearing mutations at two distant loci was sheared under controlled conditions, to give mostly half molecules. The two markers were transferred independently in a transformation test. This indicates that they were located on different DNA halves, a situation which could not always obtain if the molecules were permuted. In this context, we should note that the DNA of phage P22 is circularly permuted (Rhoades, MacHattie, and Thomas, 1968). This raises some interesting problems, not encountered with λ, in connection with the replication of this type of genome (Botstein, 1968; Rhoades and Thomas, 1968).

The λ DNA molecule has several features which have proven invaluable in studies on the relation of molecular structure to gene function and the control of gene expression. These are (1) the presence of mutually complementary 5′-terminated single-stranded ends; (2) striking differences in base composition along the length of the molecule; and (3) similar differences in base composition between the two strands. Let us consider these features in the order mentioned.

Cohesive ends. Hershey, Burgi, and Ingraham (1963) found that, when denatured λ DNA was exposed to renaturing conditions, two species were formed which sedimented more rapidly than native λ DNA in sucrose gradients. Hershey et al. proposed that the λ DNA molecule has short single-stranded ends which have complementary base sequences. One of the two new forms was seen as a circle, formed by interaction between the "cohesive ends" at opposite ends of the same molecule, while the other species was seen as a concatemer consisting of two or more molecules laid end-to-end. Electron-microscopic examination confirmed this (MacHattie and Thomas, 1964).

A good deal of information is now available about the structure of the cohesive ends. These single-stranded regions are required for biological activity of λ DNA in the Kaiser-Hogness transformation test, for "filling in" the ends, by action of purified DNA polymerase *in vitro,* destroys infectivity (Strack and Kaiser, 1965). However, treatment with exonuclease III from *E. coli* restores infectivity. Since this enzyme removes mononucleotide units from 3′ termini of double-stranded DNA, one can conclude that the cohesive ends are 5′-terminated. Moreover, from the extent of exonuclease III action necessary to give maximal restoration of activity, one can calculate that the total number of nucleotides in both sticky ends is about 40 per molecule (Kaiser and Wu, 1968). Further, one can conclude that each end has the same number of nucleotides, since hydrogen-bonded circles can be covalently closed *in vitro* through the action of DNA ligase alone, a situation which can only obtain if there are no gaps in the hydrogen-bonded structure (Wu and Kaiser, 1968; see Figure 9–1). Finally, use of *E. coli* DNA polymerase as a reagent has allowed determination of the terminal nucleotides on the sticky ends as well as parts of the sequences (Wu and Kaiser, 1967; Kaiser and Wu, 1968; see Figure 9–1).

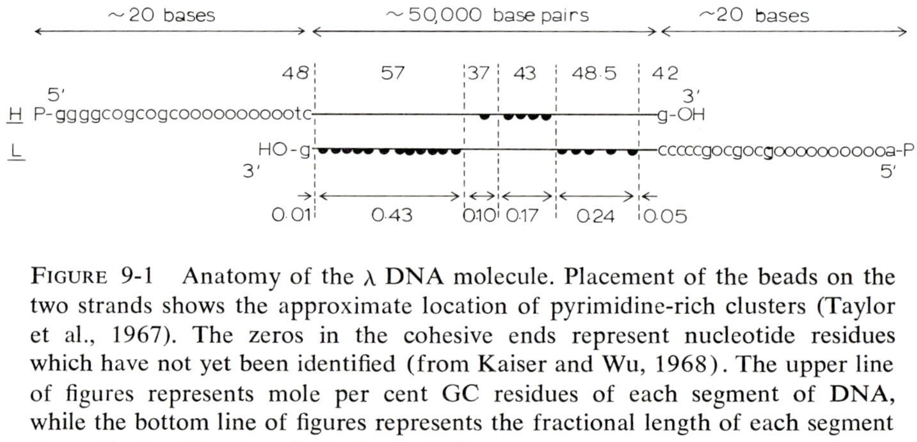

FIGURE 9-1 Anatomy of the λ DNA molecule. Placement of the beads on the two strands shows the approximate location of pyrimidine-rich clusters (Taylor et al., 1967). The zeros in the cohesive ends represent nucleotide residues which have not yet been identified (from Kaiser and Wu, 1968). The upper line of figures represents mole per cent GC residues of each segment of DNA, while the bottom line of figures represents the fractional length of each segment (from Skalka, Burgi, and Hershey, 1968).

Although the functions of the cohesive ends were not immediately clear, there is now ample evidence that circular molecules, bonded end-to-end, participate both in replication and in prophage integration (p. 238); see Chapter 10). If this is so, then one might expect all temperate phages to bear sticky ends, and to the extent to which this has been examined, they do (for list see Kaiser and Wu, 1968). In general temperate coliphages which have been examined fall into two classes: the λ family and the P2 family. Phages of the λ family, including φ 80, 21, 424, 434, and 82, can recombine with each other, and can act as helper phages for each other in the Kaiser-Hogness transformation test, while the other class, which includes P2 and 186, can also recombine with each other but not with members of the λ family. Interestingly, the cohesive ends of phages within either family will cohere with one another, but there is no reaction between the sticky ends of two phages from different families (Baldwin et al., 1966; Wang and Schwartz, 1967; Mandel and Berg, 1968).

Segmented base composition. The second unusual feature of the λ DNA molecule is the fact that the base composition is not uniform along the length of the molecule. This originally came to light when Hershey and his colleagues (Hershey, 1964; Hershey, Burgi, and Davern, 1965) found that the half-molecules of λ DNA produced under conditions of controlled shear could be separated in an equilibrium density gradient—a clear indication that one half is relatively rich in guanine-cytosine pairs (GC), while the other half has a higher content of adenine-thymine (AT) pairs. This finding, in conjunction with the availability of a biological test for λ DNA, allowed the λ genetic map to be oriented with respect to the DNA molecule (Hogness and Simmons, 1964). The GC-rich half corresponds to the left-hand portion of the genetic map as it is usually drawn (see next section). Therefore, we speak of the GC-rich left half (55 percent GC) and the AT-rich right half (55 percent AT). It should be observed that another technique has also been used to prove the colinearity of the DNA molecule and the genetic map (Meselson and Weigle, 1961; Meselson, 1964). In mixed infections between density-labeled and normal parental phage, the recombinant progeny had intermediate densities, with the values being those expected on the basis of colinearity. This type of experiment also helped to prove that recombination occurs by a breakage-reunion process (see Chapter 7).

Further fragmentation of the λ DNA molecule has revealed a greater extent of segmentation with respect to base composition and has also permitted a positioning of many genes on the DNA molecule. In the latter study Hogness et al. (1967) separated fragments of different sizes in a sucrose gradient and measured the ability of each fragment to contribute marker pairs in the transformation test. The larger the fragment, the more distant were the two farthest markers which could be borne together. Thus, the physical map corresponds to the genetic map, insofar as it has been tested. In further studies on the segmental distribution of nucleotides, Skalka, Burgi, and Hershey (1968) separated fragments of different sizes by equilibrium density centrifugation in cesium sulfate gradients of mercuric complexes. Mercury ions combine preferentially with AT-rich regions, and this greatly increases the resolving power of isopycnic separations. Skalka et al. found six rather sharply defined segments whose relative GC contents varied between 37 and 57 percent. Interestingly, the very low GC fragment is located in the middle of the molecule, in an area thought to be involved in a recombinational event which orcurs during prophage insertion (see Chapter 10). Other temperate coliphages are also segmentalized with respect to nucleotide composition (Falkow and Cowie, 1968; Skalka, 1969).

The general pattern of segmentation of the DNA of the λ-related phage φ 80 is similar to that of λ, except that the AT-rich central fragment is not present (Skalka, 1969).

DNA strand separation. Although the densities of the two strands of λ DNA differ slightly (Hogness et al., 1967), the extent of the difference is not sufficient to permit a preparative separation in a single step. However, Hradecna and Szybalski (1967) found that the synthetic polynucleotide poly I, G, which contains inosinic and guanylic acids, binds preferentially to one strand, permitting essentially quantitative strand separation. In accord with the terminology developed for T7 DNA in Chapter 8, the two strands are called H and L, or *r* and *l,* respectively (see Figure 9–1). The terms H and L refer only to the densities of the pure strands, for the L, or lighter, strand actually binds more poly I, G and therefore assumes a greater density during strand separation.

The binding of poly I, G was also used to study the distribution of cytosine-rich clusters. In contrast to the situation in T7, Hradecna and Szybalski found such clusters on both strands, as did Doerfler and Hogness (1968a). Since, as we shall see, both strands are used for transcription *in vivo* at different times during phage replication, this should not be surprising in terms of the idea that C-rich regions are sites of initiation of transcription. However, more recent work in this area (Champoux, Egan, and Hogness, 1968) suggests that a C-rich cluster is not necessary for initiation, but may be necessary for chain termination.

Strand Separation and the Genetic Map

The capability of separating strands of λ DNA has permitted the use of several methods for determining the orientation of genes, i.e., identification of the template strand and the direction of transcription *in vivo.* We shall postpone discussion of this work until we have had an opportunity to briefly cover the genetics of this virus.

The ability to separate strands of λ DNA has also permitted a new approach to the physical mapping of viral genes (Davis and Davidson, 1968; Westmoreland, Szybalski, and Ris, 1969). The method involves electron microscopic examination of artificially formed heteroduplex molecules. These are formed by denaturing DNAs from wild-type phage and a deletion mutant. The strands are separated preparatively and the H strand from one phage is renatured with the L strand from the other phage. The resulting heteroduplex molecule can form a double-helical structure everywhere except at the site of the deletion. At this site the unpaired material from the wild-type phage forms a "bush," or collapsed single-stranded region. With appropriate techniques one can measure the position and length of this region by electron microscopy. By this method, the extent of genetic homology among the related phages λ, 82, and 434 has been determined (Simon and Davidson, 1969). Moreover, a slight modification of the method, involving the use of transducing phages (see p. 228), has permitted the construction of a complete physical map of the λ genome (Hradecna and Szybalski, 1969; Davis and Parkinson, 1969).

GENES AND GENE FUNCTIONS IN λ

λ Genes and the λ Map

Plaque-type mutations. As with the T-even phages, early work on λ genetics utilized plaque-type mutations whose biochemical bases were obscure. These included host-range (*h*), minute-plaque (*mi*), and clear-plaque (*c*) mutations. The latter class of mutations have been of great value in furthering our understanding of lysogeny, since, as it turned out, the clear-plaque phenotype results from an inability to establish lysogeny. Thus, every cell which is infected on a plate undergoes a cycle of lytic phage growth. Clear-plaque mutants fall into three cistrons, c_{I}, c_{II}, and c_{III} (Kaiser, 1957; Kaiser and Jacob, 1957). Thus, at least three gene products are required for the establishment and maintenance of lysogeny. Early evidence that the one of these gene products might code for a repressor of the Jacob-Monod type came from the isolation of temperature-sensitive c_{I} mutants (Jacob, Sussman, and Monod, 1962). Lysogens containing such mutants as prophage are induced when the temperature is raised, suggesting that the gene product is a protein, and that the mutant protein has greater heat-sensitivity than the wild-type repressor. As we shall see later, there is now direct evidence that the repressor is a protein.

Conditional lethal mutations. The beginning of a physiological genetics for λ came when Campbell (1961) defined 18 new genes in terms of conditional lethal mutations: suppressor-sensitive (*sus*), temperature-sensitive (*ts,* originally called *hs*), and pH-sensitive. As we mentioned in Chapter 5, *sus* mutants are ambers. The genes were named A, B, C . . . R, and fell on a linear map, with A at the left and R on the right (see Figure 9–2). As we shall see below, the λ genome contains far more clustering with regard to function than does that of T4. In general, the genes on the left arm are concerned with maturation and virus assembly, while those to the right control early functions, such as DNA replication. In the interior of the map are genes governing regulatory functions, including the *c* genes, and another region which is involved in the establishment and maintenance of lysogeny. An exception to this clustering is the R gene, located near the right-hand extremity, which codes for the phage lysozyme (delCampillo-Campbell and Campbell, 1965; Naha, 1966), which, as in T4, is considered a late protein.

Since the work of Campbell, five new genes have been defined by conditional lethal mutations (Harris et al., 1967; Mount et al., 1968; Parkinson, 1968; Goldberg and Howe, 1969). Four of these genes, namely, U, V, W, and Z, are located in the left arm and are involved in virus assembly.

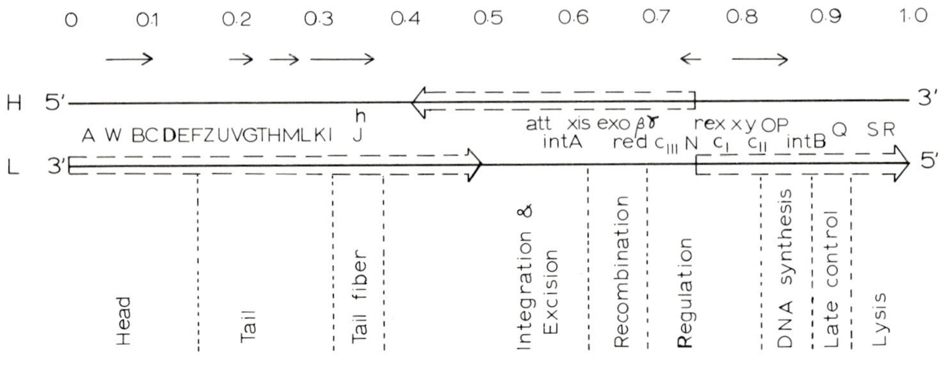

FIGURE 9-2 The λ genetic map. The upper row of numbers represents fractional distance along the map (or along the DNA molecule). The solid arrows indicate the extent and orientation of units of coordinate gene expression. The dashed arrows indicate directions of transcription. Gene functions are indicated below the map.

Gene S, which maps just to the left of R, is involved in lysis. Mutants in this gene are unable to lyse their hosts spontaneously, even though ample lysozyme (the R gene product) is formed in abundance. The biochemical basis for this phenomenon is not yet known. Interestingly, phage growth does not cease at the end of the normal latent period with a gene S mutant, as it does in R. Thus, cells infected with S mutants accumulate very high levels of phage, which can be recovered by artificial lysis of the cells (Harris et al., 1967). The S gene seems to be similar to the *t* gene in T4, studied by Josslin (1970).

Deletion mutations. Some types of λ mutations do not correspond to any similar situation in T4 or the other virulent phages, inasmuch as they are related to the ability of the phage to lysogenize. An example is the λb2 mutation (Kellenberger, Zichichi, and Weigle, 1961). This was originally isolated on the basis of its decreased density in a CsCl gradient, as related to wild-type λ. Subsequent study showed that λb2 has 18 percent less DNA than wild-type phage. Apparently, the deleted material includes the specific homology region necessary for prophage integration, for λb2 cannot establish lysogeny. However, it can establish immunity, and, if a lytic response does not occur after infection, the chromosome remains in the host in a nonreplicating condition, and it is passed to only one of the daughters at cell division. Thus, it is eventually diluted out of the population.

Defective lysogens. Another type of mutant includes the so-called λ-defective lysogens. These are mutations which occur in the prophage after mutagenic treatment of a lysogen (Jacob, Fuerst, and Wollman, 1957). Such mutants are absolutely, as opposed to conditionally, lethal. They cannot replicate after induction unless a helper phage is present to provide the needed wild-type function. At least one gene, the so-called *x* region, has been defined in studies on one such defective lysogen, λt11 (Eisen et al., 1966). As we shall see shortly, the *x* region plays an essential role in the regulation of early functions. An adjacent site, the *y* region, defined by the *c*17 mutation, appears to be a promoter region, or initiation site for RNA polymerase (Pereira da Silva and Jacob, 1967). There is now excellent evidence, from studies of RNA synthesis *in vitro,* that this is so (Roberts, 1969a).

Early proteins and their structural genes. Searches in various laboratories for λ-specific early enzymes have revealed clear-cut evidence for only one, an exonuclease which cleaves 5′-mononucleotides from 5′-termini (Korn and Weissbach, 1963; Little, 1967), although there is suggestive evidence for the existence of an endonuclease (Boyce et al., 1969; Schuster and Weissbach, 1968a) and a tRNA methylase (Wainfan, 1968; Wainfan and Visser, 1969). Radding (1966) found that defective lysogens bearing a λt11 prophage produced very high levels of exonuclease after inducing treatments. Since this defective lysogen also makes no DNA after induction, and since the purified exonuclease is indistinguishable from the wild-type enzyme, Radding (1966) concluded that faulty regulation in the t11 mutant is responsible for overproduction of the exonuclease. At the same time, Radding and Shreffler (1966) found that another protein, the so-called β-protein, is subject to the same kind of control as the exonuclease. The β-protein was discovered by immunological means, and its enzymatic function is unknown, although it appears to be involved in vegetative recombination. However, by means of recombination between λ and φ80, which produces similar but distinguishable exonuclease and β-protein, Radding, Szpirer, and Thomas (1967) were able to map the structural genes for these two proteins on the λ chromosome. The genes are called *exo* and β, respectively, since they do not correspond to any of the lettered genes in the sets of Campbell and of Parkinson. These genes appear to be nonessential for phage replication (Manly, Signer, and Radding, 1969).

It is interesting that the λ-related phages 434, 80, and 21 all induce an exonuclease very similar to the λ enzyme (Pricer and Weissbach, 1967), while the unrelated phages 186, 299, and the defective phage carried by *E. coli* 15 do not (Schuster, Breitman, and Weissbach, 1967).

Integration-negative mutations. As we shall see in Chapter 10, prophage insertion almost certainly involves a recombinational event, or crossing-over, occurring at or near a site of specific interaction between the λ and host chromosomes. In Chapter 7, we discussed the participation of enzymes, and possibly other protcins (such as the gene 32 product) in T4 recombination. If integrative recombination in λ occurs by a similar mechanism, and if it is directed by phage-coded enzymes, then there should exist mutants in the genes for these enzymes. Such mutants would exhibit the same integration-negative phenotype as λb2. However, at least some of them would be expected to be point mutants, altered in a single nucleotide pair, rather than the massive deletion involved in the λb2 mutation. Such mutants have, indeed, been found (Zissler, 1967; Gingery and Echols, 1967; Gottesman and Yarmolinsky, 1968). Most of these map at a site (*int* A) immediately to the right of the λ attachment region, although some of the mutants mapped at a second site (*int* B) which is located between genes P and Q. Thus, at least two phage-coded proteins appear to be involved in integrative recombination.

A related class of mutants is defective, not in integration, but in prophage excision (Guarneros and Echols, 1970; Kaiser and Masuda, 1970). These mutants map just to the right of *int,* in what has been termed the *xis* region.

Recombination-deficient mutations. As we shall see later, the integrative recombination system is *site-specific,* i.e., it can only promote crossing-over at the site on the λ genome at which prophage insertion normally occurs (Weil and Signer, 1968; Echols, Gingery, and Moore, 1968). However, recombination between many different pairs of λ markers ("vegetative recombination") can occur normally, even in a recombination-deficient host (Brooks and Clark, 1967). This suggests that the phage codes for a vegetative recombination system, as well as the above-discussed integrative recombination system. Again, if such a system does exist, it should be possible to isolate mutants deficient in that system, and this also has been done (Franklin, 1967; Signer and Weil, 1968a; Echols and Gingery, 1968). Such mutants, which are termed *red*−, to distinguish them from *rec*−, the term for recombination-deficient bacteria, map at sites just to the right of *int* A. These sites are the structural genes for exonuclease and β-protein (Signer et al., 1968; Radding, 1970). Thus, the major roles of these two proteins appear to be participation in vegetative recombination. Recombination between two *red*−mutants in a *rec*+ host is about 10 percent of normal, suggesting that the bacterial recombination system can function to a certain extent in promoting phage recombination.

In addition to the *red*− mutants, Zissler and Signer (1970) have described a new class of recombination-deficient mutants, the so-called *fec* mutants. These strains cannot grow in certain recombination-deficient bacterial hosts. They are double mutants containing mutations in one of the *red* genes and in a new gene, called γ.

The *ter* function. Evidence has been presented for still another λ function related to recombination, the *ter* (from termini) function (Mousset and Thomas, 1969). Maturation of λ DNA would have to involve some way of generating the mutually complementary cohesive ends. Presumably this could be accomplished by means of a nuclease which can induce single-strand breaks in opposite strands of a circular or concatenated duplex, 20 nucleotides apart (see Figure 9–3). If this is so, then one should

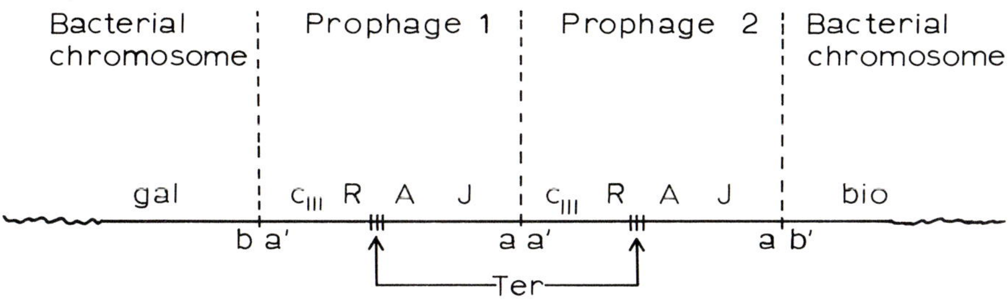

FIGURE 9-3 Action of the *ter* function in excising one complete λ genome from a double lysogen.

be able to excise a vegetative λ genome from a double lysogen containing two λ prophages in tandem (Figure 9–3). This could presumably be induced by superinfection with a *heteroimmune* phage, *i.e.,* one which responds to a different immunity substance from that of the prophage and can, therefore, replicate. Presumably, expression of the *ter* system in the superinfecting phage would lead to excision of one complete vegetative λ chromosome. (Note that two prophages must be present in tandem, because the gene order in the prophage is a circular permutation of that in phage particles. We shall discuss the basis for this in Chapter 10.) By using suitably marked prophage and superinfecting phage, Mousset and Thomas showed that this does occur, even in the absence of the phage *int* or *red* systems, or the bacterial *rec* system. Thus, the *ter* system seems to be specified by a separate gene in λ. However, ter^- mutants have not yet been isolated, nor has it been shown whether the *ter* gene is not the same as one of the lettered genes of Campbell (1961) or of Parkinson (1968). Presumably, the *ter* function is essential for replication, and one would expect a random isolation of many conditional lethal mutations to turn up some mutants defective in this function.

T4 *r*II exclusion-deficient mutants. It will be recalled from Chapter 8 that *r*II mutants of T4 cannot replicate in *E. coli* strains lysogenic for λ. For some time this was felt to reflect some metabolic consequence of the presence of the λ repressor, since the gene coding for repressor was the only phage gene known to be expressed during lysogeny. Moreover, λi^{434}, a hybrid of λ and phage 434 which contains almost all of the λ genome except for a small area which includes the 434 immunity region did not show this property of *r*II exclusion (Kaiser and Jacob, 1957). However, two groups have recently isolated λ mutants which are deficient in *r*II exclusion, even though they still exhibit the turbid-plaque phenotype characteristic of a functioning *c*I gene (Howard, 1967; Ravin and Poluchina, 1968). These mutants, termed *rex*⁻, map immediately to the left of the c_I gene. Since λ prophages would be expected to encounter T4 *r*II mutants but rarely in natural habitats, the true biological function of the *rex* gene is unknown.

The Vegetative Map

The genetic map as it exists at the time of writing was presented in Figure 9–2. Assuming that the *h, mi,* and *ter* functions are coded for by genes in the sets of Campbell and/or Parkinson, the map in its present state contains about 35 genes, including the b2 region, which may contain unrecognized genes other than the λ attachment site (cf. Tonegawa and Hayashi, 1968). Since, as we stated previously, the coding capacity of the λ genome is about 40 proteins, the amount of this capacity which has been at least partially identified in terms of function is between 80 and 90 percent.

We should not leave a discussion of temperate phage genetics without pointing out that the genetic map of P22, which contains circularly permuted base sequences in its genome, is itself circular (Gough and Levine, 1968).

Transduction

An aspect of temperate phage biology which has found widespread applications in bacterial genetics is transduction (Zinder and Lederberg, 1952), a process in which genetic information is transferred from one bacterium to another by a temperate bacteriophage. There are two types of transduction, general and specialized. Specialized transduction is seen in the case of λ and related phages, which become integrated into a specific site on the bacterial chromosome. These sites can be mapped by standard bacterial genetic techniques (cf. Hayes, 1968; Jacob and Wollman, 1961). In the case of λ, the site of prophage insertion is between the operons controlling the utilization of galactose and the synthesis of biotin. Thus, we say that the λ prophage is inserted between *gal* and *bio* on the *E. coli* genetic map, as shown in Figure 9–4. This figure shows also the attachment sites for several other temperate coliphages.

As we have mentioned, an early event following induction of a lysogen is excision of the prophage. Occasional errors occur in this process, such that the region excised includes a small amount of bacterial DNA, with a corresponding piece of phage DNA deleted. If the portion of bacterial

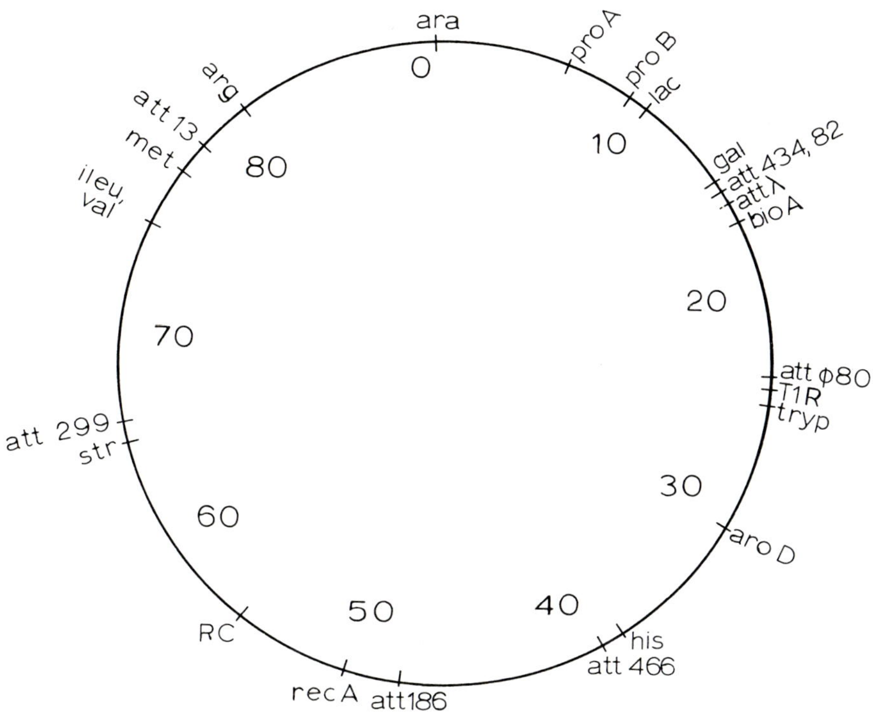

FIGURE 9-4 A partial genetic map of *E. coli* K12, showing attachment sites for some temperate phages (adapted from Taylor and Trotter, 1967). The numbers inside the circle indicate the time of transfer (in minutes) in interrupted mating experiments.

DNA is located to the left of the prophage, it may include some or all of the genes of the *gal* operon, with a compensatory loss of some of the λ genes at the right end of the prophage map. This incomplete phage chromosome then serves as the template for replication, such that essentially all of the phage progeny issuing from that cell bear the genes for galactose utilization and have lost some phage genes. Reinfection of a gal^- cell with such a phage, under conditions allowing lysogenization, can confer upon that cell the ability to ferment galactose, since the necessary genes become inserted into the bacterial chromosome. Such a transducing phage is defective in its own replication, since it lacks some essential phage genes. Such a phage is called λ*dg* (for defective-galactose), and can only be propagated in the presence of a helper phage, which supplies the missing gene functions.

Similarly, the genes to the right of the λ prophage can be transduced, with the genes governing biotin synthesis being transferred by λ to recipient cells. In some cases the phage genes which are deleted do not include any which are essential for replication. Phages which transduce *bio* are called λ*bio* or λ*dbio,* depending upon whether or not they are defective (Kayajanian, 1968). The term λ*pbio* (plaque-forming *bio*) is sometimes applied to the nondefective *bio* transducing phages.

Because the λ prophage is located adjacent to *gal* and *bio* on the *E. coli* map, these are the only genetic regions which can be transduced. There is, however, another type of temperate phage, exemplified by Pl, which has no specific attachment site. In fact, Pl is not integrated into the host chromosome at all, but is maintained inside the cell as an extrachromosomal genetic entity, subject to the same control of replication as is the bacterial chromosome (Ikeda and Tomizawa, 1968). Pl can transduce any bacterial marker (general transduction), albeit at low frequency, with the probability of a given marker being found in a Pl lysate being between 10^{-5} and 10^{-7} (Lennox, 1955). This probably results from occasional packing of a segment of bacterial DNA into a phage coat. This presents a valuable method for genetic mapping in bacterial systems with no known possibilities for transformation or bacterial mating. Since the phage chromosome (60 million molecular weight) is far smaller than the bacterial chromosome, the frequency with which a Pl particle cotransduces two bacterial genes provides a measure of their distance apart on the bacterial chromosome. Since as much as 2 percent of the *E. coli* chromosome can be transduced by a single phage particle, it is probable that at least some transducing phages contain only bacterial DNA, wrapped in a phage protein coat (Ikeda and Tomizawa, 1965).

An interesting aspect of Pl lysogens is the fact that intracellular phage DNA is circular (Ikeda and Tomizawa, 1968). In this respect, the Pl prophage resembles other extrachromosomal genetic elements, or plasmids, such as colicinogenic factors (Roth and Helinski, 1967) and, probably, sex factors (Freifelder and Freifelder, 1968; Hayes, 1968). However, the genetic map, insofar as it has been elucidated, is linear (Scott, 1968). Since DNA from phage particles is also linear, this would imply that opening of the intracellular rings during phage replication occurs at a specific point.

No discussion of transduction would be complete without a mention of its use in a *tour de force* which has been considered as the first isolation of a gene in pure form (Shapiro et al., 1969). What was actually isolated is a pure segment of the lactose (*lac*) operon in *E. coli,* containing the structural gene for β-galactosidase (the *z* gene), the operator site (*o*), and the promoter region (*p*), with parts of the adjacent genes in the *lac* operon. It may be recalled that the order of genes in the *lac* operon is *i, p, o, z, y, a* (cf. Beckwith, 1967). What was done was to use a sex factor bearing the *lac* operon (a so-called F-*lac* episome) and, with the use of certain genetic manipulations, force its integration into a bacterial chromosome, not at the usual site for the *lac* operon, but at a region adjacent to the attachment site for λ. After a derivative of this strain had been lysogenized with λ, it was induced, and a nondefective phage which bore *lac* operon DNA was isolated. Next, by use of a different sex factor and more manipulations, the *lac* operon was inserted into the *E. coli* chromosome next to the attachment site for the λ-related phage, φ80, in the opposite orientation from that of the first strain (of the two strands in the *lac* operon, only one "makes sense" genetically; this template strand is oriented clockwise in one strain, counterclockwise in the other). The second strain was lysogenized with φ80, and a *lac*-transducing phage was isolated. Because the F-*lac* episomes were inserted in opposite orientations, we now expect that, if one strand of the operon is located on the H strand of the λ strain, the complementary strand will be located on the H strand of the φ80 strain. Because there is considerable homology between λ and φ80, we would expect no homology between their separated H strands except that between the complementary strands of *lac* operon DNA. Thus, a four-tailed structure is formed, as seen in Figure 9–5. The single-stranded regions, which contain mostly unpaired phage DNA, are digested away with an exonuclease specific for single-stranded DNA, leaving pure *lac* operon DNA. Undoubtedly, by the time this book is published, this material will have been used for several important experiments on the molecular nature of genetic control.

Position and Orientation of λ Genes

As mentioned on page 222, the ability to separate strands of λ DNA has allowed several approaches to the physical mapping of the phage chromosome. The questions to be asked here are: (1) What is the physical position of a gene, and how does it correspond to the position of the same gene on a linkage map? and (2) Is the template strand for a given gene on the H or L strand, and in what direction is the gene transcribed? We have already mentioned one technique for physical mapping, i.e., formation of heteroduplexes between two λ strains one of which contains a deletion, or both of which contain regions of nonhomology. The technique of Westmoreland, Szybalski, and Ris (1969) permits measurement of the length of both the double- and single-stranded regions of the heteroduplex, when viewed in the electron microscope. An example of this technique is shown in Figure 9–6. Wild-type λ DNA was hybridized with DNA from the double deletion mutant λ*b*2*b*5, which carries the immunity region for phage 21 at the site of the b5 deletion. The heteroduplex shows the unhybridized *b*2 region and the region of nonhomology between the λ and 21 immunity regions. The physical positions and lengths of these regions of nonhomology agree quite well with the same results indicated by genetic mapping.

An extension of the above technique involves the use, not of deletion mutants, but of transducing phages (Hradecna and Szybalski, 1969; Davis

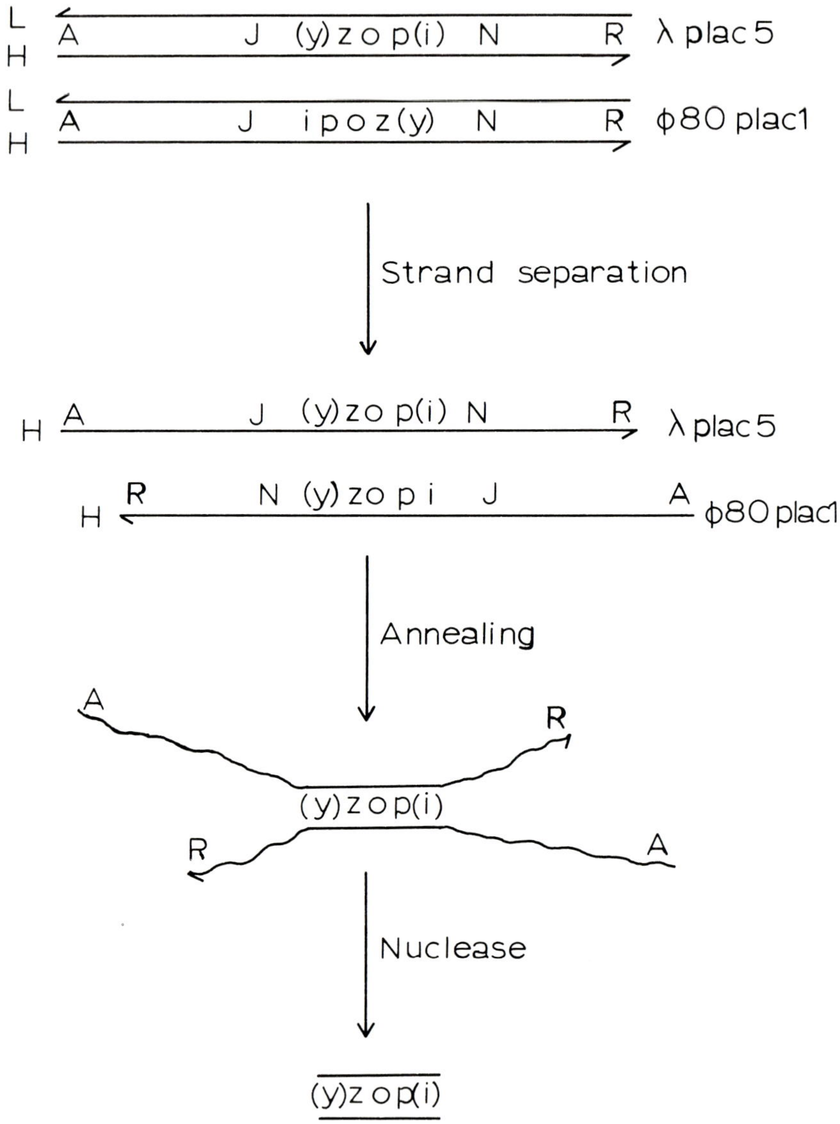

FIGURE 9-5 A rationale for isolation of *lac* operon DNA (adapted from Shapiro et al., 1969). The genes whose symbols are enclosed in parentheses are present in incomplete form. The arrow points indicate 3′ termini.

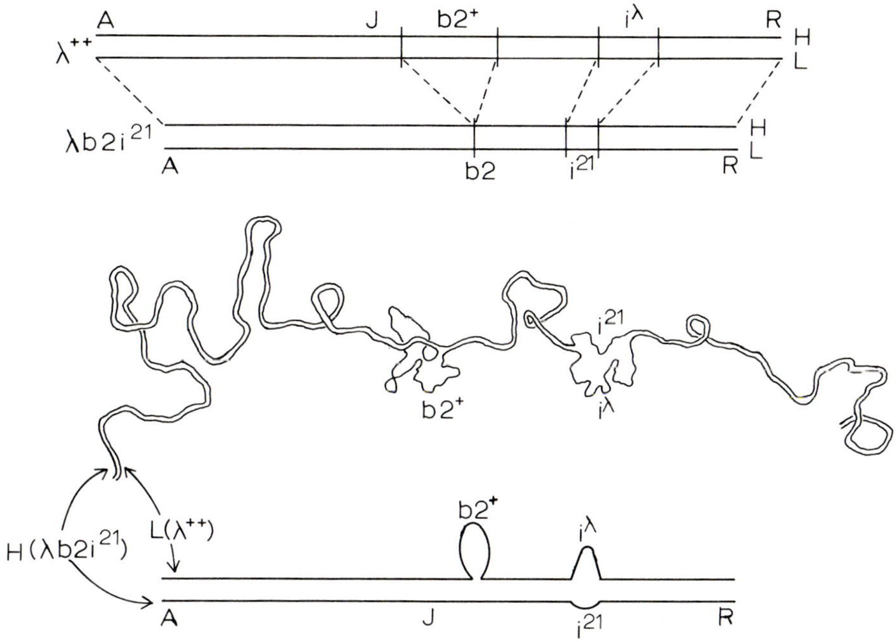

FIGURE 9-6 Physical mapping by formation of λ heteroduplexes, as described by Westmoreland et al. (1969). The upper part of the drawing indicates the genotypes of the two strains used in constructing the heteroduplex. The middle part is a tracing of an electron micrograph of one such heteroduplex, and the bottom part presents an interpretation of the results, based on measurements of strand lengths in the heteroduplex.

and Parkinson, 1969). In transducing phages, the bacterial genes are located internally. Genetic mapping had shown that, whereas the left end of bacterial DNA in *gal*-transducing phages was variable, the right end was invariably located at a specific site, at the right end of the *b*2 region (the *att*⁻ region of Figure 9–2). Similarly, the bacterial material in all *bio*-transducing phages has an invariant left end, even though the right ends are variable. The invariant left end is the same as the invariant right end in *gal*-transducing phages. Physical mapping confirms this, a fact which has important implications for the mechanism of prophage insertion and excision, as we shall see in Chapter 10.

Three methods have been used to determine the orientation, or direction of transcription, of λ genes. Transcription proceeds by addition of nucleotide units to free 3′-hydroxyl groups at the ends of growing RNA chains (Bremer et al., 1965; Maitra and Hurwitz, 1965). This RNA is presumed to be antiparallel to the DNA template strand. The λ DNA molecule is usually drawn so that the direction of the H strand is 3′ to 5′, and that of the L strand is 5′ to 3′ (see Figure 9–1). Therefore, since the direction of transcription is 3′ to 5′, we would expect that, if the H strand serves as template for a given gene, the direction of transcription is to the left, and, if the L strand is the template strand, the direction of transcription is to the right.

Studies on gene polarity have shown that the left arm of the λ genome contains four units of coordinate gene expression, all oriented rightward (Parkinson, 1968; Naha, 1968). For example, genes W, B, and C form one such unit. A nonsense mutation in gene W abolishes expression of genes B and C, indicating that this unit is transcribed to the right. Since the genes of the left arm are concerned with the late functions of virus assembly, one can infer that these late genes are transcribed from the L strand. A similar unit of rightward transcription has been seen in the "early" genes c_{II} and O (Eisen et al., 1966; Pereira da Silva and Jacob, 1967). However, the very early-acting gene N has been shown directly to be oriented leftward (Doerfler and Hogness, 1968b). Heteroduplex molecules containing one strand from wild-type λ and the complementary strand from a *sus* N mutant were tested for gene N activity in the Kaiser-Hogness transformation test, under conditions preventing the excision and repair of the mismatched sections. Only the heteroduplex containing wild-type H and *sus* N L was active. Thus, the H strand of wild-type λ is the one which serves as the template strand for N and, therefore, gene N is oriented leftward.

Hybridization between pulse-labeled λ mRNA and the separated strands of well-mapped λ *bio* strains bearing substitutions between c_{III} and *int* show that this region is transcribed leftward as well (Kumar et al., 1969). Thus, there now exist two known regions of the λ DNA molecule for which L serves as the template strand, and one for which H is the template, as shown in Figure 9–2. This has important implications for the control of transcription during viral development, as we shall see later.

Functions of λ Genes

So far we have had relatively little to say about the functions of individual λ genes, beyond the statement that some genes are involved in regulation, some in virus assembly, some in integration, and so forth. In this section, I wish to explore further the functions of genes involved in virus replication, reserving discussion of the genes involved in lysogeny until Chapter 10.

Late genes. As we have seen, genes A through J on the left arm are concerned with virus assembly, and gene R, on the right arm, is the structural gene for lysozyme, a late protein (Protass and Korn, 1966). The defective phenotypes of mutants in genes A through J in the original Campbell set have been characterized by electron microscopy (Kemp, Howatson, and Siminovitch, 1968) and in an *in vitro* assembly test (Weigle, 1966). Mutants in genes A through F are presumed to be defective in head formation, since they can donate complete tails *in vitro*. On the other hand, mutants in genes G through J can serve as head donors, indicating a defect in tail formation. Gene J has been identified as the structural gene for the major tail antigen (Dove, 1966; Buchwald and Siminovitch, 1969), and this is probably the same as the *h* marker.

We have discussed in Chapter 7 the apparent relationship between head formation in T4 and the condensation and maturation of phage DNA. A similar situation exists in λ. Mutants in genes C through J can synthesize a DNA under restrictive conditions which is infective in the Kaiser-Hogness test (Dove, 1966), while mutants in the head-defective genes A through F make a DNA which is noninfective. Work of Mackinlay and Kaiser (1969) suggests that these mutants can form a circular species of replicating DNA (see p. 238), but cannot convert this to mature linear DNA with cohesive ends.

Early genes and regulatory genes. The earliest study of the functions of these genes was that of Brooks (1965), who examined the ability of various *sus* mutants to lysogenize a nonpermissive host and to form a vegetative gene pool after uv induction of the lysogen. Her data indicated that genes N, O, and P control essential steps in DNA replication, while genes Q, R, and B acted later in the vegetative cycle. Mutants in genes N, O, and P are defective, not only in DNA replication but in late functions as well, including synthesis of tail-fiber antigen and lysozyme (Dove, 1966), and late mRNA species (Joyner et al., 1966; Skalka, Butler, and Echols, 1967). Thus, as in T4, DNA replication is required for expression of late genes. In further analogy to the T4 situation, DNA synthesis is not a sufficient condition for late gene expression, since mutants in gene Q can synthesize DNA, but not late proteins (Joyner et al., 1966). In this respect, Q seems to be analogous to the maturation genes, 33 and 55, of T4. However, in contrast to the T4 situation, the block in late mRNA synthesis is not complete in the failure of genes N, O, P, or Q to act, although the block in DNA synthesis in mutants in genes N, O, and P was complete within the limits of sensitivity of the methods used, i.e., less than 4 percent of normal (although another report states that a prophage bearing a *sus* N mutation can make phage DNA after induction [Eisen, Pereira da Silva, and Jacob, 1968]). Another point of contrast with T4 is the fact that early proteins are not oversynthesized in the absence of DNA replication by N, O, or P mutants. However, the early proteins exonuclease and β-protein are both synthesized in excess following induction of the defective lysogen λt11 (gene *x;* Radding and Shreffler, 1966; Eisen et al., 1966). Coupled with this anomaly is a failure of DNA synthesis after induction of λtll, suggesting that gene *x* controls both the turnoff of some early genes and the turn-on of genes necessary for DNA replication (Eisen et al., 1966; Joyner et al., 1966). Interestingly, λtll over-produces early mRNA, as well as early proteins (Skalka, Butler, and Echols, 1967).

Recently, Pero (1970) has described a new gene, called *tof,* whose product is involved in the turn-off of exonuclease synthesis at the proper time in phage development. The *tof* gene, which is located in the immunity region of λ, specifies a product which seems to be similar to a repressor. The *tof* gene is probably the same as *cro,* another recently-described gene whose product is involved in the regulation of immunity, possibly *via* control of c_I gene expression (Eisen et al., 1970; Szybalski et al., 1970).

Although very little is known about the roles of genes O and P in promoting DNA replication, most interest has been focused upon gene N, since this seems to affect the largest number of functions, including, as we have seen, DNA synthesis and late mRNA synthesis, but also early mRNA synthesis (Konrad, 1968; Kourilsky et al., 1968), and the synthesis of the early protein exonuclease (Radding, 1964). This latter fact suggested that N might be the structural gene for exonuclease and that formation of this enzyme is prerequisite to all of the other steps just mentioned. However, the finding that β-protein is also not formed by gene N mutants (Radding and Shreffler, 1966) raised the possibility that N is a regulatory gene, necessary for expression of other early functions and all late functions. Genetic studies (Radding and Echols, 1968) have borne this out.

Another function controlled by gene N has recently been described (Boyce et al., 1969). These workers found that superinfecting λ DNA, which is converted in stable λ lysogens to covalently closed circular molecules (see p. 239), is opened to give linear molecules following prophage induction. However, this does not occur in the absence of an active gene N product, although the products of genes, O, P, and Q are not required. This suggests that gene N might control an endonuclease, which would carry out the ring opening. Since, as we shall see in Chapter 10, it is probable that a freshly excised prophage exists as a covalent circle, this could possibly explain the pleitrophic effect of gene N. Suppose, for example, that a noncircular template is required for transcription of other genes. Of course, the present state of knowledge does not allow us to tell whether the presumed endonuclease is the product of gene N or whether it is simply subject to regulation by that gene, as are the other early proteins. In fact, preliminary evidence suggests that the endonuclease might be controlled by gene P (Schuster and Weissbach, 1968b). It seems likely that genes N and Q act by altering the template specificity of RNA polymerase, as does gene 1 in T7, but this has not yet been proven biochemically. (Note added in proof.) Very recent work (Szybalski et al., 1970) confirms that the gene Q product is an initiation factor for RNA polymerase, but indicates that gene N codes for an antiterminator protein, whose effect is to extend transcription both leftward, past gene N, and rightward, past genes *x* and P.

Butler and Echols (1970) have reported upon the properties of a new λ mutation, called *byp,* which bypasses the normal gene N requirement for late protein synthesis. Properties of this mutation, which maps between genes P and Q, suggest that gene N normally has two functional targets essential to phage growth: DNA replication and gene Q. Hopkins (1970) has independently published similar findings.

Table 9–1 summarizes the data we have been discussing on defective phenotypes associated with early λ genes.

TABLE 9–1
DEFECTIVE PHENOTYPES OF EARLY λ GENES

Mutant Gene	Covalent Ring Opening	Early mRNA and Protein	DNA	Late mRNA and Protein
none	+	+	+	+
N	−	−	+(?)	−
O	+	+	−	−
P	+	+	−	−
Q	+	+	+	−
x(λt11)	unknown	excess	−	−

Enzymatic Basis for Host-Induced Modification and Restriction

In Chapter 6 we discussed the general phenomenon of host-induced modification and restriction (or host-controlled variation), and we presented some of the enzymology of the T-even phage system, where phage DNA must be glucosylated in order to avoid breakdown in a subsequent host. The best-characterized modification and restriction system in temperate phage biology involves the two *E. coli* strains C and K (for a complete discussion of the genetics of this and other systems for host-controlled variation, see Hayes, 1968). Phage λ which has been grown by a single cycle of growth in strain K (λ · K) can infect and grow in either strain K or strain C. However, λ which has been grown in strain (λ · C) cannot grow in strain K, but only in strain C. *E. coli* K is said to be a restrictive host for λ · C; in this host the DNA of λ · C breaks down, and very few genetic functions are expressed (Terzi, 1968). However, passage of λ through *E. coli* K allows some structural modification of the phage DNA which prevents its restriction in a subsequent passage through K. Strong circumstantial evidence suggested that the modification reaction was a methylation of DNA (Arber, 1965; Wood, 1966). Presumably, the enzyme responsible for restriction would not recognize DNA with the methylation pattern brought about by the K system. As we saw in Chapter 8, there is now direct evidence for the involvement of methylation, obtained by using uv-irradiated T3 to destroy intracellular *S*-adenosylmethionine.

What is the restriction enzyme? Presumably it is a nuclease, since restricted DNA breaks down, and it must have sufficient specificity to distinguish between λ · K DNA and that of λ · C. Just such an enzyme has been isolated from an endonuclease I-deficient strain of *E. coli* K 12 (Meselson and Yuan, 1968). The enzyme induces double-strand cleavages in DNA from λ · C, but not that of λ · K. The enzyme is unique among known nucleases in that it requires both ATP and S-adenosylmethionine. The latter requirement is of special interest in light of the role of methylation in the modification reaction. The S-adenosylmethionine does not participate as a substrate in the reaction, for it is not cleaved, nor are methyl groups transferred to the DNA. By contrast, ATP is hydrolyzed to ADP, and in stoichiometric excess over the number of double-strand breaks induced.

Another restriction system exists in *E. coli* strains lysogenic for the general transducing phage P1. These strains restrict DNA from λ which has been grown in a strain not lysogenic for P1. Meselson and Yuan discovered a similar nuclease in P1 lysogens. Although the cofactor requirements for this enzyme, which they call endonuclease III · P are the same as those of the λ · C-restricting enzyme (which they call endonuclease III · C), the substrate specificities are different. Endonuclease III · P attacks DNA from any λ preparation which has not been grown in a Pl lysogen. Thus, the enzymatic bases for these systems of host-controlled variation are beginning to be understood, although much remains to be learned about the molecular details of the enzymes involved.

REPLICATION OF TEMPERATE PHAGES

DNA Replication

Because the λ DNA molecule has been so well characterized and because relatively few phage gene products are required for its replication, λ would appear to present a most attractive system for studying DNA replication, particularly when compared to T4. An additional advantage is the relatively small size of the molecule, which allows its ready isolation in undegraded form.

There has been considerable disagreement over whether the replication of λ DNA involves circular or concatenated intermediates, and present evidence would appear to indicate that both forms participate. Dove and Weigle (1965) showed that λ DNA molecules extracted from cells early during phage multiplication are noninfective in the Kaiser-Hogness transformation test, but that activity is regained later in the latent period. This suggested that λ DNA is converted to a different structure which participates in replication, but that maturation of DNA molecules late in the infectious cycle generates infective progeny molecules. The existence of cohesive ends on the λ DNA molecule suggested that the molecule could circularize through base-pairing interactions between the cohesive ends of the same molecule. Indeed, Bode and Kaiser (1965a) had shown that the DNA of phage used to superinfect immune lysogens was converted to two species which sedimented more rapidly through sucrose gradients that mature DNA. One form (species II) had the properties of a double-stranded circular molecule, while the other form (species I), which sedimented nearly twice as rapidly as linear DNA due to its greater compactness, behaved as though it were a superhelical circular modecule; in other words, a twisted circle. Young and Sinsheimer (1964) also presented evidence that parental DNA in cells undergoing phage replication could be isolated as circular molecules.

Ogawa and Tomizawa (1967a) probed the structural relationship between the two apparently circular forms of λ DNA, by observing the change in sedimentation properties of species I, following decay of incorporated P^{32} of high specific activity (recall that such decay can induce both single-and double-strand breaks). They calculated from their results that the formation of one single-strand break creates a molecule behaving like species II. This suggested that species II is a partly open circular form, covalently closed in one strand, but with a nick in the other which prevents superhelical twisting, while species I is covalently closed in both strands.

Circumstantial evidence for the involvement of circles was presented by Young and Sinsheimer (1967), who found that both circular forms of λ DNA, while inactive in the Kaiser-Hogness test, were infective in spheroplasts. This jibed with the properties of intracellular λ DNA at early stages of replication or lysogenization, which was also infective only in spheroplasts. The reason for this differential behavior in *in vitro* tests of biological activity is still unknown, but it may be related either to transport into the recipient cell or spheroplast, or else to the necessity, in the Kaiser-Hogness test, of recombination of the test DNA into the genome of the helper virus. The latter process may require cohesive ends, which, of course are not available as single-stranded regions in the circular forms.

The identification of species I and II as closed and open circles, respectively, has been confirmed by direct examination in the electron microscope (Kiger, Young, and Sinsheimer, 1968; Bode and MacHattie, 1968). The number of twists in the closed circles is strongly dependent upon the ionic strength. So far, the number of supercoils per DNA molecule *in vivo* has not been determined with precision, although an approach to the question has been reported (Hinton and Bode, 1969).

The studies cited above, as well as others, clearly indicate that λ DNA can exist intracellularly in circular forms. But what are the metabolic roles of these forms? Are they replicative intermediates, or do they participate in lysogenization? As we shall see in Chapter 10, there is good evidence that a covalently closed circular DNA molecule must be formed for prophage insertion to occur. It may also be the earliest form of excised prophage DNA after induction, since Lipton and Weissbach (1966) reported that circular DNA molecules are formed after lysogenic induction. Moreover, the study of Bode and Kaiser, as well as others, was carried out with superinfected immune lysogens, cells which permit no DNA replication (Wolf and Meselson, 1963; Green et al., 1967). The possibility existed that λ DNA molecules which cannot be used—as in superinfection of an immune lysogen—are converted to circles, which are metabolically stable. Indeed, Weissbach and Salzman (1967) proposed that circles are by-products of the replicative process, perhaps serving as templates for transcription of certain classes of genes.

At the same time, evidence was accumulating for the participation of a concatenated species in λ DNA replication, similar to that described by Frankel (1966b) for T4. Smith and Skalka (1966) reported that much of replicating λ DNA is in a form which sediments more rapidly than mature λ DNA, either in neutral or alkaline gradients. Other workers found similar evidence for the existence of long strands in replicating λ DNA (Salzman and Weissbach, 1967; Young and Sinsheimer, 1968; Kiger and Sinsheimer, 1969). In the latter study, no DNA strands of greater than two phage-equivalent units in length were seen, and the authors proposed a "bicyclic" model of DNA replication. However, electron microscopy has revealed concatemeric structures which are up to five phage-equivalents in length (Weissbach, Bartl, and Salzman, 1968). Structures of greater than 10 phage-equivalents in length have been seen by radioautography (Carter, Shaw, and Smith, 1969).

Weissbach, Bartl, and Salzman (1968) and Tomizawa and Ogawa (1968) have both used density-labeling techniques for isolation of replicating λ DNA free of bacterial DNA. Such molecules seen early in infection appear in the electron microscope mostly as double-forked circular structures, of the type expected if a circular molecule were serving as a template for replication. Concatenated structures are seen later, including circular molecules of greater than 1 phage-equivalent unit in length. Similar results have been seen in pulse-chase experiments involving characterization by sucrose gradient centrifugation (Carter, Shaw, and Smith, 1969). In this study, the formation of circles was seen in the first 15 minutes of infection. These circles do not become incorporated into phage progeny. Rather, the long-stranded intermediate, which is formed after 15 minutes, serves as the major precursor of progeny DNA. Thus, we have a situation in which at least the first round of replication occurs on a circular template, with later rounds occurring on a long-stranded template. This is reminiscent of the rolling circle model of DNA replication, which was discussed on p. 146. However, there seems to be no evidence that one arm of the replicating structure becomes "anchored" during the first round, or otherwise peeled off as replication occurs. Indeed, the density-labeling technique of Tomizawa and Ogawa (1968) has permitted isolation of DNA molecules which are probably in the second round of replication. Such preparations also show double-forked circles, the same as seen during the first round. Tomizawa and Ogawa believe that concatemers are created from circular structures by combination. In this context, it would be interesting to see what intracellular polynucleotide species are formed during replication of recombination-deficient phage mutants—perhaps in a recombination-deficient host.

Having established that λ DNA replicative intermediates include both circular and concatenated structures, let us turn to the question of the origin and direction of replication. Three types of approaches have independently established that the λ genome has a unique site for initiation of replication, located on the right half, and that the direction of replication is from right to left. Makover (1968) hybridized early-labeled replicating λ DNA with segments of the λ DNA molecule prepared as described by Skalka et al. (1968). Strongest binding was seen with the segment of 48.5 percent GC (see Figure 9–1), suggesting that the initiation site lies within this region. Tomizawa and Ogawa (1968) hybridized DNA at various stages of replication with right and left halves and confirmed Makover's conclusion, but also established the direction of replication. LePecq and Baldwin (1968) reached the same conclusions by examining the density distribution of molecules formed following a shift from bromodeoxyuridine-containing medium to one containing thymidine. In their paper, LePecq and Baldwin cited unpublished genetic studies of Dove, which tend to localize the site of initiation between genes O and c_{II}. They also mention studies in progress for further characterization of these partially BUDR-labeled molecules which are formed during the transition from BUDR to thymidine medium. The technique takes advantage of the fact that BUDR-containing DNA melts at lower temperature than normal DNA. Thus, one can melt out the BUDR-containing regions at some intermediate temperature and measure the lengths of single- and double-stranded regions in the electron microscope with some precision.

A final aspect of λ DNA replication which we should mention involves the participation of bacterial functions in the process. Little is known about this at present, but Naha (1968b) has isolated an *E. coli* mutant unable to support λ replication, and Fangman and Feiss (1969) have shown that a temperature-dependent DNA-defective mutant of *E. coli* will not support λ DNA replication at a nonpermissive temperature, although prophage excision and other phage functions occur normally. Since the nature of the defect in cellular DNA synthesis has not been clarified for this mutant, one cannot yet identify the step which is defective for λ replication. However, it is of interest that the block does not affect T4 replication. Thus, as might be expected, the larger virus T4 is somewhat more autonomous in its replication than is λ. Of course, the mere fact that λ can lysogenize its host and transduce bacterial genes might suggest that its metabolic fate is more closely tied to that of the host than is that of the more virulent T4. However, this generalization should not be carried too far, particularly in light of the isolation, in the author's laboratory, of *E. coli* mutants unable to support T-even phage growth (Mathews, 1970b). However, unlike the mutants defective in λ replication, which we have just been discussing, these new T-even-tolerant mutants do permit a limited amount of phage DNA replication.

Control of Transcription

Because of the extensive degree to which the λ genome has been mapped, both physically and genetically, and because of the various possibilities for fractionation of the λ DNA molecule, it has been possible to undertake highly detailed analysis of the transcription of phage genes *in vivo,* both in lysogeny and during phage replication. The basic approach has been to fractionate λ DNA either physically, by separation into strands or segments, or genetically, by use of hybrid phages, transducing phages, or deletion mutants. These DNA fractions are then used to measure hybridization of phage-specific mRNAs formed under various conditions. Because of the relatively large number of regulatory genes known to exist in λ, this area of phage biology has very rapidly become extremely complex. Space does not permit a detailed review of this field, nor does a detailed review seem justified in light of the rapidity with which new information is appearing. The serious reader is referred to Szybalski et al. (1969) for a recent—and formidable—discussion of the control of transcription in temperate phage development.

There is now convincing evidence that the lysogenic state is maintained through regulatory mechanisms operating at the level of transcription, as postulated by Jacob and Monod (1961). The only λ genes known to be expressed in lysogeny are the c_{I} gene, which, as we shall see in Chapter 10, is the structural gene for the repressor, and *rex,* whose gene product is involved in exclusion of T4 *r*II mutants. The repressor binds to operators located on both strands of the λ DNA molecule and prevents transcription of other blocks of genes, notably gene N, whose action is required for subsequent steps in phage replication, and region *x* through gene P, whose action is necessary for DNA replication. One might anticipate that very little λ-specific mRNA is formed during lysogeny, as expected from the small number of phage genes transcribed. Indeed, of the pulse-labeled RNA in an uninduced λ lysogen, only 0.03–0.06 percent hybridizes with λ DNA (Taylor, Hradecna, and Szybalski, 1967; Cohen and Hurwitz, 1968; Bear and Skalka, 1969). Hybridization of this low proportion of phage-specific RNA occurs primarily with the H strand, indicating that the genes expressed in lysogeny are oriented to the left (Taylor et al., 1967), although Cohen and Hurwitz (1968) have reported considerable prophage mRNA hybridizing with the L strand as well. At any rate, this material hybridzes specifically with that segment of λ DNA having a GC content of 43 percent. As can be seen from Figure 9–1, this region is located on the inner right quarter of the DNA molecule, or about the same site on the genetic map as those genes known to be functioning in lysogeny.

As predicted from the Jacob Monod scheme of regulation, the replication of λ phage after induction involves a derepression of λ-specific mRNA synthesis. The proportion of pulse-labeled RNA which will bind to λ DNA increases some tenfold during the early period after induction, and another tenfold at later times (Taylor et al., 1967; Cohen and Hurwitz, 1968). Even so, this proportion never exceeds 12 percent of the total mRNA. Thus, in contrast to the complete shut-off of bacterial functions seen in infection by the virulent T-even phages, we see that, even during active λ phage replication, the majority of RNA molecules being made are of host origin. This includes all of the major classes of cellular RNA (Terzi and Levinthal, 1967).

A recent study of the effects of λ infection upon gene expression in the host (Cohen and Chang, 1970) indicates that a virus-coded protein, whose structural gene maps near *exo* and *β*, is directly involved in decreasing bacterial RNA and protein synthesis.

As we saw when we discussed the orientation of λ genes, there are two regions of the λ genome which are transcribed to the right (from the L strand) and one region, including gene N, which is transcribed to the left (from the H strand). One can infer from this that λ replication must be controlled in large part through the operation of strand-selection mechanisms. There must be in addition, however, specific initiation sites (or promoters) for blocks of genes which are expressed at different times. Consider, for example, the righthand extremity of the λ genome. This includes both "early" genes, such as O and P, and the "late" genes S and R. It also includes gene Q, which we have already characterized as an early gene whose action is prerequisite to the expression of late functions. Moreover, there is evidence from studies of the defective lysogen λt11, that the early genes *x* through O form a single unit of transcription (Eisen et al., 1966), although this region might also include gene P (Szybalski et al., 1969). Thus, this single arm of the genome, all of which is transcribed to the right, would appear to contain at least three separate units of transcription: the *x*-P region, controlled by gene *x;* the region including gene Q, whose transcription is activated by the product of gene N; and genes S and R, whose transcription requires the product of gene Q.

In the preceding discussion, it may not be strictly correct to refer to *x* as a gene. This was originally described as a region on the λ genome, defined in terms of the λt11 mutation, which might be an operator region, a termination site, or some other element which is not, strictly speaking, a "gene," with an identifiable gene product. However, Oppenheim, Neubauer, and Calef (1970) have recently suggested, by means of a rather involved argument, that *x* is a gene and that it codes for an "antirepressor," which abolishes the action of the λ repressor early in vegetative growth. The action of this presumed antirepressor is still obscure. However, it may involve either repression of repressor synthesis, or inactivation of repressor by binding to it.

Let us now see how analysis of the λ-specific RNAs formed under various conditions of infection have helped to establish the current picture of regulation. RNA formed early during λ replication hybridizes primarily with the right half of the λ DNA molecule (Skalka, 1966), and with the H strand (Taylor et al., 1967; Cohen and Hurwitz, 1968), as expected from the fact that the early-acting gene N is transcribed to the left. As replication proceeds, more of the RNA begins to be transcribed from the L strand, and from both halves, such that, late in infection, 90 percent of the phage-specific mRNA hybridizes with the L strand. This is also expected from the fact that the late genes A through J, on the left arm, and S and R, on the right arm, are all transcribed to the right. Studies on transcription in cells infected with various *sus* mutants confirm what we have already discussed in terms of the functions of these genes. For example, gene N mutants form no RNA which can hybridize to the left arm (Skalka, Butler, and Echols, 1967), as expected from the fact that very few functions, either early or late, are expressed in the absence of a functional gene N product. Moreover, induction of the defective lysogen λt11 (gene *x*) gives the same qualitative result, but the total amount of right-arm RNA is severalfold higher than normal.

As stated earlier, we know very little about the biochemical functions of the regulatory genes. However, it does appear likely that they alter the specificity of RNA polymerase sufficiently to allow transcription of different classes of phage genes, for example, by forming altered sigma factors. In this context, we should note that studies with rifampicin yield the same result which we earlier discussed for T4, namely, that the *E. coli* RNA polymerase molecule, or at least that part involved in rifampicin binding, is involved in the transcription of all phage genes (Takeda et al., 1969; see also Chesterton and Green, 1968; Gariglio, Roodman, and Green, 1969).

We have mentioned the existence of several units of transcription in the λ genome: the c_I-*rex* region transcribed in lysogeny, the region on the H strand which includes gene N, the *x*-P region, the four polarized units of gene expression on the left arm, and so forth. The entire number of separate transcription units is not known, but the total number of RNA polymerase-binding sites on the λ DNA molecule has been reported to be between 10 and 17 (Nakano and Sakaguchi, 1969). One doesn't know whether each of these sites represents an initiation site for transcription.

However, Taylor et al. (1967) have pointed out that the distribution of cytosine-rich clusters on the segments and strands of the λ DNA molecule corresponds closely with those regions known to serve as template strands. It will be interesting to see whether the distribution of RNA polymerase binding sites corresponds as well.

Additional questions relating to the control of transcription include the following: (1) Is prophage mRNA transcribed from the H strand, or from both strands? The results of Taylor et al. (1967) and of Cohen and Hurwitz (1968) seem to be in conflict on this point. (2) The precise extents of the two regions under direct control of the immunity region have not been defined. The leftward-oriented region includes gene N, but there is a suggestion (Bode and Kaiser, 1965b) that it also includes gene c_{III}. The rightward-oriented region includes genes *x, y*, c_{II}, and O (Pereira da Silva and Jacob, 1967). Does it also include gene P, as indicated by Szybalski et al. (1969)? The work of Packman and Sly (1968) also indicates that genes O and P are controlled jointly (see p. 259). (3) The functions controlled by gene N, as well as the biochemical nature of the gene N product, are not yet well understood. Clearly, the formation of the early proteins exonuclease and β-protein are controlled by this gene, as is the activation of gene Q. It is now generally agreed that the *x*-O region is under direct repressor control. However, it has been suggested (Szybalski *et al,* 1969) that the rate of transcription of this region is considerably increased in the presence of a functional gene N product. Moreover, it is not known whether or not gene P is controlled directly by gene N (cf. Dambly, Couturier, and Thomas, 1968). Finally, we must try to resolve conflicting results on the effects of N mutations on DNA synthesis. Eisen, Sly and Siminovitch (1968) have tried to explain these in terms of the methods used in different laboratories to induce *sus* N lysogens. The results of Joyner et al. (1966), in which N mutations blocked phage DNA synthesis, were obtained in lysogens induced with mitomycin C, while their own results, and those of Eisen, Pereira da Silva, and Jacob (1968), were obtained by heat induction of a lysogen bearing a thermolabile repressor. If gene N controls the formation of an endonuclease needed to detach the prophage from the bacterial chromsome (Boyce et al., 1969), and if the gene N product is not needed for DNA replication, one might expect the undetached prophage to direct some DNA replication, perhaps directed from an initiation site on the bacterial chromosome. In the case of a mitomycin-induced culture, however, one might expect chromosomal DNA replication to be blocked, explaining the results of Joyner et al. (1966).

So far, much of the concern with enzymological aspects of transcriptional control has been expressed at the level of initiation. However, termination factors must be important as well. Kourilsky et al. (1968) analyzed the size distribution of early λ mRNA and found five more-or-less

Briefly, Campbell proposed that the DNA circularizes, and that a specific region on the λ chromosome (aa′) becomes associated with a specific region of the bacterial chromosome (bb′). A single reciprocal recombinational event then ocurs at this site, resulting in linear insertion of the phage chromosome into that of the bacterium, as shown in Figure 10--1.

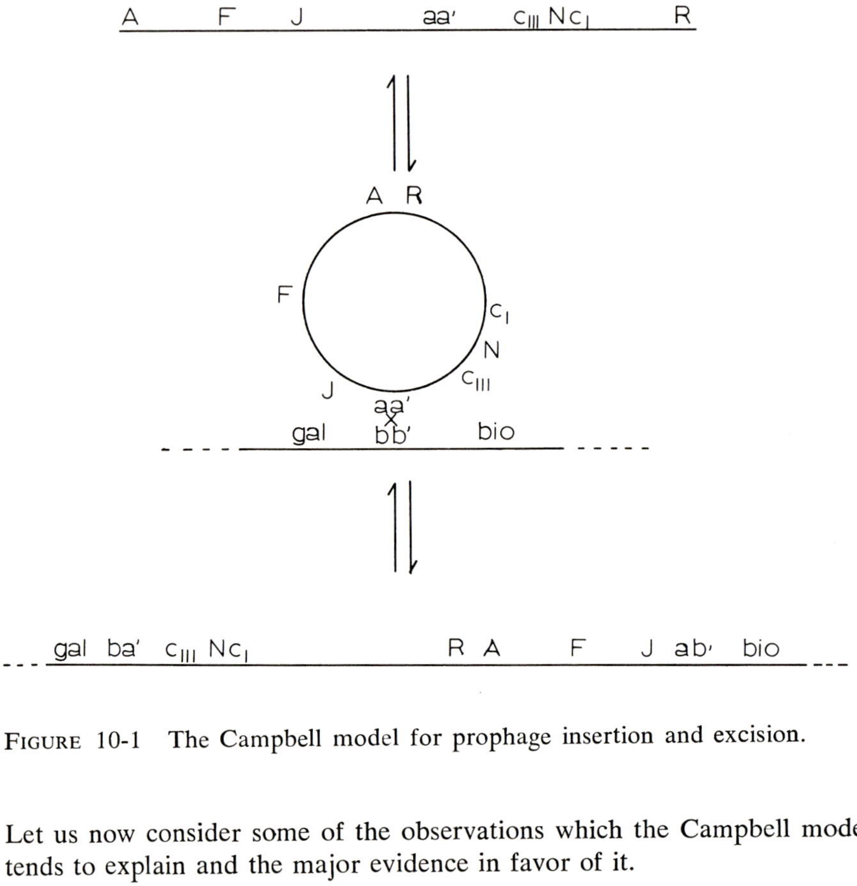

FIGURE 10-1 The Campbell model for prophage insertion and excision.

Let us now consider some of the observations which the Campbell model tends to explain and the major evidence in favor of it.

An explanation for Prophage Excision and Specialized Transduction

Campbell proposed that the excision of prophage after induction occurs simply by a reversal of the integration mechanism. Although, as we shall see, this is not exactly the case, the idea does explain how specialized transduction could occur. An occasional faulty circularization of the prophage could result in the excised region containing some bacterial genes from either end, as shown in Figure 10–2. The model also predicts

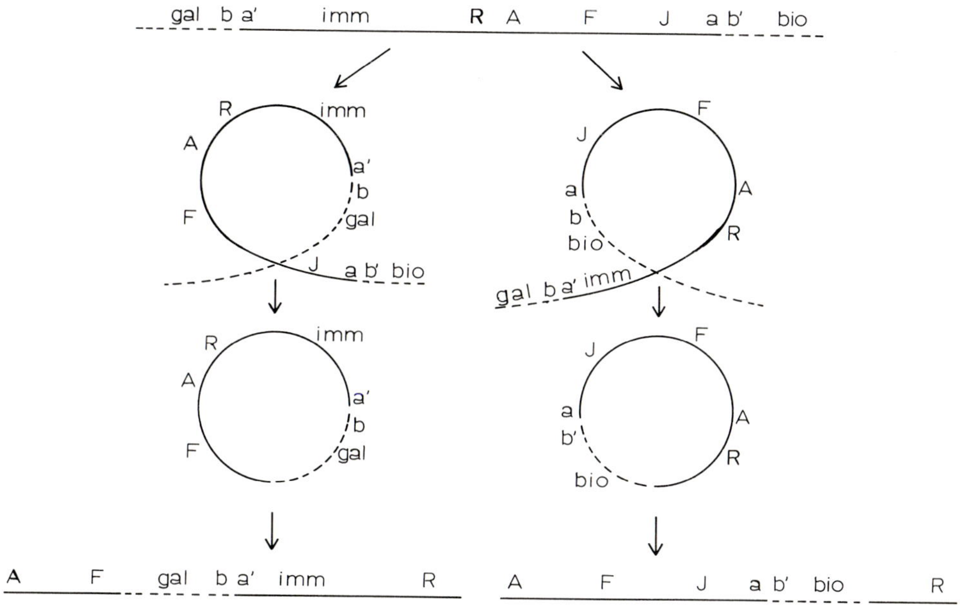

FIGURE 10-2 Transduction by λ of the *gal* or *bio* markers as explained by the Campbell model. Solid lines represent viral DNA, and dashed lines represent bacterial DNA.

that these transducing phage will bear their bacterial genes in internal positions of the phage genome, as we have already seen they do.

As we have already mentioned, not all λ*bio*'s are defective. The Campbell model would predict that all *bio*-transducing phages would have some material deleted immediately to the right of the attachment region, including *int, xis,* and *red,* which are not necessary for vegetative growth. However, transduction of more bacterial genes would be expected to delete the essential gene N, and this is what has been observed (Fuerst, 1966; Kayajanian, 1968). Moreover, all *gal*-transducing phages would lack the essential gene J, and maybe more genes, explaining why they are all defective.

As expected from the Campbell model, other temperate phages can transduce bacterial genes adjacent to their sites of integrations (cf. Signer, 1968).

Gene Order in Lysogeny

As can be seen in Figure 10–1, the Campbell model predicts that the order of phage genes in the prophage will be a circular permutation of the gene order in the vegetative map and in the DNA molecule. Indeed, the model was proposed partly to explain data which indicated that this is so. This permutation arises simply because the sites of circularization and attachment to the bacterial chromosome are not identical. Presumably, the site of circularization is the cohesive ends, which, as we have already seen, can come together, leading to the formation of circular molecules. We have seen how the attachment site was located physically, and we shall see presently how it was located genetically.

Various genetic methods have been used to determine the prophage gene order, notably P1 transduction of phage markers and closely linked bacterial markers (Rothman, 1965). This study also showed that cotransduction of the *gal* and *bio* markers is greatly reduced in λ lysogens relative to nonlysogenic strains, as expected from the idea that linear insertion would increase the physical distance between these markers.

Formation of Double Lysogens

Superinfection of a λ lysogen with λ leads to the formation of doubly lysogenic strains (i.e., carrying two prophages per chromosome), but this occurs with very low frequency. Presumably this is a rare event because the superinfected cell is immune, such that the phage-coded machinery needed for integrative recombination (cf. below) cannot be synthesized. However, since the Campbell model predicts that the initial recognition of a phage for its attachment site ocurs by base sequence homology, one would expect that the rare formation of a double lysogen could occur at any site along the prophage, because homology is not confined to the relatively limited attachment site. Indeed, when Calef et al. (1965) mapped the genes of double lysogens by P1 transduction, the results indicated that the second prophage was inserted linearly into the first.

The frequency of formation of double lysogens can be increased greatly in heteroimmune superinfection, i.e., where the superinfecting phage is insensitive to the immunity substance of the prophage. For example, in superinfection of a λ lysogen with the hybrid λi^{434}, which does not respond to the λ repressor (see p. 255), the superinfecting genome can direct all of the functions it needs to carry out prophage integration. This presents an opportunity to ask an important question about lysogeny, namely, which is the most important factor in determining where attachment will occur? The Campbell model states that recognition occurs by base sequence homology. However, recent genetic studies (Guerrini, 1969; Weisberg and Gottesman, 1969; Signer et al., 1969) suggest that the λ attachment regions on the phage and bacterial chromosomes are not completely homologous. Moreover, we have already mentioned the existence of site-specific machinery for integrative recombination (see also p. 252). If a superinfecting genome is given a "choice" of attachment sites, will it choose on the basis of sequence homology or specificity of the integration enzyme? Zissler and Campbell (1969) have established experimental conditions for asking this question. A bacterial strain diploid for the attachment region was constructed and made singly lysogenic for λi^{434}. Now this strain was superinfected with λ, and the question was asked: will the λ prophage be inserted in the *cis* position (i.e., on the already-occupied chromosome, where there is extensive sequence homology), or *trans* (a site which "looks" more like the site of action for the recombination enzyme)? Most of the superinfecting genomes were integrated in the *trans* position. Moreover, when the same experiment was carried out under conditions of homoimmune superinfection, all of the rare double lysogens produced bore the secondary prophage in the *cis* position. Thus, in the absence of the site-specific recombination system, those rare lysogenizations which do occur depend upon sequence homology, but this is not the "natural" situation.

Properties of λb2

In Chapter 9 we mentioned the existence of λb2, a deletion mutation in which 17 percent of the wild-type λ DNA is lost. This mutant behaves as though it has lost the attachment site needed to establish contact with the bacterial chromosome. However, λb2 does carry an active immunity region, so that repression can be established. When this occurs, the λb2 chromosome is maintained in a nondividing state and is transferred to one daughter cell at cell division, being transmitted independently from the bacterial chromosomes (Yarmolinsky and Korn, 1968).

Although much of the support for the Campbell model from λb2 has come from the assumption that this mutant lacks the λ attachment site, Fischer-Fantuzzi (1967) has presented evidence suggesting that part of this site is retained. She found that λb2 can be relatively efficiently integrated in a strain bearing a λ cryptic prophage. This is an incompletely characterized prophage deletion in which the c_{I} and N genes, among others, are missing. Thus, the prophage cannot maintain immunity, but it is also unable to replicate. Fischer-Fantuzzi found that, whereas wild-type λ could integrate either to the left or right of the cryptic prophage, λb2 could integrate only to the right, and never within the cryptic prophage. Thus, λb2 appears to retain enough of the attachment site to recognize that part of the cryptic prophage site at one end but not the other. This raises the question of the functional nature of the rest of the b2 region. As we have mentioned, λb2 contains 17 percent less DNA than wild-type phage, yet the only known function which is missing is part of the prophage attachment site (even the *int* gene is still present, as shown by Fischer-Fantuzzi, as well as others). Recently, Tonegawa and Hayashi (1968) have demonstrated that mRNA complementary to the b2 region is produced during normal phage replication, and, moreover, that this material is subject to temporal control. Early in infection b2-specific RNA hybridizes only with the L strand, whereas both strands of the b2 region are transcribed late in infection. Thus, it would appear that additional genes, of unknown function, are expressed in this region.

Site-Specific Recombination by the int System

In Chapter 9 we mentioned the existence of two λ-specific enzyme systems promoting recombination: the *red* system, which promotes general vegetative recombination, and the *int* system, which promotes *site-specific* recombination, only at the attachment site (Echols, Gingery, and Moore, 1968; Weil and Signer, 1968; Signer and Weil, 1968b). The *int* system can promote vegetative recombination as well, but only at the normal prophage attachment site. A cross between int^{+} red^{-} parents in a rec^{-} bacterial host can produce recombinants, but only between markers located on either side of the attachment site. For example, recombination between J and *c* markers occurs with about 27 percent of the frequency observed when all three recombination systems are present, while a similar cross between *c* and R markers gives less than 1 percent of normal recombination. To account for complexities of the *int* system, particularly the data of Fischer-Fantuzzi (1967) and the results of b2 × b2 crosses, Signer and Weil (1968b) have proposed that the enzyme specified by the *int* system functions similarly to that presumed for the *ter* system (see p. 227), in producing a pair of staggered nicks, such that recombination can occur by reassortment of double-overlap structures. To date, little enzymology has been done with the *int* system. What needs to be accomplished is first, the isolation of the protein product of the *int* system, and also the isolation of both phage and bacterial DNA enriched for the attachment sites, such that the process can be observed *in vitro*.

Since the *int* system is designed to function under conditions promoting lysogeny over lytic growth, one might expect it to be controlled differently from genes involved in virus replication. However, Signer (1970) has found that, insofar as tested, control of the *int* system is the same as that affecting other early genes. For example, double *sus* N mutants are unable to integrate, but can become established as plasmids. In other words, a state akin to lysogeny can be brought about, even though the *int* system does not function, a clear indication that *int* is under the control of gene N.

As one might expect from the Campbell model, the *int* system is involved in prophage excision, as well as integrative recombination (Gingery and Echols, 1968; Gottesman and Yarmolinsky, 1968b). Lysogens can, by various treatments, be "cured," i.e., rid of prophage without loss of viability. Frequencies of curing with int^{-} mutants are far lower than those seen with wild-type phage (in these experiments the int^{-} phage were lysogenized by complementation with wild-type phage).

A gene in phage P22 responsible for prophage integration has been recognized on the basis of integration-negative (termed L) mutants (Smith and Levine, 1967). This phage has the interesting property of being both a general and a specialized transducing phage. General transduction predominates under conditions provoking a lytic response to infection, whereas induction of a P22 lysogen leads primarily to specialized transduction of the *pro*A and *pro*C genes (for proline synthesis) adjacent to the prophage. The role of the L gene in prophage excision has been probed by Smith (1968). Induction of an L mutant lysogen results in the formation of defective phage, most of which is bacterial in origin, deriving from the *proC* side of the prophage insertion site.

Fate of Phage DNA in Lysogeny

So far, most of the evidence we have cited in favor of the Campbell model has consisted of genetic evidence which, however elegant, may be capable of other interpretations. We have discussed the fact that intracellular λ DNA molecules can be recovered as circles, a finding which must have provided considerable gratification to Dr. Campbell, who presented his model before circular molecules had been demonstrated. What else do we know about the fate of the DNA of virus destined to become prophage? This kind of question has been rather difficult to approach experimentally, since the extent of phage DNA synthesis preceding lysogenization was not known. In others words, one could label parental phage DNA but one had no assurance that the input molecules were the ones which became prophage. Moreover, since bacterial DNA synthesis was presumed to continue during the period preceding lysogenization, it was difficult to label intracellular phage DNA specifically, in order to follow its fate. Smith and Levine (1965) devised a means for separating phage and bacterial DNA in the P22 system, and were able to show that considerable DNA replication precedes the establishment of lysogeny. However, two groups have now shown that in the λ system, at least a portion of input phage DNA can become associated with bacterial DNA in the absence of replication. Hoffman and Rubenstein (1968a, b) infected cells with H^3- and BUDR-labeled phage under conditions favoring lysogeny. Parental label was found in heavy, hybrid, and light density classes, indicating that some replication had occurred. Some of the parental label was associated with DNA which behaved as though it were linear and of high molecular weight. In some of this material the density was not altered from the light position, even though it contained the parental phage density label. This, of course, is expected if the dense or hybrid prophage is incorporated into a much larger DNA molecule of light density, such that the contribution of the phage DNA to the overall density is not appreciable. Hodgetts and Rubenstein (1968) obtained similar evidence for the association of bacterial and parental phage DNA. Phage labeled with BUDR was used to lysogenize light bacteria. The culture was superinfected with P1. The density distribution of *gal*-transducing particles in the P1 lysate led to the conclusion that the *gal* marker is specifically associated with prophage DNA, as expected if the latter is inserted into the bacterial chromosome.

Menninger et al. (1968) have performed an elegant experiment which shows that some parental DNA can be lysogenized and at least one strand of this recovered intact. Parental phage DNA in this experiment was labeled both with a density label and a host-induced modification property. The phage were used to infect an Hfr (male) bacterial strain. The chromosome of this strain was transferred to a non-immune T6-resistant recipient. Once prophage has been transferred in this system, phage development is initiated by zygotic induction (Jacob and Wollman, 1961). At the same time the male bacteria were destroyed with T6 phage, to prevent replication of any phage chromosomes which had not been transferred with the bacterial chromosome. A significant proportion of the phage in the lysate retained both the density and host modification labels. Since DNA replication occurs simultaneously with mating, such that only one of the transferred strands existed prior to mating, we can only conclude that at least one of the two strands of parental DNA has remained intact through lysogenization, detachment, and replication. However, this represents, to me at any rate, some of the most direct evidence in favor of the Campbell model. Indeed, although no one has yet been so brash as to say, "The Campbell model is proved," there certainly seems to be no reasonable alternative to its acceptance. As a final piece of direct evidence in support of the Campbell model, we might cite a recent experiment of Freifelder and Meselson (1970). These authors determined the rate at which circular *E. coli* sex factor DNA is converted to linear molecules by x-irradiation. Sex factors containing a λ prophage are converted more slowly, and the magnitude of the difference in rates, plus the absence of circular molecules from the product, can only be interpreted on the basis of the assumption that prophage DNA is inserted linearly into the sex factor DNA.

MAINTENANCE OF LYSOGENY

As we saw in Chapter 9, clear-plaque mutants of λ fall into three cistrons: c_{I}, c_{II}, and c_{III}. Mutation in any of these genes can lead to failure of the infecting phage to establish lysogeny. Studies on complementation between *c* mutants and c^{+} strains (Kaiser, 1957) indicated that the c_{II} and c_{III} genes act early after infection, probably providing some transient repression of replicative functions while lysogeny is being established. However, the products of these genes are not needed for maintenance of a stable lysogenic state. This has been shown even more directly in P22 (Levine and Smith, 1964), where, by the use of shiftup experiments with temperature-sensitive *c* mutants, it has been possible to estimate the time during which each gene must act in the establishment of lysogeny. One early-acting gene in P22, the c_{1} gene, was shown to act between 7 and 11 minutes after infection. Moreover, one effect of the action of this gene was shown to be an inhibition of bacterial DNA synthesis.

A detailed study of the functions controlled by the c_{II} and c_{III} genes (McMacken et al., 1970) indicates that the primary effect of these genes is directed against late gene transcription. These authors suggest that the c_{II} and c_{III} gene products may act as repressors for late genes, and they also present some evidence in favor of the idea that these gene products somehow augment the effect of the c_I gene.

The c_I Gene and the Nature of Repression

Although relatively little is known about the biochemistry of the c_{II} and c_{III} gene products and their functions in λ, a great deal is known about the functioning of the c_I gene. Most of our early information on the c_I gene product came from various experiments suggesting that this gene controls the formation of a diffusible substance which could bind to a phage chromosome and somehow prevent its replication. The phenomenon of zygotic induction, which we mentioned in the previous section, was in accord with this idea. Presumably, when a chromosome containing a prophage is transferred from an immune to a sensitive cell, it is moving into a new cytoplasm where this substance is not present. At this point, lytic growth can ensue. Further information was provided through the construction of various hybrid phages bearing the immunity region of one phage and the remainder of the genome from another phage (Kaiser and Jacob, 1957). An example is phage λi^{434}, which we have already mentioned. This phage bears the immunity region of 434, but the rest of the genes are those of λ. The immunity region in this hybrid includes the c_I gene, as do the immunity regions of all similar hybrids involving λ. We say that λ and λi^{434} are heteroimmune. In other words, λ, but not 434, can undergo lytic growth in cells lysogenic for λi^{434}, showing that the immunity region of the prophage determines the immunity pattern expressed. Moreover, λ, but not λi^{434}, can replicate in cells lysogenic for 434, showing that immunity is directed against the immunity region of the superinfecting phage.

In their classic paper on genetic regulation of protein synthesis, Jacob and Monod (1961) proposed that the c_I gene directs the formation of a cytoplasmic repressor, which binds to a specific operator site on the phage genome and prevents mRNA formation. As such, the λ repressor was postulated as being formally similar to the *lac* repressor, which was presumed to control the lactose operon in *E. coli.* One difference, of course, was that the *lac* operon contains but three structural genes, all of which are under direct control of the repressor, while the λ genome contains many genes, only a few of which are under repressor control. However those few which are, including genes N and *x,* must be expressed if the entire phage genome is to be "derepressed." Therefore, we can say that virtually all of the λ genes are controlled, either directly, or indirectly, by the repressor.

A key element of the Jacob-Monod formulation was that repressors, by binding to specific operator sites, prevent transcription of other genes, presumably by blocking initiation of RNA synthesis. We have already seen that the formation of λ-specific mRNA is very low during lysogeny and that it increases manyfold during vegetative growth. Thus, phage induction represents a true derepression.

Several years elapsed between publication of the model of Jacob and Monod and the first direct demonstrations of the existence of repressors as protein products of regulatory genes. In the meantime, indirect evidence was accumulating, both on the nature of the λ repressor, and on its site of action. The existence of temperature-sensitive c_{I} mutants was in accord with the idea that the product of this gene is a protein. Presumably the induction which occurs upon heating cells lysogenic for such mutants occurs because of inactivation of a heat-sensitive repressor protein molecule. The existence of *amber* mutants in the c_{I} gene also indicated that the gene product is a protein. Moreover, the repressor appeared to be growth-stable under laboratory cell culture conditions. Ogawa and Tomizawa (1967b) determined immunity in a bacterial population for several generations after infection by λb2. Immunity was maintained long after the non-integrating, non-replicating λb2 chromosomes had been diluted out by bacterial growth.

Jacob and Monod proposed that the primary target of genetic repression is transcription. In the case of λ, two primary sites of repressor blockage would be transcription of the early-acting N gene and the *x*-P region. We have already discussed the work on λ-specific mRNA metabolism which is consistent with this hypothesis. Genetic data (Thomas, 1966) are also in accord with the idea of gene N transcription as a primary target for repressor action. However, there is also evidence that the λ repressor can act to block DNA replication directly, even in the presence of all known functions needed for phage replication. Thomas and Bertani (1964) found that in superinfection of a λ lysogen by the immunity-insensitive phage λi^{434}, very few infectious λ are recovered in the progeny. Thus, even though the superinfecting phage provides all necessary functions needed for its own replication, and presumably that of the λ prophage, only the superinfecting genome replicates. In confirmation of this finding, Green et al. (1967) have found that, of those few λ particles which do appear in the final lysate under such conditions, essentially all contain DNA which has not replicated during the superinfection. However, the immunity region, as we have seen, includes genes other than c_{I}, and there is no assurance that the product of c_{I} is actually the agent which directly blocks DNA synthesis in heteroimmune superinfection. Moreover, interpretation of these experiments requires the assumption that all of the gene products required for DNA synthesis can diffuse fom their sites of formation and act upon other genomes, and this assumption has not yet been rigorously proven. Finally, any conclusions drawn about the nature of c_I control over replication must take into account the existence of "replication inhibition constitutive" (ri^C) mutants (Dove, 1968). These mutants replicate, in spite of immunity, in mixed infection of a lysogen with a heteroimmune phage. As suggested by Dove (1969b), this may involve altered control of the c_{II}-O region, which must be transcribed early in lytic growth, but which also probably contains the site for initiation of replication.

The existence of the λ repressor was directly demonstrated for the first time when Ptashne (1967a) isolated a protein product of the c_I gene, by means of an ingenious double-labeling technique. Bacteria were lysogenized with a non-inducible λ prophage and irradiated heavily with ultraviolet light, to reduce bacterial protein synthesis to a very low level (the noninducible prophage, which presumably binds repressor irreversibly, was used to prevent induction when the cells were irradiated). The culture was then divided in half, and one part infected with a c_I *amber* mutant in the presence of C^{14}-leucine, to label any proteins formed after infection. The other half was infected with a λ strain bearing a wild-type c_I gene, in the presence of H^3-leucine. To maximize repressor formation relative to that of other phage-coded proteins, both superinfecting phages also bore mutations in gene N, such that most phage proteins would not be formed. The two cultures were then mixed, and the presumed repressor was isolated on the basis of the assumption that it would be the only protein present which was labeled with H^3 but not with C^{14}. One such protein was found and shown to have a molecular weight of approximately 30,000.

If the above protein is really the repressor, and if it functions in the manner proposed by Jacob and Monod, then it should bind specifically to the immunity region of λ DNA. By cosedimentation of labeled repressor and DNA through sucrose gradients, Ptashne (1967b) verified this prediction. The repressor binds to λ DNA but not to that of λi^{434}. Moreover, the repressor will bind only to native, not to denatured, DNA. Finally, Ptashne showed that the repressor formed by an ind^- mutant (noninducible prophage) has different properties from those of wild-type phage, confirming that the *ind* mutation produces an altered repressor, one which is not inactivated under conditions normally leading to induction.

Using similar techniques, Pirrotta and Patshne (1969) have isolated the 434 phage repressor. As expected, this binds specifically to λi^{434} DNA, but not to that of λ. The 434 repressor is of slightly lower molecular weight than the λ repressor and somewhat more basic.

The effect of the λ repressor has been tested in a partially purified *in vitro* RNA-synthesizing system (Echols, Pilarski, and Cheng, 1968). As expected, the repressor specifically inhibits transcription of λ DNA, but not that of *E. coli.* Moreover, Hayward and Green (1969) have shown, by analyzing complexes containing λ DNA and *E. coli* RNA polymerase from both repressed and derepressed lysogens, that one effect of repressor is to prevent binding of RNA polymerase to DNA sites under repressor control. To my knowledge, no one has yet performed similar experiments on DNA synthesis *in vitro,* to test the idea that the repressor also has a direct effect on DNA replication.

Virulence: Loss of Repression

In our discussion of the regulation of λ gene transcription, we described two operons which seem to be under direct control of the c_I gene: a leftward-oriented region including gene N, and the rightward-oriented region which includes genes c_{II} and O. Since these are oppositely oriented, one would presume that they are controlled by separate operators. Ptashne and Hopkins (1968) have explored the question of the number of operators by use of a mutant called λ*vir*. This is a virulent mutant, described by Jacob and Wollman (1954), which can overcome λ immunity and replicate in λ lysogens. The genetic structure of λ*vir* is complex, including three mutations (v_1, v_2, and v_3) in the immunity region. Jacob and Monod (1961) proposed that λ*vir* bears mutations in its operator site(s) which weaken or prevent binding of the repressor. As such, λ*vir* is formally comparable to the operator-constitutive mutations in the *lac* operon, where the structural genes of the operon were presumed to be expressed constitutively because the operator could not bind the *lac* repressor. In confirmation of this idea, Ptashne and Hopkins found that labeled λ repressor does not bind to the DNA of λ*vir in vitro*. Moreover, at least two of the individual mutations in λ*vir* lead to decreased affinity of DNA for the repressor. Thus, there would appear to be at least two operators on the genome to which the repressor binds. From this and from the chromosomal locations of the mutations (see Figure 10–3), it seems likely that they are the operators for the N and c_{II}-P operons.

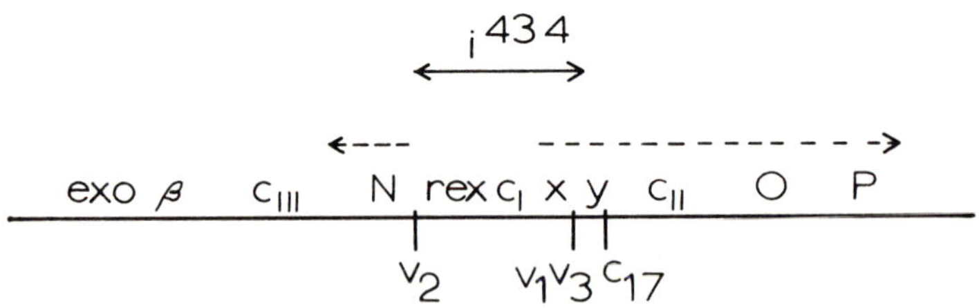

FIGURE 10-3 The λ immunity region, showing the locations of mutations which lead to virulence. The double arrow at the top indicates the extent of the substituted region in λi^{434}, and the dashed arrows represent orientations and extents of regions under direct control by the c_I gene.

Another type of virulence is presented by double mutants containing a c_I mutation and the *c*17 mutation. As we discussed on p. 225, this latter mutation, which maps in the *y* region, is considered to define a promoter site for the c_{II}-P operon. Packman and Sly (1968) found that the *c*17 mutation leads to constitutive phage DNA replication and host cell killing. However, phage replication does not occur unless a c_I mutation is also present, and in this case replication is virulent. Sly and Rabideau (1969) explain this by saying that the constitutive DNA replication oversaturates the capacity of the prophage-directed repressor to bind to all of the DNA molecules present. Thus, the *c*17 mutation appears to generate an altered promoter region which leads to uncontrolled expression of genes O and P and, therefore, constitutive DNA replication, even though other phage functions are repressed. In its effect it is similar to the v_1v_3 mutation described by Ptashne and Hopkins although the v_1v_3 mutation bears more formal resemblance to an operator-constitutive mutation. Sly and Rabideau pointed out differences between these mutations in the patterns of virulence observed. Moreover, the *c*17 mutation is outside the λ immunity region. It can be crossed into 434, where the v_1v_3 mutation cannot.

PROPHAGE INDUCTION

As we discussed earlier in this chapter, and in Chapter 9, vegetative phage development can occur in a lysogen when the level of active repressor becomes too low to maintain lysogeny. This happens spontaneously with low frequency, explaining why plaques are seen when lysogenic bacteria are plated on a sensitive indicator strain. The frequency with which lysogeny is broken can be increased to the point, in some cases, where virtually every cell in the culture undergoes lytic growth. In the case of induction of a thermolabile lysogen by heating, it is easy to comprehend how this comes about; if the repressor is heat-labile, it becomes inactivated, and in essentially all cells. The mechanism of induction by other agents—such as ultraviolet irradiation, thymine starvation, mitomycin C, or treatment with nitrogen mustards—is considerably more complex. All of these agents interfere with nucleic acid metabolism in some way. How do these treatments inactivate repressors? There is evidence that uv light as an inducing agent does not interact directly with repressor molecules. For example, the transfer of a chromosome from irradiated, nonlysogenic male bacterial to nonirradiated, lysogenic females, leads to induction in the recipient (Borek and Ryan, 1958). Since repressors are not present in the donor, the transfer of some other material must have caused induction.

The above would suggest that some altered intracellular condition leads to the events promoting vegetative growth. These altered conditions may pertain to some aspect of the bacterial system for repair of uv damage, for repair-defective bacterial mutants are also defective in induction of a $\phi 80$ prophage (Takebe et al., 1967). A slightly more specific hypothesis has been put forth by Goldthwait and Jacob (1964). These authors isolated a thermosensitive bacterial mutant in which induction of a λ prophage occurs when the temperature is raised to 40°. The presence of certain agents, including hadacidin (a purine antagonist) or guanine and cytosine, prevent this induction, while other agents, including adenine, reverse this effect and promote induction (Kirby, Jacob, and Goldthwait, 1967). This led to the suggestion that some alteration in the distribution of low-molecular-weight nucleic acid components might be responsible for induction. Suppose, for example, that one particular nucleotide could combine with repressor, possibly by allosteric interactions, and change its conformation so that it could no longer bind to operator sites. It should be emphasized that there is no direct evidence for this hypothesis; no measurements of pool sizes of various compounds have yet been carried out with reference to this model. Indeed, it is not even known for certain that the repressor protein is the agent whose properties change after treatment with uv or related agents. Perhaps the structure of the prophage is altered by environmental conditions after induction in such a way that repressor will no longer bind. Certainly the mechanism by which immunity is lost after prophage induction remains one of the major problems of temperate phage biology. Whatever cellular events occur are interrelated with other major cellular functions, such as uv repair, as we have just seen, and also with recombination (Brooks and Clark, 1967) and cell division. In the latter context, Kirby et al. found that when their bacterial mutant was treated in the nonlysogenic state under conditions which would promote induction if a prophage were present, there was a gross disturbance of cell division, with long, filamentous cells being formed.

A feature of the Campbell model is that induction of a prophage occurs by a reversal of the process used to promote its integration. Thus, the prophage should be physically excised from its site of insertion. Ptashne (1965) was able to show that this occurs in at least one case, by superinfecting a λ lysogen which had been grown for several generations in dense medium with the heteroimmune phage λi^{434}, in light medium. Some particles in the lysate had a density consistent with the idea that they contained a full complement of unreplicated DNA. One would, of course, like to know whether the same result would be seen following uv induction of a density-labeled prophage. Of course, interpretation of such an experiment would be difficult because the excised prophage could be expected to undergo replication and recombination, both of which would alter its density.

Also in line with the idea of induction as reversal of insertion are the facts, previously noted, that *int*⁻ mutants are defective in this process, and also genetic studies on the need for unmodified attachment sites to be present (Franklin, Dove, and Yanofsky, 1965). These workers selected for deletion mutants lacking the T1 resistance marker, located immediately to the right of the site of insertion of a $\phi 80i^{\lambda}$ prophage. Some of these deletions apparently spanned the attachment site at the right end of the prophage, a site necessary for prophage excision, for many of the deletion mutants were defective in this process, even though no other known phage genes were deleted.

There are two aspects of prophage excision which indicate that it does not occur by an exact reversal of integration. First is the existence of *xis* mutants, different from int^- mutants, which are specifically defective in excision (Guarneros and Echols, 1970). Dove (1970) has rationalized the existence of *xis* mutations by proposing that the integrated state for prophage is thermodynamically more stable than the nonintegrated state. Thus, prophage insertion is an exergonic reaction, which can be catalyzed by the *int* product alone, while the *xis* product is seen as a type of energy-coupling device which drives the excision reaction in the energetically unfavorable direction.

The second point of difference between insertion and excision derives from two lines of evidence indicating that in most cases the ends of the bacterial chromosome are not rejoined after excision. A reversal of the steps of Campbell's model would regenerate an intact chromosome. This does occur on occasion, for lysogenic cells can become cured of prophage, with low frequency—i.e., rid of prophage under conditions where the virus does not replicate, such that the cell retains its viability (cf. Cross and Lieb, 1967). However, in most cases of prophage induction the cell dies because one effect of phage growth is lysis of the cell. However, Eisen, Siminovitch, and Mohide (1968) found that cell death usually occurs even when the prophage is defective and cannot replicate. Lethality is directed, not against the cell, but against the intracellular structure bearing the prophage. When the defective prophage was borne on an F-*gal* episome, an extrachromosomal genetic determinant, the cell was not killed after induction, but the episome was lost from the cell (Sly, Eisen, and Siminovitch, 1968). These studies suggest that the destruction of functional capacity, either of a chromosome or an episome, could result from the formation of an unrepaired break. This was tested by measuring the linkage of the *gal* and *bio* markers after zygotic induction of a λ prophage (recall that this occurs following transfer of a lysogenized chromosome to a nonimmune cell by conjugation). Linkage between these markers was destroyed following induction of an N mutant, decreased fivefold with an O mutant, and restored with P or *x* mutants (Eisen, Tallan, and Siminovitch, 1968). Of course, this type of experiment cannot be done with wild-type phage, since replication would destroy the recipient cell. However, the results do suggest that the chromosome is broken in induction of a wild-type prophage, and that some phage genes take part in the process. On the other hand, Oppenheim (1969) has found that transient heat induction of thermolabile prophage leads to prophage excision, with the ends of the chromosomes being rejoined. These events occur early, before the detached prophage has been committed to phage replication.

11

Bacteriophages with Single-stranded DNA

Even though we have encountered some oddities in viral DNA base composition, all of the phages which we have encountered thus far are similar to all cellular life forms in that their genetic material is double-stranded DNA. In this chapter we shall discuss a group of minute coliphages whose chromosomes—as isolated from mature virions—contain only single-stranded DNA. These are also the first phages we shall have encountered which do not contain a distinct head and tail, but, rather, are symmetrical in structure. The small amount of genetic information borne by these viruses has made them attractive systems for the study of a number of biological problems.

GENERAL PROPERTIES

There are two major classes of single-stranded DNA phages about which significant biochemical information is available: the spherical phages, including ϕx174 and S13, and the filamentous phages, including fd, fl, and M13. These phages all infect *E. coli,* although there are many small DNA phages known which infect other bacterial hosts. By far the best-studied of the single-stranded DNA phages is ϕx174, although there is a rapidly-expanding literature describing the others. Indeed, although the first filamentous phage was not isolated until 1960 (Loeb, 1960), a recent review of this class of viruses alone includes 205 references (Marvin and Hohn, 1969). The general field of the small DNA phages has also been reviewed recently (Ray, 1968).

The spherical phages ϕx174 and S13, which are actually icosahedral (Maclean and Hall, 1962; Edgell, Hutchison, and Sinsheimer, 1969), are related to one another immunologically, as are the filamentous phages fd, fl, and M13. Although the structures of these two classes of phage are strikingly different, there are close similarities among all five phages, both in amount of DNA per virion and in mode of DNA replication. All of these phages have circular, single-stranded DNA molecules with molecular weights of the order of 1.5–2 $\times$ 10^6. This is sufficient genetic information, in a single-stranded template, to code for between six and ten proteins of average size. For ϕx174 and for S13 the total number of genes thus far described is eight. Thus, the goal of completely correlating genes with gene functions seems well within range for these viruses. On the other hand, a complete understanding of the mode of replication of these phages will require more knowledge of bacterial physiology than we presently possess.

The spherical phages are minute by any standard, with diameters of about 250 Å, whereas the filamentous phages are several times longer than T-even phages. In modes of attachment to bacteria and release from infected cells, we also encounter striking differences among these classes of viruses. The infective cycle with ϕx174 or S13 is similar to those seen with other virulent phages. Following phage attachment to a site on the cell wall, the infected cell undergoes an eclipse of infectivity, and the cycle terminates with simultaneous cell lysis and release of progeny. On the other hand, the filamentous coliphages infect only male strains of *E. coli,*

TABLE 11—1
GENERAL PROPERTIES OF SINGLE-STRANDED DNA PHAGES[a]

	Spherical Phages	*Filamentous Phages*
Physical dimensions	250 Å (diameter)	50 $\times$ 8000 Å
Particle weight, daltons	6.2 $\times$ 10^6	11.3 $\times$ 10^6
DNA content, daltons	1.7 $\times$ 10^6	2 $\times$ 10^6
Number of protein components	4[b]	2[c]
Susceptible strains	Male or female	Male only
Killing of host	Yes	No
Mode of release	Lysis	Extrusion through cell wall

[a] The quantitative data are approximate for each class of phages. Unless otherwise indicated, data were compiled from Ray (1968).
[a] Edgell, Hutchinson, and Sinsheimer, 1969.
[c] Henry and Pratt, 1969.

where they attach to the tips of the sex pili (Caro and Schnös, 1966), little hairs which protrude from the surface of male bacteria. The mode of viral release is unusual in that, instead of lysing their hosts, the filamentous phage are able to parasitize without killing (Hoffmann-Berling and Mazé, 1964; Hofschneider and Preuss, 1963). Cells continue to grow and divide, at only a slightly reduced rate, and progeny phage are extruded into the medium through the cell wall.

Table 11–1 summarizes some of the major properties of the spherical and filamentous coliphages.

STRUCTURAL COMPONENTS AND ADSORPTION

The single-strandedness of ϕx174 DNA was recognized following the demonstration that its nucleotide composition could not be explained from the base-pairing rules of the Watson-Crick structure (Sinsheimer, 1959). The DNA of ϕx174 contains about 25 percent adenine, 33 percent thymine, 24 percent guanine, and 19 percent cytosine. Base compositions of other phages in this class are similar. The circularity of ϕx174 DNA was first shown by its resistance to exonucleolytic attack (Fiers and Sinsheimer, 1962), and by zone sedimentation studies. A useful property of the DNAs of these phages is infectivity in spheroplasts, even in the absence of a helper phage. Circularity is a necessary condition for infectivity (Burton and Sinsheimer, 1965). As one might expect for single-stranded DNA molecules, these phages, and their chromosomes, are far more sensitive to radiation damage than are duplex molecules (cf. Ray, 1968).

Although the structure of the ϕx174 virion appears deceptively simple, its very simplicity poses some rather complex questions. For example, the structure of the virion tells us very little about the mechanism by which viral DNA penetrates through the cell wall. We do know that the virion contains four detectable proteins (Edgell, Hutchison, and Sinsheimer, 1969; Gelfand and Hayashi, 1969; Burgess and Denhardt, 1969). As mentioned above, the capsid is icosahedral. From each of the 12 vertices protrudes a projection, or spike. As we shall see below, these spikes bear the attachment specificity of the phage. Of the four proteins in the virion, three are components of the spikes, while the fourth is the major capsid protein, as determined following the removal of the spikes by urea treatment.

Adsorption of ϕx174 is not sufficient to guarantee penetration of the DNA. Dalgarno and Sinsheimer (1968) have isolated a class of temperature-sensitive mutants which can replicate at a restrictive temperature, 40°, but cannot initiate new cycles of infection. Under these conditions phage adsorbs but virus cannot penetrate. However, penetrability can be regained by incubation of the phage at 30°, suggesting that the mutation confers a reversible conformational change upon a component of the capsid, probably a component of the spikes. Apparently wild-type phages can also undergo conformational changes *in vitro,* for Bleichrodt and Van Abkoude (1967) demonstrated two interconvertible forms of ϕx174 which differ in heat sensitivity and in adsorption characteristics.

The great bulk of the capsid of M13, as an example of a filamentous phage, is formed from a single protein. However, there is also a minor protein, the so-called A protein (Henry and Pratt, 1969; Pratt, Tzagoloff, and Beaudoin, 1969). This protein forms 0.5 percent of the total virion protein, and, from its estimated molecular weight (70,000), one can calculate that it is present to the extent of only one molecule per virion. Genetic studies indicate that it is required for adsorption. It is very probably located at one end of the molecule, which is in agreement with the observation that the initial attachment of filamentous phages to bacterial sex pili is tip-to-tip (Caro and Schnös, 1966).

Sex pili form the conjugation tubes through which transferred chromosomal DNA is passed during bacterial mating (cf. Hayes, 1968). Little is known regarding the precise mechanism whereby these pili serve as the agents of entry for filamentous phage into an infected cell. Marvin and Hohn (1969) have proposed that phage adsorption triggers a retraction of the pilus, which pulls it into the cell, at a site where the cell wall adheres to the membrane. A distinctive feature of filamentous phage penetration is the fact that, in infection by fd, both viral DNA and protein enter the infected cell (Trenkner, Bonhoeffer, and Gierer, 1967). In this respect, these phages are similar to certain animal viruses which are engulfed by the cells they infect, and, once inside, become "uncoated" before the genome can express itself. A similar uncoating must occur in fd-infected cells, for it can be shown by use of appropriate density labels and radioactive labels that parental DNA is extruded with newly-formed protein, and some newly formed DNA is extruded in association with parental protein.

THE INFECTIVE PROCESS

Effects on the Host Cell

As we stated above, infection of bacteria with filamentous phage is not lethal to the host. Bacterial synthetic processes continue at near-normal rates. In one study of the effects of infection with M13, Brown and Dowell (1968a) found that the most significant change appeared to be a lengthening of the division time of infected cells by about 25 percent.

By contrast, infection with ϕx174 is invariably lethal to the host. In an early study, Rueckert and Zillig (1962) reported that infection with this virus had little or no effect on bacterial protein or nucleic acid synthesis. Later, work by Lindqvist and Sinsheimer (1967a) showed that bacterial DNA synthesis is inhibited at 12 to 14 minutes after infection, and that synthesis of a viral protein is necessary for this to occur. These studies were carried out with a lysis-defective phage mutant, in which the latent period is considerably extended (Hutchison and Sinsheimer, 1966). This led to an explanation of the disagreement with the work of Rueckert and Zillig, for, in the early study, lysis was beginning to occur at about the same time as the onset of DNA inhibition, thus obscuring this latter effect. Lindqvist and Sinsheimer found, in agreement with Rueckert and Zillig, that there is no appreciable affect of ϕx174 infection on cellular RNA or protein synthesis. However, Palchoudhury and Poddar (1968) have presented data which they interpret as evidence for a virus-induced inhibition of the synthesis of the 23S component of ribosomal RNA. However, since inspection of the sucrose gradient profiles in this latter study shows (1) poor resolution of the 16S and 23S components, and (2) extensive breakdown of all classes of RNA, presumably during preparation, I find it difficult to accept their conclusions.

More recently, Stone (1970) has confirmed the results of Lindqvist and Sinsheimer. In addition, he has demonstrated a general inhibition of *E. coli* macromolecular synthesis which occurs at high multiplicities of infection.

None of the known genes, either in ϕx174 or in S13 seem to be directly involved in the shutoff of host DNA synthesis (Ishiwa and Tessman, 1968).

DNA Replication

As one might expect, the replication of a single-stranded DNA occurs by a quite different route than those we have encountered for replication of double-stranded viral DNAs. Among the single-stranded DNA phages, the replication of ϕx174 DNA is the best understood, thanks principally to the prodigious output of Sinsheimer and his colleagues. Most of what we say about DNA replication will be with reference to this phage, although we shall point out a few distinctive features of the replication of other DNAs in this class of viruses.

Following injection of ϕx174 DNA into a host cell, we recognize three stages in its replication: (1) attachment of parental DNA to a site on the membrane, where it primes the synthesis of a complementary strand, to form a two-stranded circular replicative intermediate (RF); this is accomplished with bacterial enzymes; (2) replication of this "parental RF" to give "progeny RF" molecules, which are released into the cytoplasm; this process requires phage-coded protein and is semiconservative, although the parental DNA molecule (the parental "plus" strand) remains bound to the membrane site and exchanges "minus" strands at each round of replication; (3) displacement of plus strands from the progeny RF molecules, by synthesis of new plus strand with simultaneous packaging of the displaced material into phage particles. Free single-stranded (SS) DNA is not detectable in infected cells. This latter process, which can be considered "asymmetric semiconservative" replication, occurs by a process very similar to that proposed in the rolling circle model of Gilbert and Dressler (1968).

The replicative form. This was first described by Sinsheimer et al.

(1962). Like mature viral DNA, the RF was infective to spheroplasts. The material had several properties expected of a double-stranded circular molecule, including lowered density relative to mature DNA, increased resistance of its infectivity to ultraviolet light, and resistance to exonucleolytic attack (Burton and Sinsheimer, 1963). Electron microscopic examination of RF DNA confirmed its circular nature (Kleinschmidt, Burton, and Sinsheimer, 1963; Chandler et al., 1964). Electron microscopy has also shown that a small percentage of RF molecules are multiple-length rings (Rush et al., 1967). Like the circular DNA molecules found in cells infected by λ, RF DNA exists in two forms: a rapidly-sedimenting, covalently closed, twisted circle, (RF I), and a less rapidly-sedimenting open circle with one closed strand and one open (RF II; Roth and Hayashi, 1966). Half of the RF II molecules have an open plus strand and a closed minus strand, and half have the converse structure (Knippers, Komano, and Sinsheimer, 1968).

The RF DNA can be fractionated into its component DNA strands by brief treatment with deoxyribonuclease, followed by resolution in an equilibrium density gradient. Resolution can be enhanced by inclusion of a density label in one of the strands. By use of preparative separation of RF strands, it has been possible to show that minus strands, as well as plus strands, are infective in spheroplasts (Siegel and Hayashi, 1967; Rüst and Sinsheimer, 1967).

Genetics of single-stranded DNA phages. As we stated earlier, phages in this group contain sufficient DNA to code for between six and ten proteins of average size. So far, eight cistrons have been defined, both for φx174 and for S13, on the basis of complementation tests with conditional lethal mutants. Sinsheimer's group has numbered these genes in φx174 with Roman numerals: I, II, . . . VIII (Hutchison and Sinsheimer, 1966; Sinsheimer, Hutchison, and Lindqvist, 1967; Knippers et al., 1969). The defective functions associated with mutations in each gene are listed in Table 11–2. The ability of these mutants to direct steps in DNA replication was tested in cells pretreated with mitomycin C. As shown by Lindqvist and Sinsheimer (1967b), such cells do not incorporate appreciable isotope into bacterial DNA, but are unhindered with respect to their ability to support phage replication. As one might expect, some gene products are required for replication of RF DNA, while others are involved in the replication of SS DNA. However, no phage mutants have yet been isolated which are unable to form the parental RF structure. This suggests that the parental RF is formed by bacterial enzymes. Although the enzymes involved in this process have not yet been identified, one might expect *a priori* that DNA polymerase could synthesize a minus strand, using a plus strand as a template, and that the resulting open ring structure could be closed *in vitro* by the action of DNA ligase. In fact, purified DNA ligase and DNA polymerase from *E. coli* have been used to form such structures

TABLE 11–2
GENES AND GENE FUNCTIONS IN φx174

Gene	*Function*	*References*
I	Lysis	Hutchison and Sinheimer (1966)
II	Spike component determining attachment specificity	Sinsheimer *et al* (1967) Dalgarno and Sinsheimer (1968)
III	Spike component concerned with DNA injection and needed for SS DNA synthesis	Lindqvist and Sinsheimer (1967b) Knippers and Sinsheimer (1968a) Edgell *et al* (1969)
IV	Coat protein component	Sinsheimer *et al* (1967)
V	SS DNA synthesis	Sinsheimer *et al* (1967)
VI	Chloramphenicol-resistant protein involved in progeny RF replication	Lindqvist and Sinsheimer (1967b) Levine and Sinsheimer (1968; 1969a, b)
VII	Coat protein component	Levine and Sinsheimer (1969b)
VIII	SS DNA synthesis	Knippers, Razin, Davis, and Sinsheimer (1969)

in vitro, with viral plus strands as the template, and the newly formed minus strands have been shown to be biologically active (Goulian, Kornberg, and Sinsheimer, 1967). This, of course, does not prove that the same enzymes are used *in vivo.* In fact, Steinberg and Denhardt (1968) have shown that a temperature-sensitive, DNA-defective *E. coli* mutant is unable to support conversion of parental SS DNA to parental RF at a restrictive temperature, even though DNA polymerase is active at that temperature.

As can be seen from Table 11–2, mutants in gene III, which codes for a viral structural component, are also unable to synthesize single-stranded DNA. This fact, plus the absence of free intracellular SS DNA in infected cells (Sinsheimer et al., 1962), led to the idea that single-stranded DNA is packaged into virions at essentially the same time as it is formed. Siegel and Hayashi (1969) have also studied a φx174 coat protein mutant which is defective in DNA synthesis. In this case, linear single-stranded DNA and noninfectious phage particles containing either linear or circular DNA are formed. Because of the different nomenclature for the mutants of Siegel and Hayashi, it is difficult to identify the gene in the Sinsheimer set with which this corresponds.

Table 11–2 also shows that gene I mutants are defective in host cell lysis. The mechanism by which ϕx174-infected bacteria lyse is obscure, since no enzymatically active lysozyme has yet been detected in extracts of cells infected with wild-type phage.

Finally, although a linkage map for ϕx174 has not been published, such a map is in existence for the closely related phage S13 (Baker and Tessman, 1967). As one might expect for a virus whose DNA is circular, the map is also circular. In addition, it contains at least one unit of polarized transcription—cistrons IIIa and IIIb, both of which determine structures of phage coat components (Tessman, Kumar, and Tessman, 1967). Amber mutations in gene IIIa abolish the activity of IIIb. The eight known genes in ϕx174 are homologous to counterpart genes in S13, as shown by complementation tests between the two phages, and the electrophoretic patterns of the phage-induced proteins are very similar for the two phages (Jeng et al., 1970).

Eight complementation groups have been recognized in the filamentous phages, but most of them have not yet been characterized (Marvin and Hohn, 1969). Unlike ϕx174, mutations in either of the two M13 genes for coat proteins have no effect on DNA synthesis (Pratt et al., 1969). An interesting aspect of the genetics of filamentous phage is the existence of "heterozygotes." Upon examination, these proved to be particles twice as long as normal phage filaments, containing two molecules of DNA (Salivar, Henry, and Pratt, 1967). Apparently, these arise as the result of some error in phage assembly.

Replication of RF DNA. In their initial characterization of RF DNA in ϕx174-infected cells, Sinsheimer et al. (1962) found that, once formed from parental DNA, the RF replicates several times more. By use of appropriate density-labeling experiments, Denhardt and Sinsheimer (1965a) found that replication of the parental RF is semiconservative and that the conserved subunit is a single polynucleotide strand. Moreover, the material constituting progeny SS DNA was derived from the medium, not from the parental RF. This finding was in agreement with earlier work (Sinsheimer, 1961; Kozinski, 1961b) which showed that labeled parental ϕx174 DNA is not incorporated into progeny. The suggestion raised from these observations, namely, that parental RF plays some special role in DNA replication which cannot be fulfilled by progeny RF, was extended by further work of Denhardt and Sinsheimer (1965b, c). First, when cells were infected in cold medium with P^{32}-labeled phage of high specific activity, the plaque-forming ability of the infected complexes was lost with decay of the isotope, even though substantial RF DNA replication had occurred prior to harvesting of the cells. Similarly, experiments involving the inactivation of infected cells by ultraviolet light supported the idea of a special role for the RF containing parental DNA.

What is this "special role" played by the parental RF? Genetic evidence suggested that it was a particular intracellular site, to which RF must attach in order to replicate, and the number of which is limited in any bacterial cell (Yarus and Sinsheimer, 1967). When cells in which infection was synchronized by prior starvation for 90 minutes were multiply infected with several different ϕx174 strains, the progeny from single bursts rarely contained more than one genotype. On the other hand, unstarved cells were found to support the growth of up to four parental genotypes. Since the number of genomes which could replicate was larger in unstarved than in starved cells, this suggested that unstarved cells contained a greater number of the "sites" at which replication was presumed to occur.

From all that we have said in previous chapters about the involvement of bacterial membranes in the control of DNA replication, it seemed reasonable to expect that the site through which the parental RF acts is a membrane attachment site. In line with this expectation, Knippers and Sinsheimer (1968b) found that in gently lysed cells, parental DNA is attached to a cellular component which sediments rapidly in sucrose gradients. Although the forces involved in the attachment were not explored in great detail, it was found that parental DNA could be released by detergents, high pH, or digestion by pronase (a proteolytic enzyme). DNA attached to the site replicated semiconservatively. Replicated DNA, which was released from the site, did not replicate further in the cytoplasm. Most of the RF bound to the rapidly-sedimenting material was the open RF II form, whereas all of the cytoplasmic RF was the supercoiled RF I.

What happens to parental DNA molecules which are unable to attach at the essential site because it is already occupied by another molecule? The failure of Sinsheimer (1961) or Kozinski (1961b) to observe parental-to-progeny transfer of label would suggest that all of the parental molecules attach to sites which make them unavailable for packaging. However, these experiments were carried out at low multiplicity. When infections are carried out at higher multiplicities, such that the essential sites are flooded, then parental DNA is found in the progeny (Knippers, Salivar, Newbold, and Sinsheimer, 1969). This label does not derive from membrane-associated parental DNA, but from cytoplasmic DNA molecules, which pass through the double-stranded RF form. Thus, there appears to be no inherent structural difference between RF containing parental DNA and progeny RF, since both can be packaged into virions. Rather, there is only a functional difference, depending upon whether or not the RF molecule becomes attached to the membrane. In general agreement with this conclusion, Stone (1967) has found that in a multiply infected cell, only one of the parental DNA molecules undergoes replication, although all form RF molecules.

Whereas Knippers et al. found that parental DNA which is transferred to progeny is incorporated as intact molecules, Matsubara et al. (1967a) reported that parental DNA which does not serve as a template for replication is broken down and its nucleotides reutilized. Moreover, these authors concluded, on the basis of a rather involved triple-density-label experiment, that progeny RF molecules as well as parental RFs replicate semiconservatively. The reasons for these discrepancies with the results from Sinsheimer's laboratory are not clear.

The number of possible membrane binding sites for ϕx174 DNA, and also DNA from phage λ, has been explored by Salivar and Sinsheimer (1969), who devised a simple test involving Millipore filters for measuring the binding of labeled DNA to membrane. By titration of cells with labeled phage, Salivar and Sinsheimer estimated that the total number of binding sites for ϕx174 DNA is at least 60 to 80. There is a similar number of sites for λ DNA, but these are not homologous with the ϕx174 DNA sites. Even though the number of potential DNA-binding sites is quite large, Salivar and Sinsheimer confirmed previous results from the same laboratory, namely, that the number of parental RF molecules per cell which can actually undergo semiconservative replication is only one to three, depending upon the state of the cell. Presumably all of the DNA-binding sites are identical, at least before infection, and some modification occurs after binding of one or a few DNA molecules in a cell to inactivate the other sites. Otherwise, singly infected cells would rarely yield progeny. Left unanswered, however, is the question of why progeny RF molecules, and excess parental RFs, remain free in the cytoplasm. Why, in view of the large excess of potential (if nonfunctional) binding sites, do these RF molecules remain unbound?

What is the enzymatic machinery responsible for replication of the parental RF? Both phage-coded and bacterial functions are involved. The initial studies on RF replication (Sinsheimer et al., 1962) showed that this process could occur in the presence of chloramphenicol. Although this would seem to indicate that no phage gene products are involved, Lindqvist and Sinsheimer (1967b) found that phage bearing mutations in cistron VI were unable to replicate the parental RF. Subsequent investigation (Levine and Sinsheimer, 1968) showed that the synthesis of the gene VI protein is "chloramphenicol-resistant." The protein is formed in the presence of 30 μg/ml of chloramphenicol, but not 100 μg/ml. Similar chloramphenicol-resistant proteins have been detected in uninfected *E. coli,* where they are believed to participate in the control of DNA replication (Lark, 1966), and from *E. coli* infected with phage λ (Levine and Sinsheimer, 1969c).

Levine and Sinsheimer (1969a, b) have further explored the role of the gene VI protein, both by use of temperature-sensitive mutants and by direct isolation of the protein. The work with *ts* mutants indicates that active gene VI protein must be present continuously in order for RF synthesis to proceed. RF DNA synthesis stops abruptly following a temperature shiftup, although the formation of single stranded DNA is relatively unaffected, except to the extent that the rate of SS DNA synthesis is a function of the size of the intracellular RF pool at the time of shiftup. However, continuous synthesis of gene VI protein is not required, for RF formation can continue long after the formation of the gene VI protein has been blocked by the addition of 150 μg/ml of chloramphenicol.

Levine and Sinsheimer (1969b) have devised a double-label chromatographic method for isolation and assay of the gene VI protein. The protein is formed with about the same kinetics as is the RF, and when it first appears in infected cells, it is found attached to the membrane. However, it is not required for attachment of parental RF to the membrane. The gene VI protein is present in cells in slight molar excess over RF DNA molecules (160 molecules of protein vs. 80–120 RF molecules). In asking about the biochemical role of the gene VI protein, Levine and Sinsheimer found that purified fractions possessed no nuclease or DNA polymerase activity. The suggestion was raised that the protein might be a denaturing agent, which binds to DNA and causes it to unwind slightly in the region of the initiation site. Moreover, if the binding site for gene VI protein were on the minus strand, and the exposed region of plus strand became bound to the membrane, this could explain why semiconservative replication of the parental RF always releases minus strands into the cytoplasm, rather than parental plus strands. The suggestion that this protein acts as a denaturing agent is reminiscent of similar suggestions which have been advanced for gene 32 of T4 (Alberts et al., 1968). In this context, it would be interesting to know whether synthesis of the gene 32 product is also chloramphenicol-resistant.

By studying DNA replication in chloramphenicol-treated cells, Knippers, Whalley, and Sinsheimer (1969) have been able to concentrate specifically upon the replication of parental RF molecules. This process involves the covalent addition of nucleotides to the initially circular template. Electron microscopy of partially purified replicative intermediates shows some circular structures with linear tails. Inasmuch as the replication process generates two double-stranded circles, it is difficult to build a model based upon such an intermediate unless the tip of the tail becomes reattached to the circle.

As mentioned earlier, some bacterial functions are also required for semiconservative replication of the parental RF. Irradiation of cells with ultraviolet light prior to infection abolishes the ability of phage to direct progeny RF formation (Matsubara et al., 1967b). Moreover, Denhardt, Dressler, and Hathaway (1967) have isolated bacterial mutants which are unable to complete a cycle of phage replication. The step blocked in these mutants, which are termed REP$^-$, is the formation of progeny RF from parental RF. A possible clue to the identity of the defective gene product is provided from the fact that these mutants are also defective in genetic recombination.

With regard to the enzymatic machinery involved in ϕx174 DNA replication, there is a report of increased DNA polymerase activity in cells infected by this phage (Huh and Helleiner, 1968). However, this report has been questioned (Grunau and Daniel, 1968).

Synthesis of single-stranded DNA. As we mentioned earlier, single-stranded DNA is formed by a process of asymmetric displacement, in which the plus strand of an RF molecule is peeled off and packaged concomitantly with synthesis of a new plus strand on the RF template. Evidence for this process was originally obtained in pulse-chase experiments, even though it was not possible to detect free single-stranded DNA in infected cells. Lindqvist and Sinsheimer (1968) found that, although net synthesis of RF ceases at 15–20 minutes after infection, RF DNA (particularly RF II) is labeled by pulses of H^3-thymidine administered late in infection, implying that RF DNA is turned over. Moreover, the label incorporated into RF II during a pulse was subsequently incorporated into SS DNA during a chase, confirming that RF DNA is a precursor to SS DNA. These studies were extended by Knippers, Komano, and Sinsheimer (1968) and by Komano, Knippers, and Sinsheimer (1968), who showed that the pulse-labeled RF material which is chased into SS DNA is exclusively plus strand material, while the label remaining in RF DNA molecules is exclusively minus strand material.

Application of the rolling circle model of DNA replication to this asymmetric displacement process would lead to the prediction that nascent plus strands are covalently attached to displaced plus strands, at least during the replication process. This in turn leads to the expectation that one could detect, during the period of SS DNA replication, plus strands of greater than one phage-equivalent unit in length. Because the ϕx174 DNA molecule is so small, the displacement reaction is quite rapid. Therefore, such long plus strands were not detected until very short pulse-labeling periods (5 to 10 seconds), in cells infected at low temperature, were used (Dressler and Denhardt, 1968; Sinsheimer, Knippers, and Komano, 1968). Presumably a transient intermediate in the displacement reaction would be a duplex circular molecule with a single-stranded tail (see Figure 11–1). Such structures have indeed been characterized, by sucrose gradient analysis (Denhardt and Burgess, 1968); by electron microscopy (Knippers, Razin, Davis, and Sinsheimer, 1969); and by chromatography on benzoylated, naphthoylated DEAE-cellulose, a technique which resolves polynucleotides on the basis of differences in secondary structure (Sedat, Kelly, and Sinsheimer, 1967; Knippers et al., 1969).

The overall process of ϕx174 DNA replication, based upon what we have discussed to this point, is presented in Figure 11–1. The molecular nature of the displacement reaction has been probed in considerable detail by Knippers et al. (1969), and some of their conclusions are summarized in Figure 11–2, while Figure 11–3 shows some electron micrographs which helped to establish these conclusions. Briefly, these workers found that the

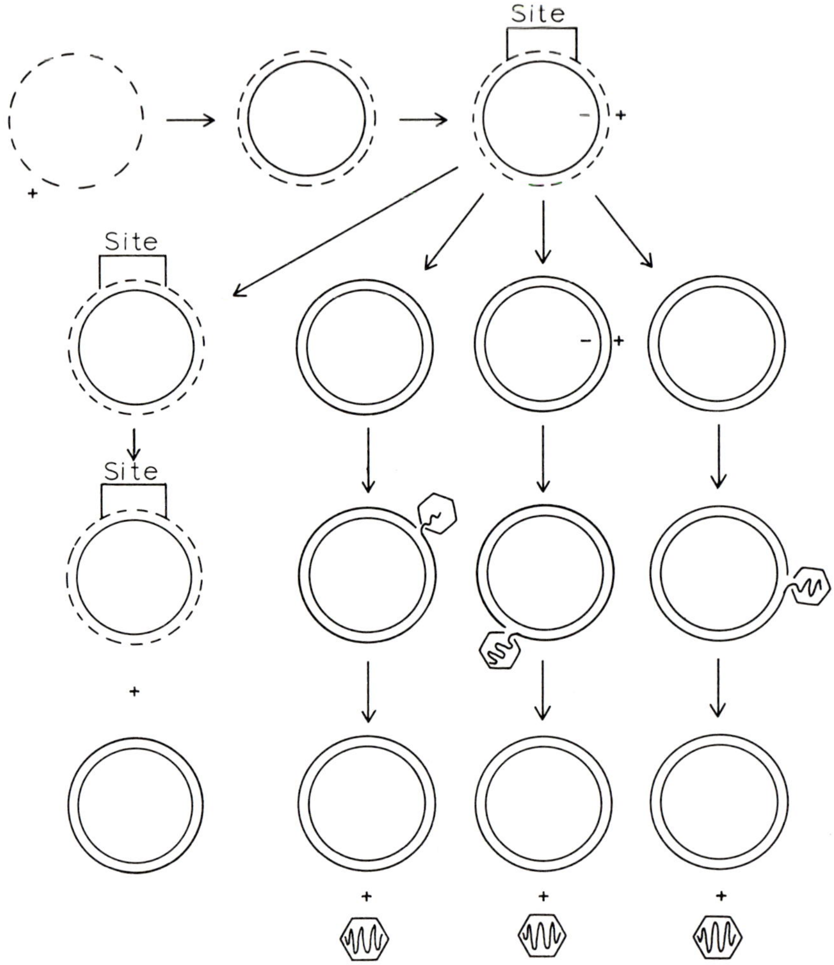

FIGURE 11-1 Replication of ϕx174 DNA. Dashed lines represent parental material, and solid lines represent DNA synthesized after infection .

RF species which serves as the most direct precursor of SS DNA is an RF II species, with a nick on the plus strand between a 3′-hydroxy-terminated deoxyguanosine residue and a 5′-hydroxy-terminated deoxycytidine residue. The displacement reaction occurs by covalent addition of material to the 3′ end of the plus strand, with concomitant displacement of the 5′ terminus. Interestingly, the 5′ terminal residue of the displaced plus strand changes from deoxycytidine to deoxyguanosine. Whether this occurs by addition or cleavage is not clear. A possible function for this replacement reaction, which could lead to closure of the single-stranded ring, is shown in Figure 11–2.

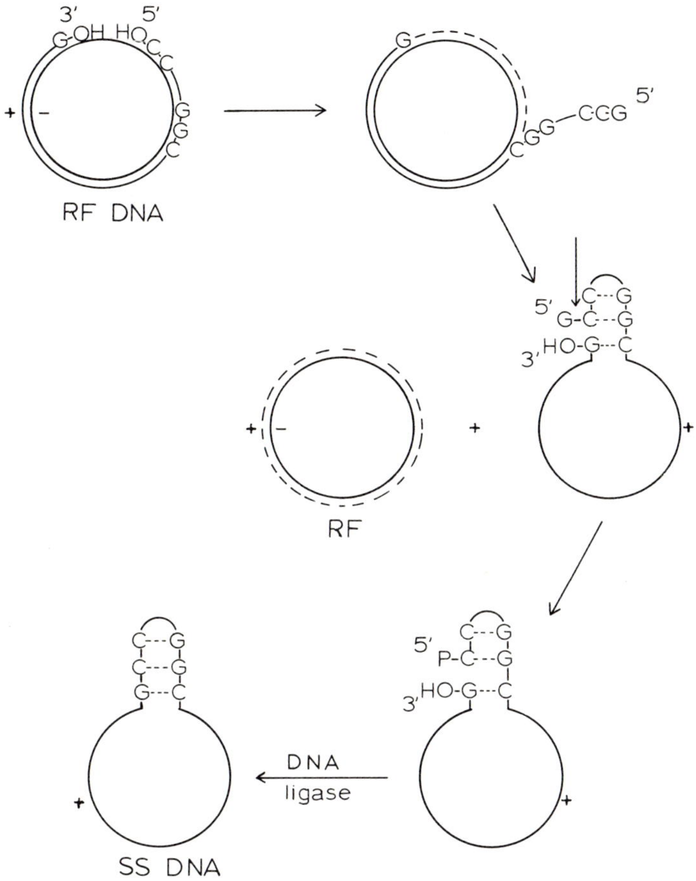

FIGURE 11-2 A mechanism for synthesis of single-stranded DNA from an RF DNA template, as proposed by Knippers, et al., (1969). The dashed line represents DNA synthesized during the indicated round of replication, while the solid lines represent DNA already formed before initiation of this round of synthesis.

This important paper of Knippers et al. showed also that the displacement reaction occurs, not at the membrane, but in the cytoplasm, as expected from earlier work. As also expected from earlier work, coat protein was shown to be associated with the SS DNA-producing complexes.

Another approach to the mechanism of SS DNA synthesis is being carried out in Denhardt's laboratory (Denhardt and Burgess, 1968; Denhardt, 1968). These workers have developed an *in vitro* DNA-synthesizing system, part of whose product can be shown to be SS DNA, by means of DNA-DNA hybridization tests. Much other material is also formed, but

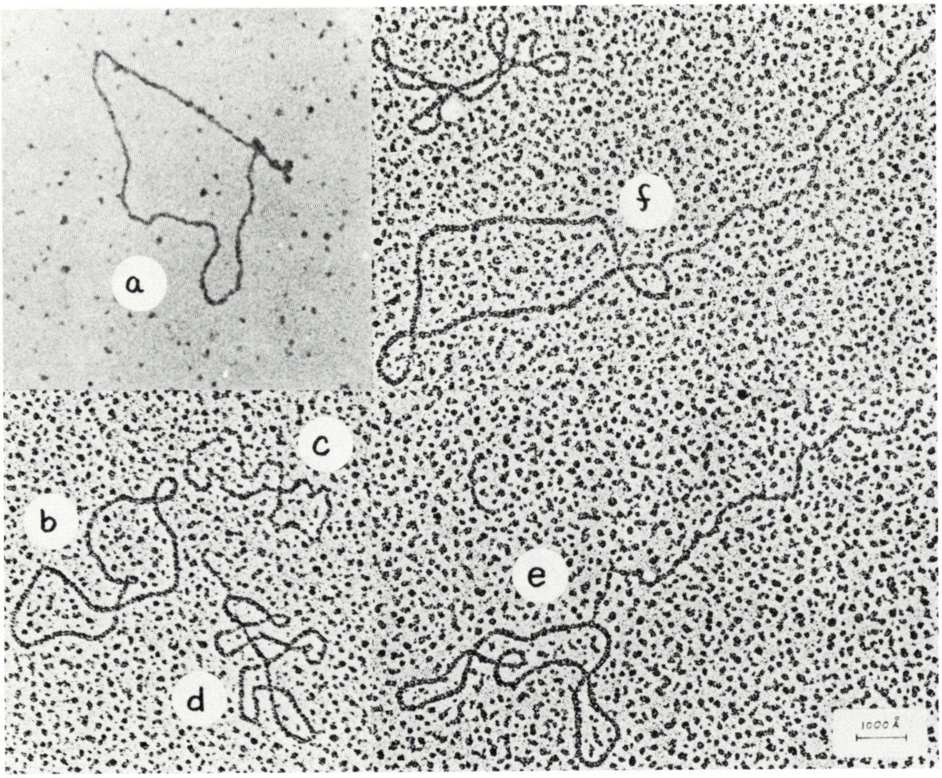

FIGURE 11-3 Electron micrographs of double-stranded ϕx RF DNA molecules in the process of synthesizing single-stranded DNA (from Knippers et al., 1969. I am grateful to Dr. Robert L. Sinsheimer and to Academic Press © for permission to reproduce this photograph).

Denhardt believes that this represents repair of bacterial DNA which has been damaged in preparation of the extracts. Interestingly, addition of antibody to purified *E. coli* DNA polymerase yields only partial inhibition of SS DNA synthesis, suggesting that, if this enzyme is involved in the process, it is not the only enzyme. However, the extent of inhibition of the polymerase reaction per se by the antibody was not checked, and it is conceivable that the antigen-antibody complex could still have some enzymatic activity.

Further light has been shed on the role of DNA polymerase in a recent paper by Knippers and Strätling (1970). These workers have prepared an *in vitro* system for ϕx174 DNA synthesis, using a membrane fraction from phage-infected cells of the DNA polymeraseless mutant of de Lucia and Cairns (1969). Their results suggest that DNA replication, both bacterial and ϕx174-induced, is catalyzed by a membrane-bound enzyme system distinct from the *E. coli* DNA polymerase. Similar conclusions have been drawn by Smith, Schaller, and Bonhoeffer (1970).

Among the filamentous phages, DNA replication has been studied most extensively in M13. Whereas the basic features are similar to ϕx174 DNA synthesis, there are some striking differences. One does see a circular RF formed by synthesis of a new minus strand, followed by its replication, and, finally, formation of SS DNA from the RF (Ray, Bscheider, and Hofschneider, 1966; Pratt and Erdahl, 1968; Brown and Dowell, 1968b; Ray, 1969; Ray and Schekman, 1969; Mechler and Arber, 1969). However, the following specific differences are noted: (1) M13 contains no chloramphenicol-resistant gene functions needed for DNA replication; (2) only two known genes have any effect on DNA replication; (3) intact parental DNA can be recovered from progeny phage particles; and (4) a pool of uncoated single-stranded DNA exists in infected cells, before being packaged.

Transcription of Phage Genes

ϕx174 was one of the first phage systems used to test the specificity of transcription, particularly with respect to strandedness. Hayashi, Hayashi, and Spiegelman (1963) showed that RNA prepared from pulse-labeled infected cells hybridized only with RF DNA, and not with DNA extracted from virus particles. Thus, in this virus, transcription is totally asymmetric, with the minus strand serving as the template strand. Asymmetry in this system depends upon circularity of the RF molecule. Transcription of RF DNA *in vitro* is asymmetric, but only if the circularity of the structure is maintained (Hayashi, Hayashi, and Spiegelman, 1964). More recently Warnaar et al. (1969) have shown that a few single strand breaks in the RF DNA molecule do not interfere with asymmetry of *in vitro* transcription, so long as circularity is maintained. Either the supercoiled RF I or the open RF II forms can serve as templates for transcription *in vitro*.

Hayashi and Hayashi (1968) have explored the nature of the complex formed between newly synthesized RNA and template RF DNA during asymmetric transcription, both *in vivo* and *in vitro*. The complexes contain a ribonuclease-resistant region, which appears to be the area at which polymerization is occurring. Interestingly, this region is not labeled by S^{35}, which would suggest that a protein, such as RNA polymerase, is not involved in stabilizing the complex. An alternative possibility, namely, that such a protein does exist but contains no sulfur, cannot, of course, be ruled out.

We have seen that the RF DNA molecule containing parental DNA plays a special role in infected cells, as the template for replication. Are there also special RF molecules which serve as the sole templates for transcription? We have no answer to this at present. However, it is known that in S13 the parental RF DNA molecule can serve as the only template for

mine in its DNA, and other phages have since been shown to contain HMU. The related transducing phages PBS1 and PBS2 were shown to contain uracil substituted for thymine (Takahashi and Marmur, 1963a). A question related to this latter discovery, which seems not yet to have been asked, is whether the DNA of this phage, when in the lysogenic state, also contains uracil.

There were reports that the DNA of *subtilis* phages is glucosylated. Takahashi and Marmur (1963b) reported that PBS2 DNA contains covalently bound glucosyl residues, even though this DNA contains no hydroxymethyl groups through which they could attach. Rosenberg (1965) reported that SP8 DNA, which does contain hydroxymethyl groups is also glucosylated, and that a host range mutant of SP8 contains no glucose. Other *subtilis* phages, notably ϕe (Roscoe and Tucker, 1966) and SPO1 (Okubo, Strauss, and Stodolsky, 1964) were shown to contain no glucose. The question of glucosylation of SP8 and PBS2 DNA was extensively reinvestigated (Alegria and Kahan, 1968). These workers found very much less glucose in their DNA preparations than had the earlier workers, and they concluded that the apparent presence of glucose was due to contamination of DNA samples by a bacterial polysaccharide, such as a teichoic acid.

ENZYMOLOGY OF PHAGE-INFECTED CELLS

The enzymology involved in replacement of thymine by uracil was first discussed in a brief communication by Kahan (1963), who explored pathways of nucleotide metabolism in cells infected by PBS2. Two new enzyme activities were found: a deoxyuridylate kinase and a thymidylate phosphatase. The former enzyme is presumably involved in uracil incorporation into DNA, the latter in thymine exclusion from DNA. A second pathway for uracil incorporation has been suggested in recent work of Tomita and Takahashi (1969), who discovered a dCTP deaminase in cells infected by the related phage PBS1. This enzyme, which converts dCTP to dUTP, is noncompetitively inhibited by thymidine triphosphate. Table 12–1 shows probable pathways of pyrimidine deoxynucleotide metabolism resulting in substitution of uracil in DNA in cells infected by these two phages.

The mechanism of replacement of thymine by hydroxymethyluracil has been explored by several groups. In cells infected by three different phages —SP8, ϕe, and SP5—the mechanism of synthesis of an HMU nucleotide seems to be identical (Kahan, Kahan, and Riddle, 1964; Roscoe and Tucker, 1966; Aposhian and Tremblay, 1966; Nishihara, Chrambach, and Aposhian, 1967). A phage-specific deoxycytidylate deaminase is induced. This presumably creates a large pool of deoxyuridylate, which is acted upon by a tetrahydrofolate-dependent dUMP hydroxymethylase, analogous to the T-even phage-specific dCMP hydroxymethylase. The phosphorylation of hydroxymethyldeoxyuridylate, at least to the diphosphate level, is catalyzed by a phage-induced kinase (Kahan et al., 1964). Of interest is the fact that SP8-infected cells can incorporate the nucleoside, 5-hydrox-

TABLE 12–1
NUCLEOTIDE METABOLISM IN CELLS INFECTED BY PBS1 OR PBS2

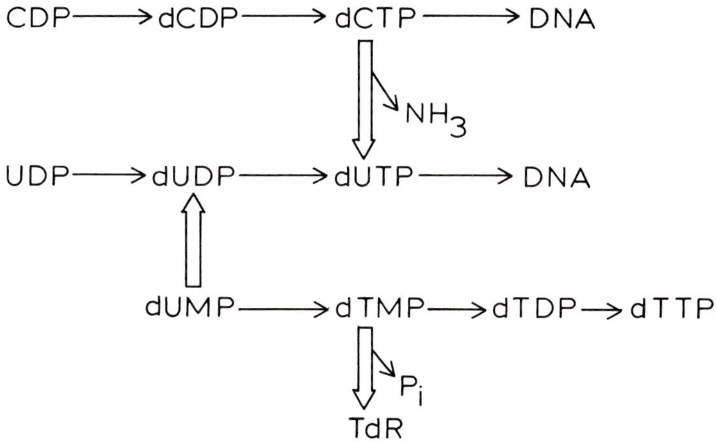

Reactions indicated by heavy arrows are catalyzed by phage-specific enzymes.

ymethyldeoxyuridine, but not the base, hydroxymethyluracil, into phage DNA (Nishihara, Friedman, and Aposhian, 1969). This is in contrast to T-even-infected cells where, as we have seen, the deoxynucleoside of 5-hydroxymethylcytosine cannot be incorporated into phage DNA.

Although the above three phages all induce the same enzyme, insofar as has been tested, for incorporation of HMU into DNA, they all have different machinery for exclusion of thymine. SP8 induces what appears to be a single enzyme capable of cleaving dTTP and dUTP to their respective monophosphates (Kahan et al., 1964), an enzyme seemingly analogous to the dCTPase-dUTPase induced by T-even phages. Phage ϕe forms a protein which acts as a specific inhibitor of thymidylate synthetase (Roscoe and Tucker, 1966; Haslam, Roscoe, and Tucker, 1967), while SP5 induces a thymidylate phosphatase, which cleaves dTMP to thymidine and orthophosphate (Aposhian and Tremblay, 1966). This latter enzyme is similar to the dTMPase induced by the uracil-substituted phages.

Newlon and Marcus (1970) have inquired about the biological significance of substitution of HMU for thymine in phage ϕe. By infecting cells at raised temperatures with phage ϕe strains bearing *ts* mutations in the genes for dTTPase or dUMP hydroxymethylase, these workers have obtained phage particles in which 10 to 25 percent of the HMU is replaced by thymine. These particles seem to be fully viable.

Pathways of nucleotide metabolism in the HMU-containing phages are indicated in Table 12–2.

TABLE 12–2
PATHWAYS OF HMU NUCLEOTIDE SYNTHESIS AND THYMINE EXCLUSION

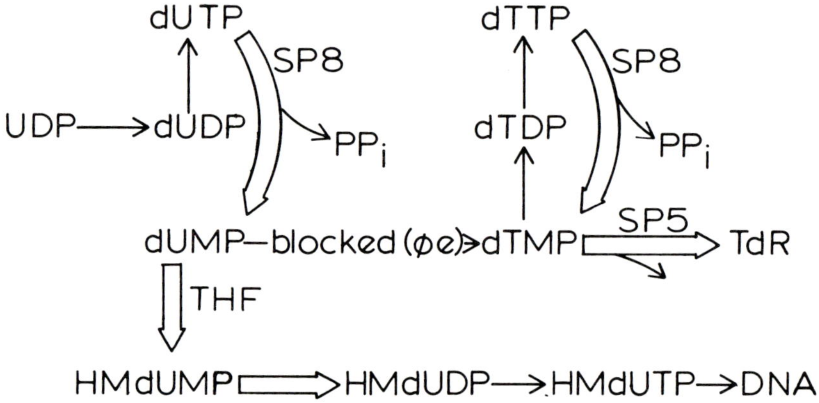

Reactions indicated by heavy arrows are catalyzed by phage-specific enzymes. Reactions specific to individual phages are identified with the name of the phage.

Another *subtilis* phage-induced enzyme which has been studied in some detail is an unusual deoxyribonuclease, found in cells infected by SP3, a phage of "normal" base composition (Trilling and Aposhian, 1965, 1968). This enzyme is neither an endonuclease, nor a true exonuclease, which cleaves terminal phosphodiester bonds. Rather, the enzyme preferentially cleaves the second phosphodiester bond in from the end, for the products of digestion are almost all dinucleotides, with about 10 percent trinucleotides, and no intermediate products of larger size can be detected.

A final enzymological aspect of interest is the acquisition of thymidylate synthetase activity by a thymine auxotroph of *B. subtilis* following infection by the temperate phage φ3 (Tucker, 1969). Since the phage is temperate, the increased activity seen after infection can be due either (a) to transduction of a bacterial gene for thymidylate synthetase, a gene closely linked to the chromosomal attachment site for phage; or (b) phage conversion, a term which denotes a metabolic change in a lysogenized cell, which results from expression of a prophage gene (cf. Robbins et al., 1965). So far, the data are insufficient to determine between these alternatives, although Tucker favors the idea of phage conversion. If this is so, then the newly-acquired thymidylate synthetase activity should be a phage gene product and should, in turn, be distinguishable from the corresponding bacterial enzyme, isolated from wild-type cells. To date, however, this acquired enzyme activity has not been characterized.

TRANSFECTION

Bacillus subtilis can undergo genetic transformation; in a particular state of growth, which is termed "competence," cells can take up DNA added to the medium, and some of the information in the added DNA can be expressed in the recipient cell. When one uses bacterial DNA to transform *B. subtilis,* an added marker must be incorporated into the bacterial genome by recombination in order to be permanently expressed. One can also use viral DNA to infect competent bacteria. Such a process is called "transfection," to distinguish it from transformation and infection, but to indicate its relationship to both processes. In transfection, a small percentage of cells which take up viral DNA undergo a lytic cycle of growth and release viable progeny (Harm and Rupert, 1963; Földes and Trautner, 1964; Green, 1964; Okubo et al., 1964; Reilly and Spizizen, 1965). Since the added DNA need not be integrated into the host chromosome in order to be expressed, this process is not the same as transformation. In principle, transfection presents an assay for biological activity of DNA which is superior to any of the systems developed for the coliphages, for it does not depend upon conversion of the cells to spheroplasts, or prior infection with a helper phage.

Phage DNAs fall into two categories with respect to their behavior in transfection assays: group I, in which plaque-forming titer is directly proportional to the concentration of added DNA (Harm and Rupert, 1963; Reilly and Spizizen, 1965), and group II, in which plaque-forming titer is proportional to the third or fourth power of DNA concentration (Okubo et al., 1964; Green, 1964). This latter result means that three or four molecules of DNA must invade a cell together in order for the cell to liberate at least one progeny particle. The best explanation of this phenomenon is that DNA molecules from group II phages are partially degraded after uptake into competent cells, and some sort of recombination process must occur between fragments of different molecules, in order to maintain at least one intact phage chromosome (Green, 1966). In line with this idea, Epstein (1967) found that ultraviolet irradiation of cells to less than 1 percent survival, immediately prior to introduction of viral DNA, enhances the activity of the DNA some one hundredfold and restores a first-order relationship between DNA concentration and plaque-forming titer. Epstein's interpretation of this result is that the ultraviolet irradiation damages the host cell DNA to the extent that the cell's complement of repair enzymes and nucleases is preoccupied with the bacterial DNA. The viral DNA slips into the cell unobserved by the cell's defense mechanisms and is able to initiate its own replication before significant degradation has occurred. This is an intriguing explanation, but as yet there is no direct evidence for it.

The differences between phages in groups I and II, which determine whether intracellular DNA is attacked, arc not clear. One would like to invoke some structural difference. The presence or absence of hydroxymethyl goups is superficially appealing, since, as we have already seen, T-even phage DNA is resistant to attack by *E. coli* nucleases by virtue of its hydroxymethylcytosine residues. However, SP82, whose DNA contains hydroxymethyluracil, is in group II. Therefore, if the basis for difference between groups I and II is structural, the critical element of structure must be some other functional group.

An approach to understanding the transformation of individual viral genes is being explored by Alegria, Kahan, and Marmur (1968). These workers have devised a sensitive test for phage-induced hydroxymethylases, based upon reversibility of the reaction. Preliminary studies of the ability of transformed cells to synthesize dUMP hydroxymethylase show, as expected, that an intact virus genome is not needed for expression of this individual gene.

SOME ASPECTS OF REGULATION

One advantage of *B. subtilis* over *E. coli* as a host for experimental phage work is that, being gram-positive, *B. subtilis* is permeable to actinomycin D. This agent, which binds to DNA and blocks RNA polymerase, is useful in studies of mRNA metabolism. Pène and Marmur (1967) have used actinomycin, and mitomycin C, in studies on the relationship between DNA replication and early and late gene expression in cells infected with 2C, a hydroxymethyluracil-containing *subtilis* phage. Examination of the kinetics of dCMP deaminase and lysozyme synthesis under various conditions showed that, as in T4, transcription of different messages occurs at different times during the latent period, and that some DNA replication is required for transcription of late genes (or at least for normal lysozyme synthesis).

Geiduschek and his colleagues have initiated an extensive study of viral gene transcription in *B. subtilis* infected with phage SPO1. As we saw in Chapter 7, this system is similar to the T4-*E. coli* system in that the host cell RNA polymerase, or at least that part which determines sensitivity to rifampicin, is used for transcription of all phage messages (Geiduschek and Sklar, 1969). There is, however, at least one aspect of this system which is so far unknown in T4. The syntheses of some of the earliest-formed mRNA species are turned off later in infection (Gage and Geiduschek, 1967), as might be expected if the template genes for these species were being repressed. Wilson and Geiduschek (1969) have isolated, from SPO1-infected cells, an agent which might well be the repressor responsible for this negative control mechanism. This material, which seems to be a protein and whose formation is inhibited by chloramphenicol, binds to undenatured DNA of SPO1 and closely related viral species. It blocks the initiation of RNA synthesis *in vitro,* but has no effect on the elongation of chains already begun. To my knowledge, no such "repressor-like" substance has yet been detected in cells infected by T-even phages.

The *B. subtilis* phage SP8 was one of the first for which it was possible to separate DNA strands preparatively, on the basis of a difference in density between the two strands (Marmur and Greenspan, 1963). Thus, this system provided an early opportunity to test the asymmetry of transcription *in vivo,* and it was found that pulse-labeled RNA could hybridize only with the lighter of the two strands. More recently, Riva et al. (1968) have found that phage SPP1, a virus whose DNA has high activity in a transfection assay, also has DNA strands differing sufficiently in density to permit preparative strand separation. The heavier strand, with a density of 1.725 g-cm^{-3}, is rich in pyrimidines (57 percent C + T), while the lighter strand, with a density of 1.713 g-cm^{-3}, is poor in pyrimidines (43 percent C +T). Transcription in cells infected with this phage is also asymmetric (Riva, 1969). Pulse-labeled RNA binds only to the pyrimidine-rich heavy strand, which is of interest in light of the proposal of Szybalski et al. (1966), which we have already discussed, that pyrimidine-rich regions provide sites of initiation and/or termination of transcription. This virus has also been used as a test system in studies on discontinuous DNA replication. As we have already mentioned in Chapter 9, pulse-labeled DNA in cells infected by phage λ binds equally well to both separated strands of λ DNA, as expected from the model proposed by Okazaki et al. (1968). Similar results were seen in cells infected by SPP1, even when the pulse-labeling period was as short as 2 seconds (Polsinelli et al., 1969).

DEFECTIVE PHAGES

Strain 168 of *B. subtilis* carries a most interesting defective phage, PBSX (also called PBSH), which seems to be intermediate, on a "biological" scale, between a bacteriocin and a normal temperate phage (Okamoto et al., 1968; Okamoto, Mudd, and Marmur, 1968; Haas and Yoshikawa, 1969a, b, c,). Phage particles are formed following induction of bacteria with mitomycin C. These particles contain well-defined structures, including hexagonal heads and contractile tails (Okamoto et al., 1968), yet they are non-infectious to any bacterial host yet studied. Susceptible cells are killed by the phage particles, but phage DNA is not injected, and, of course, phage replication inside the killed cell does not occur.

Analysis of the DNA from defective phage particles show that it is linear, double-stranded, and relatively homogenous in size, with a molecular weight of about 12 million (Okamoto et al., 1968; Haas and Yoshikawa, 1969a). It is mostly, if not completely, bacterial in origin, some having been present in cells before induction and some being synthesized after induction. "Cutting" of bacterial DNA into phage-sized pieces seems to be a phage-directed process, for it occurs quite efficiently, without detectable accumulation of fragments intermediate between bacterial and phage-sized. Moreover, reduction of bacterial DNA to phage-sized fragments is not seen when induction is carried out in the presence of chloramphenicol (Okamoto, Mudd, and Marmur, 1968).

When the frequency of various bacterial markers in DNA extracted from phage particles was analyzed by transformation, a rather surprising result was observed (Haas and Yoshikawa, 1969b). The relative frequency of all markers was about the same as seen in bacterial DNA, except for an adenine marker, *ade-16,* which is located close to the site of origin of bacterial chromosome replication. The frequency of this marker in phage-derived DNA was far higher than that in DNA extracted from bacteria. This could mean either (a) the prophage attachment site is adjacent to the *ade-16* gene on the chromosome, and this gene is detached from the bacterial chromosome along with phage DNA, or (b) some event following induction stimulates multiple initiations of chromosome replication, with this newly replicated DNA being packaged into virus particles. Haas and Yoshikawa favor the latter explanation. Measurement of the sedimentation rate of intracellular DNA bearing the *ade-16* marker after induction showed that it sediments more rapidly than DNA extracted from phage particles. However, the possibility was not ruled out that phage DNA can detach from the chromosome and replicate, with formation of a rapidly-sedimenting replicative intermediate, of the type seen in infection by many nondefective phages. Since virus-specific DNA has not yet been detected or isolated, the difficulties in determining whether a segment of DNA is viral or bacterial are apparent. In such a system, indeed, the distinction between "viral" and "bacterial" becomes definitely blurred.

An unusual aspect of viral DNA containing the *ade-16* marker is that it renatures much more readily after denaturation than viral DNA containing other bacterial markers, or bacterial DNA containing the *ade-16* marker (Haas and Yoshikawa, 1969). This is not due to extensive cross-linking in this region of the DNA molecule, of the type that might be expected after treatment with mitomycin C. The reason for this anomalous behavior and its biological significance are unknown. However, as one possible explanation, Haas and Yoshikawa postulated the existence of complementary repeating sequences in both strands, near the *ade-16* gene. Such sequences could be of the type

....ACACACAC....
....TGTGTGTG....

These could provide a site for rapid strand association, not necessarily by exact complementarity all along the chain. However, positioning of the complementary strands for renaturation could then occur by "slippage" along the above region. Whether this intriguing possibility is true remains to be seen.

Finally we should state that, although PBSX may be unusual in its status as a defective phage, it is certainly not unique. It has been known for several years that *E. coli* 15 strains harbor a defective phage. Lee, Davis, and Davidson (1970) have reported extensively upon the properties of this phage, which is similar to PBSX in that it cannot initiate productive infection in any known host. Unlike PBSX, however, phage particles formed after induction seem to contain primarily virus-specific DNA. This DNA molecule has a molecular weight of 27 million. It is terminally redundant and circularly permuted with respect to base sequence.

13

RNA Phages

It has long been known that many of the common disease-producing animal viruses, such as those causing influenza, polio, and encephalitis, as well as virtually all plant viruses, contain RNA as their genetic material. However, RNA-containing bacteriophages have only been discovered within the past decade (Loeb and Zinder, 1961). The availability of RNA phages has presented unique opportunities for gaining a complete understanding of some basic biological processes at the molecular level. Such opportunities derive first, from the fact that phage RNA replication, as studied in highly purified *in vitro* systems, is very similar to the *in vivo* process, and secondly, from the small size of the phage chromosome (about 3300 nucleotides) which makes possible a detailed comparison between chromosome structure and gene expression. Some of the most exciting current work on RNA phages is concentrated upon those parts of the chromosome which are *not* translated into protein, but which serve as signals for initiation or termination of the translation of different messages. It seems not overly brash to say that the determination of the entire nucleotide sequence of a viral RNA molecule is an attainable goal. This information, plus our understanding of the genetic code and our knowledge of the structures of viral gene products, should allow us, ultimately, to specify the function of virtually every nucleotide in the phage chromosome.

GENERAL PROPERTIES

The Virion

Among the RNA phages, biochemical interest has focused chiefly upon the coliphages, including the closely related f2, MS2, R17, and M12, and the unrelated phage, Qβ (Zinder, 1965). All of these phages are similar in structure, being tail-less and hexagonal in outline, when viewed in the electron microscope, with diameters of about 225 Å (Bradley, 1967). For some time it was thought that the virion contained but a single coat protein, of molecular weight between 14,000 and 15,000, which was present to the extent of about 150 subunits per virion. However, similar to the situation we have seen with the filamentous DNA phages, the RNA phage virion contains a single molecule of another protein, the so-called A protein, of molecular weight about 35,000, which is responsible for attachment of phage to host cells (Roberts and Steitz, 1967). That these are the only protein components of the phage was strongly suggested in this study of Roberts and Steitz, who were able to reconstitute infectious R17 phage from coat protein, A protein, and R17 RNA. The reconstitution process is species-specific, for Ling, Hung, and Overby (1969) found that, whereas MS2 could form phage-like particles with Qβ RNA, and vice versa, such heterologous particles were noninfectious.

As stated above, the RNA molecule from an RNA phage contains about 3300 nucleotides, and has a molecular weight of about one million. The molecule is single-stranded and linear (Sugiyama, 1965). Like a molecule of transfer RNA, which folds back upon itself to give a highly compact conformation, phage RNA molecules possess extensive secondary structure. In one study (Boedtker, 1967), it was shown that 73 percent of the nucleotides in the R17 RNA molecule are present in a hydrogen-bonded structure.

The complete amino acid sequences are known for the coat proteins of R17 and f2 (Weber, 1967; Weber and Konigsberg, 1967). Both proteins contain 129 amino acid residues and differ in only one position, while the structure of the MS2 coat protein is very similar. With a molecular weight of 35,000, the A protein contains about 320 amino acid residues. Since the coding ratio for protein synthesis is 3, the information for synthesis of these two proteins must constitute about 1400 nucleotides, or some 40 percent of the coding capacity of the viral genome.

How many proteins are coded for by viral RNA molecules? The currently accepted—and probably correct—answer is three. In addition to the two structural proteins, a third virus-specific protein, RNA replicase, has been detected. This is an enzyme catalyzing replication of viral RNA, a process which we shall discuss in some detail in the next section. The total number of virus-specific proteins which have been detected by gel electrophoresis of infected cell extracts is also three (Viñuela, Algranati, and

Ochoa, 1967). Moreover, phage RNAs serve as efficient messengers in *in vitro* amino acid-incorporating systems (Nathans et al., 1962; Nathans, 1965). Protein products formed in these *in vitro* systems have electrophoretic properties similar to the viral proteins formed *in vivo,* and the total number of products formed *in vitro* is three (Nathans et al., 1966; Eggen, Oeschger, and Nathans, 1967). Although the products of phage RNA-directed protein synthesis *in vitro* are very similar to the naturally occurring phage proteins, there are some chemical differences. For further discussion of this point, see Fraenkel-Conrat and Weissmann (1968).

It can reasonably be expected that each of the above three proteins would be essential for replication of infectious phage, and that amber mutants in the genes for any of the proteins would be lethal under nonsuppressed conditions. In line with this expectation, amber mutants of all RNA phages which have been examined fall into three complementation groups (Horiuchi, Lodish, and Zinder, 1966; Gussin, 1966; Tooze and Weber, 1967). Mutants in the gene for replicase are unable to initiate viral RNA replication, while coat protein mutants can make RNA but no phage particles, owing to the absence of complete coat protein, and mutants in the gene for A protein make noninfectious phage particles. It has been impossible to map these cistrons genetically, since recombination has not been detected among the RNA phages. However, as we shall see later, the physical positions of these genes are known.

RNA phage virions are extraordinarily stable. Infectivity is not lost under conditions which destroy most other viruses, including desiccation and treatment with detergents. As a result, there is a practical problem of contamination accompanying work on RNA phages, a problem which has occasionally made workers in the field unpopular with their colleagues who must share the same or adjacent laboratory facilities. There is an amusing story, no doubt apocryphal, about two distinguished scientists, Dr. X and Dr. Y. Dr. X desired a culture of a particular RNA phage, so he wrote to Dr. Y, who had originally isolated it. Dr. Y, fearing the competition, wrote to Dr. X that he had lost the strain and no longer had it in his laboratory. Skeptical, Dr. X homogenized Dr. Y's letter with buffer in a Waring blendor, plated aliquots on susceptible bacteria, and obtained plaques containing the RNA phage in question. Whereupon he wrote the mystified Dr. Y to thank him for his cooperation.

The Infective Process

Like the filamentous DNA phages, the RNA phages are sex-specific, being able to infect only male strains of bacteria. Attachment of the phage to the bacterium occurs at the sex pili (Crawford and Gesteland, 1964), but, in contrast to the situation with filamentous DNA phages, attachment occurs along the side of a pilus, rather than the tip (Caro and Schnös, 1966; Ippen and Valentine, 1966). In further contrast, the whole virion is not transported into the cell; rather, RNA is injected into the pilus in some unknown fashion and is then borne into the cell (Knolle and Kaudewitz, 1963; Valentine and Wedel, 1965). Attachment of a phage particle to a pilus is not energy-dependent, but injection and transport are, since adsorption, can occur at 0°, or with partially purified pili which have been dissociated from the bacterial surface (Valentine and Strand, 1965; Valentine and Wedel, 1965).

The sex-specificity of the RNA phages seems to be related only to the capacity of male bacteria to uncoat the viral RNA and transport it into the cell. Viral RNA is infective to spheroplasts, whether formed from male or female bacteria. Moreover, RNA phages can infect bacteria of other genera, such as *Shigella, Salmonella,* and *Proteus,* to which an *E. coli* sex factor has been transferred by conjugation (cf. Horiuchi and Adelberg, 1965).

The elucidation of distinct biochemical stages in the replication of RNA phages has not been as easy as similar studies in other phage systems, because adsorption is slow and infections are asynchronous. Nevertheless, we do know that the latent period is about 25 minutes. Lysis inhibition can be established, in which case one sees immense burst sizes, of the order of 20,000 phage per bacterium. Under these conditions, the mass of intracellular phage is so great that the virus crystallizes inside the infected cell (Schwartz and Zinder, 1963). Although this seems like a spectacular output of phage, particularly when compared to the hundredfold lower burst sizes commonly seen with T-even phages, it should be recalled that, because of the minute size of RNA phages, the actual mass of virus-specific material formed per cell is probably comparable in the two cases.

The question of the mode of regulation of RNA phage gene expression is at once simpler and more complex than the same question when asked about a DNA-based organism. One cannot invoke the existence of control mechanisms operating at the level of transcription, since the viral chromosome serves directly as the template for translation, and regulation must be expressed at this level. Obviously, some element of control must exist, for the great bulk of phage-specific protein synthesized, both *in vivo* and *in vitro,* is viral coat protein (Eggen, Oeschger, and Nathans, 1967). The earliest direct demonstration of a control mechanism operating *in vitro* was made possible by the fact that MS2 coat protein contains no histidine. In an *in vitro* system primed with viral RNA, there was a distinct lag in histidine incorporation, whereas other amino acids were incorporated with no lag (Ohtaka and Spiegelman, 1963). No such lag was seen when *E. coli* RNA was used as the *in vitro* template. This suggested that the coat protein, which would not incorporate histidine, was formed preferentially in the early period of incubation. By means of electrophoretic separation of the *in vitro* products, Eggen et al. (1967) were able to show that the synthesis of all three proteins is sequential, with coat protein made first, followed by the A protein, and, finally, the replicase. Similar sequential protein synthesis is seen *in vivo* (Nathans et al. 1966).

What is the nature of this translational control mechanism? Lodish and Zinder (1966a) presented evidence that the capsid protein itself functions as a translational repressor, inhibiting the translation of the genes for replication and A protein. This was extended by Ward, Shive, and Valentine (1967), who showed that purified f2 coat protein, when added to an *in vitro* system primed with f2 RNA led to a greater inhibition of histidine incorporation than arginine incorporation, while calf thymus histone had equal inhibitory effects on both amino acids. Ward et al. concluded that the coat protein specifically inhibited the synthesis, *in vitro,* of histidine-containing, or noncoat proteins. *In vivo* evidence for this idea has recently been obtained (Sugiyama, Stone, and Nakada, 1969). An MS2 strain bearing an amber mutation near the middle of the coat protein gene was found to synthesize an incomplete portion of the coat protein *in vitro,* as expected from the nature of the amber mutation. However, no such polypeptide could be demonstrated *in vivo* under nonsuppressed conditions. A reasonable explanation for these observations is that the preferential translation of the coat protein gene does not occur in the absence of the regulatory role expressed by a functional coat protein. Strandedness of the template may be another factor in regulation. Engelhardt, Robertson, and Zinder (1968) found that double-stranded f2 RNA, when used as template in an *in vitro* protein-synthesizing system, led to formation of only the coat protein. However, as we shall see in the next section, the intracellular existence of double-stranded RNA has not been definitely established.

In addition to its roles in virus structure and in regulation, Zinder and Lyons (1968) have proposed that the coat protein plays a third role, in promoting lysis of the host cell. This was based upon the fact that amber mutants in the coat protein gene fail to lyse their hosts under unsuppressed or poorly suppressed conditions. No effects of purified cell wall components upon cell wall constituents *in vitro* have yet been noted.

There is a report of another altered function in RNA phage-infected cells. Hung and Overby (1968) reported a modification of proline-specific tRNA of host cells following infection with Qβ. The polyctidylate-stimulated binding of prolyl-tRNA to ribosomes was decreased by two- to threefold when the tRNA was extracted from infected, rather than uninfected, cells. However, no changes in chromatographic properties were seen, as earlier reported for T4 phage-specific modifications in tRNA. Moreover, the specificity of the modification reaction was not checked except to the extent that the authors reported no effect when polyuridylate was used instead of polycytidylate. Finally, they stated, without presenting data, that the effect was specific for Qβ, but was not seen in infection by MS2.

REPLICATION OF VIRAL RNA

RNA Replication *in vivo*

The existence of RNA viruses raises questions about the storage and flow of genetic information in a nonDNA-based life form. The viral RNA must serve at once as template for its own replication and as messenger for the synthesis of viral proteins. Our biochemical knowledge of these phenomena is more completely developed for the RNA phage systems than for RNA-containing animal or plant viruses. Insofar as they have been investigated, the modes of replication of most RNA viruses seem to be basically similar, with the exception of a class of animal viruses which contains double-stranded RNA in the virion (for reviews see Erikson, 1968; Höfschneider and Haüsen, 1968; Shatkin, 1968).

It seemed reasonable *a priori* to assume that information for replication of an RNA chromosome would be transmitted in the same fashion as in replication of a DNA chromosome, i.e., by Watson-Crick base-pairing interactions. Since viral RNA is single-stranded, one would assume that a polynucleotide strand of complementary sequence would be formed in infected cells, and that this strand would then provide the template for formation of more viral RNA. But is the complementary material RNA or DNA? The suggestion had been made for certain RNA-containing animal viruses that a complementary strand of DNA is formed in infected cells. However, Doi and Spiegelman (1962) were unable to detect, in MS2 phage-infected cells, a DNA species which could hybridize with viral RNA. Moreover, inhibitors of DNA synthesis had no effect on RNA phage replication (Cooper and Zinder, 1962), nor did actinomycin D, an agent which blocks DNA-primed RNA synthesis by binding to DNA (Haywood and Sinsheimer, 1963).

The above findings strongly suggested that viral RNA does not replicate via a DNA intermediate. Direct evidence for the existence of an RNA intermediate began to accumulate with the discovery, in RNA phage-infected cells, of an RNA-dependent RNA polymerase (August et al., 1963; Weissmann et al., 1963; Haruna et al., 1963), and of linear, double-stranded virus-specific RNA molecules (Weissmann and Borst, 1963; Kelly and Sinsheimer, 1964; Fenwick, Erikson, and Franklin, 1964). Such molecules were seen as analogous to RF DNA of single-stranded DNA phages, i.e., double-stranded RNA containing one strand derived from the parent or identical to parental RNA (the plus strand) and a complementary strand (the minus strand).

Assuming that double-stranded RNA is an intermediate in RNA replication, and arguing by analogy to the mode of ϕx174 DNA replication, one might propose that plus strands are formed by a displacement reaction involving double-stranded RNA as the template. If this occurs, then one might expect to find also RNA molecules caught in the midst of the displacement reaction—in other words, double-stranded molecules with single-stranded tails. Such molecules were indeed detected and characterized as intermediates in phage replication (Fenwick et al., 1964; Granboulan and Franklin, 1966). The single-stranded tails are sensitive to attack by pancreatic ribonuclease, whereas the double-stranded portion of the molecule is not. Using this information, Kelly and Sinsheimer (1967a) asked whether the displacement reaction is conservative or nonconservative. In other words, if the reaction involves simultaneous synthesis of one plus strand and displacement of another, is the plus strand which is released the parental strand or the nascent strand? By separately labeling parental RNA and intracellular RNA, isolating the replicative intemediate, and determining which label is released during RNase digestion, Kelly and Sinsheimer concluded that both forms of replication occur. By use of density-labeling experiments, Kelly and Sinsheimer (1967b) were also able to show that both parental and progeny RNA molecules serve as templates for replication. This is in further contrast to the replication of ϕx174 DNA, where, as we have seen, the parental DNA molecules play a special role in replication by virtue of their attachment to a special site. In RNA phage infection the parental RNA must serve itself as template, not only for replication but also for translation, since labeled parental RNA can be isolated from infected cells in association with cellular polyribosomes (Godson and Sinsheimer, 1967).

The above studies, and many others which are not cited here, indicated that phage RNA replication occurs *via* the formation of double-helical RNA intermediates—the duplex replicative form (Ammann, Delius, and Hofschneider, 1964) and the so-called replicative intermediate, containing an RNA duplex with one or more single-stranded tails (Fenwick et al., 1964). Recently, however, the existence of double-helical structures as true intermediates in replication has been seriously questioned (cf. Weissmann et al., 1968). Because these later ideas have derived from extensive studies of the mechanism of RNA replication *in vitro,* let us consider this topic now and postpone for a few pages any further consideration of the mode of phage RNA replication *in vivo.*

RNA Replication in vitro

RNA phage workers are, in at least one respect, more fortunate than their colleagues who are studying viral DNA replication. Whereas the enzymes involved in DNA replication have still not been identified with certainty, there is virtually no doubt that the RNA phage-induced RNA polymerase is the enzyme catalyzing RNA replication *in vivo.* Since, as we shall see shortly, this enzyme has been highly purified and shown to catalyze the synthesis, *in vitro,* of infectious RNA molecules, one can study this process in the test tube with reasonable assurance that the results bear a close relationship to RNA replication in the infected cell.

The Qβ replicase. In the previous section, we mentioned the discovery of viral RNA-directed RNA polymerases. These enzymes, which have been variously called polymerases, RNA synthetases, or replicases, all catalyze the RNA-dependent incorporation of ribonucleoside 5′-triphosphates into polyribonucleotides, in a reaction analogous to the DNA-primed synthesis of RNA which is ubiquitous in cellular organisms. However, in contrast to the DNA-dependent RNA polymerase, which can transcribe virtually any DNA template offered it *in vitro,* the RNA-dependent enzymes are highly specific. The MS2 enzyme, for example, was found to exhibit a strong preference for MS2 RNA, as compared to RNA molecules isolated from other viruses. This species specificity is not unexpected, for a cell infected with an RNA virus contains far more host cell RNA molecules than phage RNA molecules, and it is clearly to the advantage of the virus to be able to avoid the futile replication of cellular RNA.

Of the various RNA phage-induced RNA synthetases, the Qβ enzyme offers the greatest stability, and extensive studies of the properties of this enzyme have been carried forth in the laboratories of Spiegelman, August, and Weissmann (for reviews see Spiegelman and Haruna, 1966; August et al., 1968; Weissmann, Feix, and Slor, 1968; Spiegelman et al., 1968; August, 1969; August et al., 1969). The demonstration that the Qβ enzyme is a true "replicase," i.e., that it can catalyze the replication of a viral RNA molecule, rather than the incorporation of triphosphates into some undefined polynucleotide, came from a series of brilliant experiments carried out by Spiegelman and his co-workers. Haruna and Spiegelman (1965a) purified the enzyme to the point where contaminating nuclease and polymerase activities were undetectable (a more extensive purification, giving up to 5000-fold enrichment of replicase, has since been described by Eoyang and August, 1968). Not only does this enzyme exhibit almost absolute specificity for Qβ RNA as a template, but the RNA molecules must be intact. Fragments of Qβ RNA have essentially no template activity.

The product of Qβ RNA-directed RNA synthesis *in vitro* was infective, suggesting that a true replication of the template was occurring. Moreover, when the enzyme was provided with a subsaturating level of template, the kinetics of RNA synthesis were those expected of an autocatalytic reaction, which showed that the enzyme has the same activity on the product as it does on the template (Haruna and Spiegelman, 1965b). Rigorous proof that the RNA product is indeed a faithful copy of the input template was provided by Spiegelman et al. (1965). Although the kinetics of RNA synthesis *in vitro* were identical to the kinetics of formation of infective units, one could entertain ad hoc hypotheses, such as the possibility that the enzyme increases the infectivity of the template RNA, and that the observed RNA synthesis represents the formation of extraneous, noninfectious material. To circumvent this possibility, Spiegelman et al. set up a series of identical tubes, each containing replicase, triphosphates, and magnesium ions. The first tube was primed with a small amount of Qβ RNA. After extensive synthesis had occurred, a small portion of the reaction mixture was transferred to the second tube. Incubation was again carried out, and this transfer operation was repeated 15 successive times, such that the amount of original template RNA calculated to be present in the last incubation mixtures was considerably less than one strand per tube. Yet the formation of infectious units paralleled the kinetics of RNA synthesis all the way, demonstrating conclusively that the enzyme was replicating infectious RNA molecules, which served in turn as templates for formation of further infectious RNA. Moreover, when RNA from a temperature-sensitive mutant was used as template in such a transfer experiment, the infectivity of all of the product molecules formed was also temperature-sensitive (Pace and Spiegelman, 1966). Thus, the RNA molecule, rather than some contaminant or component of the enzyme, is the true instructive agent in this extracellular replication process.

As isolated in August's laboratory, the Qβ replicase has a molecular weight of about 150,000 (Eoyang and August, 1968). However, there is an additional requirement for two macromolecular factors from the host cell, one of which seems to be a protein. Essentially the same results were obtained in Spiegelman's laboratory (Eikhom and Spiegelman, 1967; Eikhom, Stockley, and Spiegelman, 1968), except that their purification procedure results in an enzyme containing two components which can be separated by sucrose gradient centrifugation. One component, of molecular weight about 130,000, is of viral origin, whereas the lighter component, of molecular weight about 80,000, is present in uninfected bacteria —even in *E. coli* B, which does not serve as host for any known RNA phage. Presumably this lighter component contains both of the factors described by August and colleagues, but direct evidence has not been presented on this question, nor is anything yet known about the roles of these factors in cellular metabolism.

The existence of at least two components of the Qβ replicase was interesting in light of the discovery of a temperature-sensitive f2 mutant which could form double-stranded RNA at high temperature but could not replicate single-stranded RNA. Lodish and Zinder (1966b) postulated the existence of two enzyme activities, one needed for synthesis of minus strands, the other involved in plus strand formation. One could assume, *a priori,* that the phage component catalyzes one reaction and the host component alone has no activity on any phage RNA species. A consideration of the coding capacity of phage RNA would tend to rule out the existence of more than one phage-specific polypeptide as a component of RNA replicase. As we have already seen, the genes for coat protein and A protein account for some 1400 out of the 3300 or so nucleotides in an RNA genome. Even assuming that all of the remaining 1900 nucleotides code for protein (which, as we shall see, is not so), the maximum size of a polypeptide component of replicase would correspond to about 600 amino acids, or a molecular weight of 50,000. Since the viral component of the replicase has a molecular weight estimated at 130,000 or 150,000, this means that it must be composed of two or more identical subunits. One cannot, of course, rule out the possibility that the 50,000 daltons of replicase protein which can be coded for represent two or more nonidentical subunits, but these would be of small size relative to the protein subunits of other known nucleotide-polymerizing enzymes.

Mechanism of *in vitro* RNA replication. Much of the interest in the mechanism of viral RNA replication *in vitro* has focused upon the direction of chain growth and the structures of the intermediates formed. August et al. (1968) showed that the replicase binds to a site on the RNA molecule which is near the 3′ hydroxyl terminus. Zone centrifugation of RNA-replicase complexes indicated that one molecule of RNA is bound to one molecule of enzyme. If one assumes that complementary RNA molecules are antiparallel to their templates and that the enzyme begins to function at its site of binding, one would expect that the first nucleotide inserted in a nascent minus strand would give a 5′ triphosphate terminus, and, therefore, that the direction of synthesis would be from the 5′ to the 3′ end. Direct evidence for this was obtained by use of triphosphates labeled with P^{32} in the γ position. As shown in Figure 13–1, if synthesis proceeds from 5′ to 3′, some label should be permanently incorporated immediately upon initiation of synthesis, from the triphosphate corresponding to the 5′ terminus, whereas the 3′ to 5′ direction would give no permanent incorporation of label into product until the end of the reaction. As expected from the predictions of 5′ to 3′ synthesis, P^{32} from γ-labeled GTP was immediately incorporated into product. Moreover, GTP was the only one of the four nucleotides to label the product under these conditions. If synthesis proceeded from 3′ to 5′, one would expect label from each nucleotide to be transiently associated with product during synthesis, but this was not observed. Moreover, pulse-labeling studies with H^3-labeled nucleotides indicated that chains which had incorporated P^{32} into the 5′ terminus could increase in length without loss of the P^{32}. Thus, direction of chain growh in minus strand synthesis was shown to be from the 5′ to the 3′ end. This was in agreement with results obtained from Spiegelman's laboratory (Bishop, Pace, and Spiegelman, 1967). However, an unexpected result was that the 5′ terminal nucleotide of the minus strand, namely GTP, was not complementary to the 3′ hydroxyl terminus, which is adenosine. Since the penultimate residue on the plus strand is cytidylate, which is complementary to GTP, it is assumed that copying begins one residue in from the end. The significance of this is not known. However, Kamen (1969) has

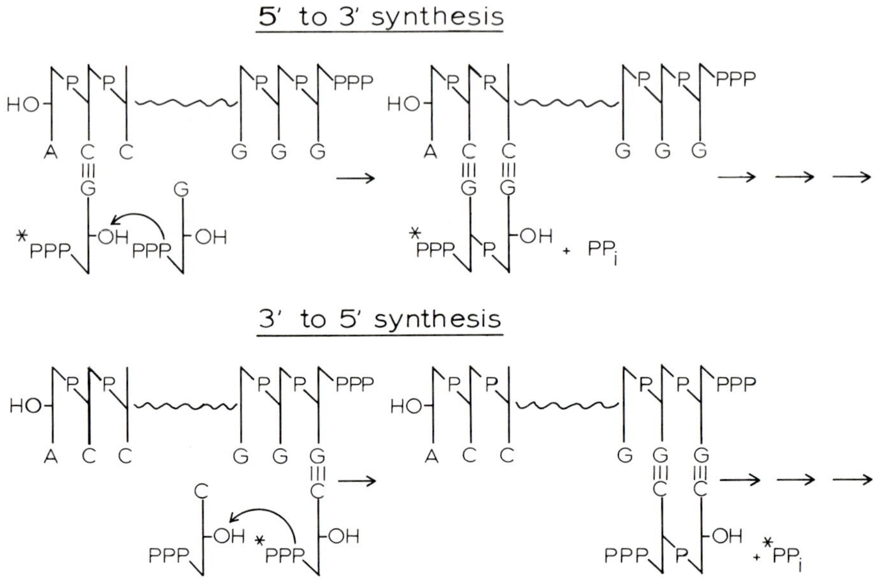

FIGURE 13-1 Use of terminal triphosphate labeling to determine the direction of viral RNA synthesis. *P represents labeled phosphorus. Note that synthesis of the minus strand from the 5′ to the 3′ terminus is initiated opposite the penultimate nucleotide residue on the plus strand. For further discussion of this point, see p. 297.

shown that the 3′ terminal adenylate can be removed from Qβ RNA without destroying its infectivity. August et al. (1969) have suggested that one role of the host factors in the Qβ replicase is to catalyze the terminal addition of a 3′ adenosine residue to the penultimate cytidylate, an event necessary for the initiation of minus strand synthesis. In line with this expectation, the factors are not required for synthesis of plus strands on a minus strand template.

Although Bishop et al. (1967) had reported that the direction of plus strand synthesis on a minus strand template is 3′ to 5′, the availability of purified minus strands (Pollet, Knolle, and Weissmann, 1967; Iglewski and Franklin, 1967) permitted a direct test of the direction of plus strand synthesis (August et al., 1968). Minus strands were obtained by annealing denatured double-stranded RNA with an excess of fragments derived from purified Qβ RNA (i.e., plus strand fragments). This gave duplex molecules consisting largely of intact minus strands and fragmented plus strands. These were purified chromatographically, denatured, and the minus strands separated from the plus strand fragments by zone sedimentation. By use of the techniques described in the previous paragraph, August et al. (1968) were able to show that the synthesis of plus strands occurs from the 5′ to the 3′ end, and a similar conclusion was reached by Spiegelman et al. (1968).

The availability of purified minus strands has permitted a number of informative experiments. This material is complementary in nucleotide composition to viral RNA (Weissmann et al., 1968). It is noninfectious, in contrast to minus strands of the single-stranded DNA phages (Iglewski and Franklin, 1967). In retrospect, this should not be surprising, since the RF DNA molecules are formed by bacterial enzymes, whereas a virus-specific enzyme must be present for replication of RNA molecules. Since, as we have already seen, viral RNA (plus strands) can serve as template *in vitro* for synthesis of viral proteins, it is clear that the information for this synthesis does not reside directly in the minus strands. For a minus strand to be infectious, it would have to be able to form plus strands, and it could not do this without first priming the synthesis of replicas. In line with this reasoning, R17 minus strands were found to be virtually inactive as a template for protein synthesis *in vitro* (W. T. Iglewski, personal communication).

The noninfectivity of minus strands allowed a determination of the length of time needed for synthesis of a complete RNA molecule (Weissmann et al., 1968). One merely incubates minus strands with replicase and determines the minimum time elapsed before infectious units can be detected in the medium. As shown in Figure 13–2, this period is about 100 seconds at 37°, which means, since the RNA molecule is something over 3000 nucleotides in length, that the replicase *in vitro* can catalyze the addition of about 30 nucleotides per second. Although the corresponding rate for *in vivo* replication has not yet been determined, this value is well within the range of observed rates of DNA-directed transcription *in vivo* (see p. 168).

In the previous section we discussed some evidence that replication *in vivo* occurs via the formation first, of a double-helical RNA molecule, next of a double-helical molecule with one or more single-stranded tails. These structures have been called, respectively, replicative form (RF) and replicative intermediate (RI; see Erikson, 1968). However, work on RNA synthesis *in vivo* has raised some question whether RF and RI are true intermediates in replication. First, purified RF and RI are inactive as templates *in vitro* (Feix et al., Mills, Pace, and Spiegelman, 1969), and secondly, RNase-resistant structures are not detectable during the synthesis of $Q\beta$ RNA *in vitro* (Weissmann et al., 1968). Recall that purified RF and RI are RNase-resistant by virtue of their double-helical structures. Almost all work on the isolation of RF and RI, as well as almost all other intracellular RNA species formed during RNA phage replication, has involved phenol extraction. Weissmann et al. found that phenol extraction of *in vitro* reaction mixtures generates polynucleotides having the same properties as RF and RI. Thus, Weissmann believes that they are artefacts of isolation. One doesn't know exactly how they would arise, but Weissmann

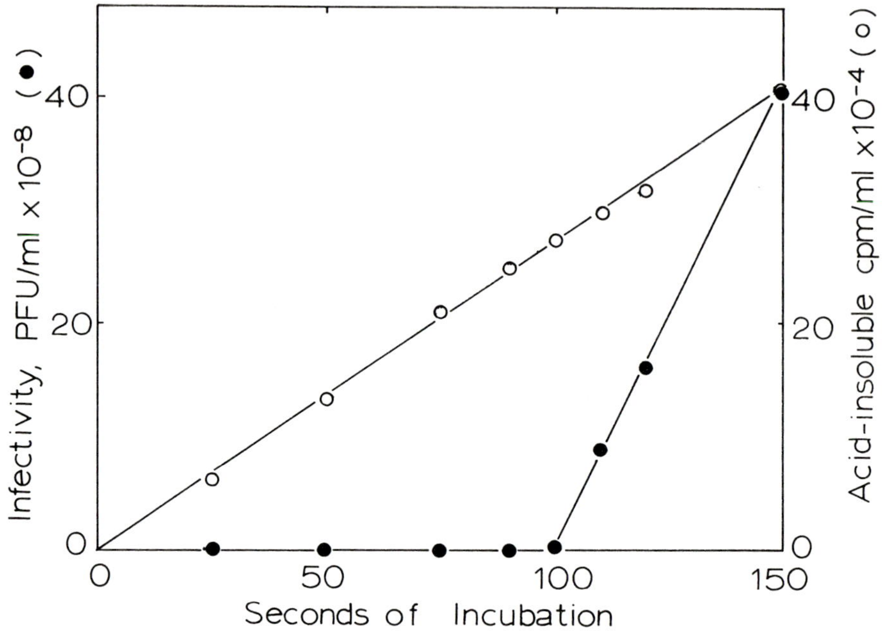

FIGURE 13-2 Synthesis of infective Qβ plus strands *in vitro* on a purified minus strand template (modified from Weissmann et al., 1968).

believes that there is a short stretch of double-strandedness at the growing point, but that the enzyme binds the template and nascent chains in such a way as to prevent them from permanently winding about one another. However, deproteinization, as in phenol extraction or treatment with detergent, allows the complementary strands to come together, with formation of double helices. The role of the enzyme in binding complementary strands so that this can occur is shown by the fact that purified plus and minus strands, when treated together with phenol, yield no RNase-resistant RNA molecules.

From the above, it would appear that the intermediates in replication *in vitro* are largely, if not completely, single-stranded. In line with this expectation, Weissmann et al., (1968) found that purified Qβ minus strands constitute a better template for synthesis of infectious RNA *in vitro* than do plus strands: the rate of synthesis is more rapid, and there is a shorter lag period, when minus strands are used. Presumably, the longer lag with plus strands reflects a requirement for the formation of complete minus strands, which are then used in turn to prime the synthesis of complete plus strands. These findings, and others, support a mechanism for *in vitro* RNA replication which involves first, the formation of a minus strand on a plus strand template, next, binding of enzyme near the 3′ terminus of the nascent minus strand, and finally, the formation of new plus strand material, with all chain elongations occurring from the 5′ to the 3′ end. Except for brief regions of hydrogen bonding near the growing point, all of the templates and products are single-stranded, as indicated in Figure 13–3. The true relationship of this scheme to the mechanism of RNA replication in the infected cell remains to be determined. However, this scheme does account for the results of Kelly and Sinsheimer (1967a), who found that the RNase-sensitive material in isolated intracellular RNA is derived both from parental and nascent RNA. One of the major problems in establishing this mechanism *in vivo* is devising a method for isolation

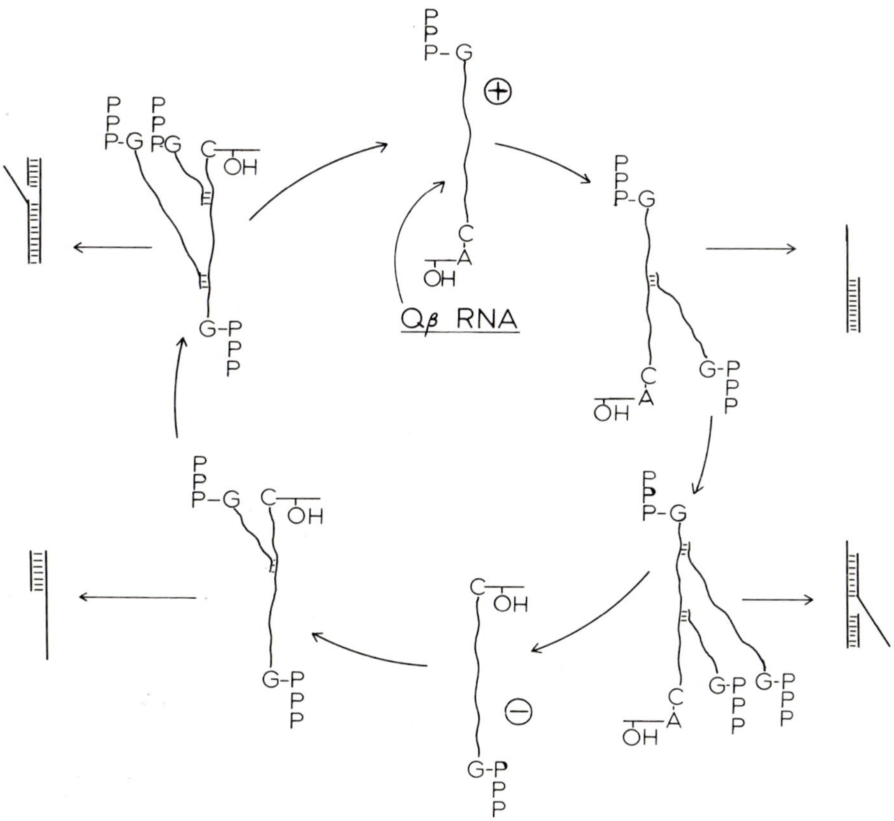

FIGURE 13-3 A scheme for phage RNA replication, as proposed by Weissmann et al., (1968). The diagrams to the right and left indicate the structures formed when the indicated complexes are subjected to phenol extraction or similar procedures.

of intracellular RNA which will not alter the secondary structure of this material.

Spiegelman and extracellular Darwinism. The demonstration of Pace and Spiegelman (1966) that RNA is the instructive agent in RNA replication *in vitro* allowed the performance of a fascinating series of experiments, in which RNA molecules were seen to evolve *in vitro,* in response to specific selection pressures. In the first such experiment Mills, Peterson, and Spiegelman (1967) asked what would happen to RNA molecules when they were subjected to a large number of serial transfers of the type earlier described (Spiegelman et al., 1965). The transfer procedure was modified slightly, such that the periods of incubation were relatively short and became even shorter as the experiment progressed. Thus, a selection pressure was applied which would favor the survival of molecules which could replicate relatively rapidly. Of course, the molecules did not need to retain functional genes for structural proteins or replicase, for the enzyme was provided, and the RNA molecules did not need to form virions in order to survive. All that really needed to be retained was a binding site for replicase and the necessary instructions for initiation and termination. During the transfer experiment, it was found that infectivity was lost after the fourth transfer, as might be expected from the loss of genes needed for viral replication but unnecessary for RNA replication *in vitro.* By the end of the 75th transfer, the RNA molecules which remained, and were still replicating, had lost some 83 percent of the material present in viral RNA, and contained only about 550 nucleotides. However, this material replicated some 15 times faster than the original Qβ RNA. Thus, it appeared clear that a dramatic series of evolutionary steps had occurred in the test tube during this experiment.

Using a different type of transfer regime, but one which also selected for fast-growing RNA molecules, Levisohn and Spiegelman (1968) isolated a second variant RNA molecule (termed V-2, to distinguish it from V-1, the RNA molecule of Mills et al. (1967)). Although the V-2 molecule is the same size as the V-1 molecule, it replicates slightly faster (doubling time of 0.403 minute, as compared to 0.456 minute for V-1). Thus, different regimes of selection lead to the evolution of different RNA molecules. Because of the obvious advantages of these small RNA molecules as objects for sequence studies, it became desirable to ask whether all of the molecules in a population were identical, at least within the range of mutation frequency. Levisohn and Spiegelman (1968) approached this question by asking whether they could clone RNA molecules, in other words, grow a population of RNA molecules from a single strand. V-2 RNA molecules were diluted to the point where there was, on the average, one molecule per 0.1 ml of solution. Aliquots of 0.1 ml were distributed to 82 tubes, which also contained replicase and radioactive triphosphates. The number of tubes in which synthesis occurred, and the distribution of radioactive counts incorporated among the tubes coincided amazingly well with predictions made on the basis of the Poisson distribution. For example, one would expect that at an average input of 1 molecule per tube, e^{-1}, or 30 tubes, would receive no RNA and incorporate no counts, and this is exactly what was observed. Among the remainder of the tubes, it was clear that many had synthesized RNA from a single molecule present initially.

Another recent publication (Levisohn and Spiegelman, 1969) describes what may be a most exciting application of the formation of variant RNA molecules. Variants of V-2 were selected for an ability to replicate rapidly in the presence of a suboptimal concentration of CTP. These variants had the same nucleotide composition and size as the original molecules, and were able to replicate efficiently in the presence of suboptimal concentrations of all four triphosphates. One possible explanation for this is the existence of altered binding to replicase which permits the complex to bind triphosphates more tightly. Another variant is one which was selected for in the presence of the triphosphate of tubercidin, an inhibitory analog of adenosine. Tubercidin triphosphate is incorporated into RNA in place of ATP. However, the variant is able to replicate efficiently in spite of this, while replication of the parental V-2 is significantly impaired. These few experiments suggest that the isolation of variant molecules might have enormous practical consequences. Consider, for example, the use in chemotherapy of an RNA virus-caused disease, of a variant RNA molecule which (1) binds more tightly to replicase than viral RNA; (2) replicates more rapidly; and (3) is selected for resistance to the effect of some other chemotherapeutic agent with which it is used in conjunction. Of course, this statement ignores numerous practical difficulties, such as transport into cells and other intracellular consequences, but it does, I believe, suggest that this approach has enormous potential.

PHAGE RNA AS A MESSENGER *IN VITRO*

The availability of phage RNAs as well-defined polycistronic messengers which can direct the production of recognizable proteins *in vitro* has been indispensable in the development of our present ideas on the nature of protein biosynthesis. For example, in Chapter 5 we referred to the use of RNA from f2 coat protein amber mutants in the demonstration that suppression of nonsense mutations involves formation of altered transfer RNA molecules. Much of the work on the mechanism of protein synthesis *in vitro* with viral RNA messengers is beyond the scope of this monograph. For this material, the reader is referred to an excellent review by Lengyel and Söll (1969). In this section I shall be concerned primarily with the role of viral RNA as messenger, and not with the nature of its interaction with ribosomes, initiation factors, etc. In other words, what does the structure of viral RNA molecules tell us about the nature of the genetic code? Of the many coding triplets which might be used in protein synthesis, which are actually used? We know that certain coding triplets can lead to polypeptide chain initiation or termination *in vitro,* but what actual signals are used for these purposes in natural messengers? What is the extent of mRNA molecules which are not translated as part of their function, and what are their functions, if any, aside from initiation and termination? Before discussing relationships between phage RNA structures and the questions raised above, let us briefly review some basic information on coding triplets which have been shown to act as chain terminators or initiators in protein synthesis *in vitro*. The reader is referred to the reviews of Garen (1968) and Lengyel and Söll (1969) for further information on these matters.

The *amber* codon UAG and the *ochre* codon UAA, when present in a polynucleotide sequence as a result of mutation, both serve as chain terminators *in vivo* and *in vitro*. In an unsuppressed *E. coli* strain there are no transfer RNA molecules which can respond to either of these triplets. Thus, the triplets cannot be translated, and growing polypeptide chains are released at the site of the mutation. Suppression involves alteration of a particular tRNA molecule such that it can respond to *amber* or *ochre* triplets. Recent studies on UGA (sometimes called *umber* or *opal*) mutants indicate that mutations generating this codon have the same results, namely chain termination *in vitro* at the site of mutation, but completion of synthesis in the presence of tRNA from suppressor strains of bacteria (Model, Webster, and Zinder, 1969).

Although the existence of three nonsense triplets gives us at least three possible signals for termination of translation at the normal end of a cistron, it wasn't known for some time whether any of these codons actually does fulfill a natural chain-terminating role. One difficulty was the existence of nonlethal suppression. If a single nonsense triplet forms a natural termination signal, then the existence of a suppressing tRNA species would be expected to stitch the adjacent ends of proteins together during translation, since the termination signal would be translated. This could in turn have deleterious consequences to the cell. However, no evidence for the existence of super-length protein molecules in suppressed bacterial strains has yet been seen. One can, of course, invoke ad hoc hypotheses concerning altered secondary structure at termination sites which lead to nonrecognition by altered tRNA, but these are not very satisfying.

A phenomenon related to the existence of nonsense mutations, and one which we discussed briefly in Chapter 5, is polarity. A nonsense mutation quite often will abolish expression, not only of the gene bearing the mutation, but of genes located distal to it. The existence of polarity is usually interpreted as evidence that a multicistronic region is transcribed as a unit and that the individual genes are translated sequentially. Presumably, the initiating signal for the entire region is located at the left end of the first gene, and, when translation ceases at the site of mutation, it cannot be reinitiated. We have seen how the existence of polarity was used to ascertain the direction of transcription of certain genes, both in T4 and in λ (pp. 87 and 234).

An aspect of polarity which has not yet been satisfactorily explained is the fact that the nearer a polar mutation to that part of the gene specifying the N-terminal amino acid, the greater is the polar effect. This general observation has been confirmed in studies on f2 RNA translation *in vitro* (Engelhardt, Webster, and Zinder, 1967). A strongly polar amber mutant in the coat protein gene was altered in the codon specifying the sixth amino acid, while a nonpolar mutant in the same gene was altered at the codon for the 70th amino acid. During *in vitro* translation, the nonpolar mutant RNA was found in polyribosomes, while the polar mutant RNA could form only monosomes. This suggested that there is only one initiation site on the f2 RNA molecule and that all three f2 genes are translated sequentially. However, as we shall see below, there is now more direct evidence for the existence of separate initiator sequences for each of the three genes in the related phage R17. There is good evidence, from studies of the tryptophan operon in *E. coli* (Morse and Yanofsky, 1969) that polarity in this system results from premature degradation of mRNA, starting at the site of the exposed nonsense triplet. However, this does not help us to understand polarity in RNA phage systems, for there is no basis for believing that phage RNA is metabolically unstable under any known conditions. For a recent review of polarity, see Zipser (1969).

With regard to polypeptide chain initiation, there is now abundant evidence that all *E. coli* proteins are synthesized with N-formyl methionine as the N-terminal amino acid. There are two species of methionine tRNA in *E. coli* cells, both of which accept methionine but not N-formylmethionine. However, only one of the two methionyl tRNA species can subsequently by formylated (by an N^{10}-formyltetrahydrofolate-dependent enzyme). This species, formylmethionyl $tRNA_F$, responds to two triplets, AUG or GUG, whereas the nonformylatable species, methionyl $tRNA_M$, responds only to AUG. The GUG codon presents some ambiguity *in vitro,* inasmuch as it also codes for valine. In principle a GUG or AUG triplet could serve as a simultaneous terminator signal for one protein and initiator signal for the next, because the formyl group of N-formylmethionine prevents peptide bond formation. However, the dual roles for the GUG and AUG codons suggest that the initiator sequences are somewhat more complex. Specifically, what factors determine whether GUG is to be read as formylmethionine or valine, or whether AUG is to be read as formylmethionine or internal methionine? Again, one can invoke *ad hoc* hypotheses such as altered secondary structure of the messenger in the vicinity of an AUG or GUG which is to function in initiation, and, as we shall see shortly, there is now some evidence that this may be a factor.

isolated and sequenced ribosomal binding sites containing the initiation regions for all three R17 genes. Unfortunately, inspection of these sequences does not immediately clarify the initiation problem. Only one of the three oligonucleotides can be visualized as existing in a double-helical loop. Moreover, Steitz cites evidence that at least two internal AUG codons in the coat protein gene exist in hydrogen-bonded structures, which seems to rule against the use of loops as specific ribosomal recognition regions.

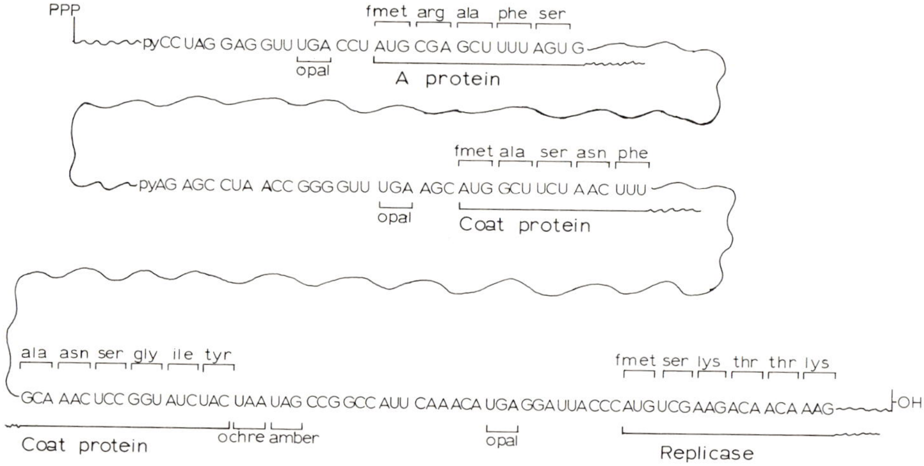

FIGURE 13-5 Partial structure of the R17 RNA molecule. This indicates the sequences of the oligonucleotide regions near the initiation site for each of the three cistrons (Steitz, 1969) and the region corresponding to the C terminus of coat protein, with the intercistronic region between the genes for coat protein and replicase (Nichols, 1970). The gene order is determined from work of Jeppesen et al., (1970).

Inspection of all three of Steitz's sequences shows, to the left of the initiating AUG codon, UGA triplets in phase with AUG and removed by several nucleotides (Figure 13–5). Thus, if these triplets participate in the termination of previous codons, it is clear that there must exist untranslated material between two adjacent codons. This in turn, plus the existence of one ribosomal binding site corresponding to the N-terminus of each gene, strongly suggests that each of the three R17 genes is translated independently. Further light on the relationship between polypeptide termination and initiation of the next chain has come from a recent paper by Nichols (1970), who has isolated and determined the sequence of an oligonucleotide spanning the entire intercistronic region between the R17 genes for coat protein and RNA replicase. Nichols partially digested P^{32}-labeled R17 RNA with T1 ribonuclease and resolved the fragments by gel kinase following infection of mutant cell lines lacking the enzyme; and (3) isolation of virus mutants unable to induce the enzyme. So far, to my knowledge, no one has succeeded in purifying an animal virus-induced enzyme to homogeneity, a task which is prerequisite to the type of rigorous demonstration of *de novo* synthesis of enzyme protein which we discussed in Chapter 4.

A quite different situation exists in cells infected with the small oncogenic DNA viruses SV40 and polyoma, and this highlights the need for caution in concluding that an increase in activity of an enzyme signifies virus-coded protein synthesis. It has been known for some years that infection by these viruses of stationary-phase cells stimulates DNA synthesis, much of which is cellular (Weil, Michel, and Ruschmann, 1965; Dulbecco, Hartwell, and Vogt, 1965). Associated with this turn-on of DNA synthesis is the activation of many enzymes of DNA synthesis, including nucleotide reductases, deoxynucleoside and deocynucleotide kinases, thymidylate synthetase, dihydrofolate reductase, deoxycytidylate deaminase, DNA polymerase, and DNA ligase (Kit, 1967; Kara and Weil, 1967; Sambrook and Shatkin, 1969). However, these viruses are quite small. The genome contains about 3×10^6 daltons of double-stranded DNA, sufficient to code for only 5–10 proteins. This is insufficient information to code for the enzymes listed above, to say nothing of viral capsid proteins or functions involved in malignant transformation (see next section). Therefore, one is forced to the conclusion that infection by these viruses increases the activity and/or synthesis of cellular enzymes of DNA metabolism, and most current evidence is in line with this idea. However, one cannot exclude the possibility that one or a few of these enzymes are virus-coded. In particular, the properties of thymidine kinase isolated from infected cells differ markedly from those of the same enzyme isolated from uninfected cells, suggesting that it is a viral gene product. However, when a cell line lacking thymidine kinase was infected with SV40, no enzyme induction was seen, indicating that the enzyme is not a viral gene product (Basilico, Matsuya, and Green, 1969). A possible explanation for this apparent anomaly is provided from some preliminary work on thymidine kinase from adenovirus-infected cells (Hatanaka, Twiddy, and Gilden, 1969). These authors presented evidence that the host kinase is modified after infection to give a more active form. This is analogous to the situation we have discussed with respect to RNA polymerase and valyl-tRNA synthetase in phage T4-infected bacteria.

Not only do animal viruses induce new enzyme activities, by the mechanisms discussed above, but some of them also contain enzymes bound to purified virions. In Chapter 3 we discussed certain enzymes, notably a lysozyme and dihydrofolate reductase, which seem to be specific components of T-even phages. It has long been known that influenza and related viruses contain a bound neuraminidase which functions in the cell-surface attachment reaction (Gottschalk, 1960). More recently, it has been found that vaccinia virions contain two bound enzymes, DNA-dependent RNA polymerase and a nucleoside triphosphatase with greatest activity toward ATP (McAuslan, 1969; Munyon et al., 1968). Perhaps even more intriguing than the question of whether these enzymes are viral gene products, which is still unsettled, is the question of their biological role as agents of virus structure. There is evidence that some early mRNA species in vaccinia-infected cells are formed before the viral DNA is uncoated, and McAuslan has proposed that this transcription is accomplished by the particle-bound RNA polymerase. With respect to the ATPase, Munyon et al. (1968) found no resemblance of this enzyme to known cellular ATPases. Of interest is the fact that vaccinia virions increase in heat stability when treated with nucleoside triphosphates, due to binding of these compounds by the virus-associated enzymes (Munyon, Mann, and Grace, 1970).

A final example of an animal virus-associated enzyme is an RNA polymerase bound to purified reovirus particles (Borsa and Graham, 1968). Inasmuch as reovirus contains double-stranded RNA, which is not found in any known cellular organisms, and since the reovirus-bound polymerase utilizes double-stranded RNA as a template *in vitro,* it is probable that the enzyme is truly a virus-specific component.

The existence of virus-coded enzymes, which presumably play essential roles in viral structure and replication, presents the basis for development of a rational chemotherapy of virus diseases. Attempts at treatment of such diseases with chemical agents have been under way for some years (see review by Hilleman, 1969). However, such attempts have met with only sporadic success. For only a very small group of agents is there any real knowledge of the chemical basis of therapeutic effectiveness. These include the primidine nucleotide antagonists 5-iododeoxyuridine and cytosine arabinoside, which have been used to treat corneal infections caused by herpes simplex, and even here the biochemical targets for these drugs are not known with certainty.

Baker (1967a) has pioneered in the development of an approach which should, in principle, allow one to design specific and irreversible inhibitors for any desired enzyme. Such agents are called "active site-directed irreversible inhibitors." One builds a molecule containing a functional group which can bind at the active site. This is attached to an apolar region which can associate with a similar apolar region on the enzyme molecule, just outside the active site. Finally, there is a group, such as a sulfonyl chloride, which can alkylate some functional group on the enzyme molecule which is just outside the apolar region. Such an agent is directed toward the active site by the first functional group, and the apolar region aligns the molecule such that the alkylating agent is brought into contact with the susceptible site on the enzyme molecule; the alkylation reaction occurs, and the enzyme is irreversibly inhibited. The specificity of this approach derives from the expectation that enzymes from different species, or different tissues of the same species, are fundamentally similar in the structure of the active site, but different in the area outside the active site, including the hydrophobic bonding region adjacent to the active site. By adjusting the size of the apolar region of the inhibitor, one can take advantage of these differences and design inhibitors which are highly species-specific, or tissue-specific. Baker has even designed inhibitors of dihydrofolate reductase which have a stronger inhibitory effect toward the enzyme from a rat hepatoma, or liver tumor, than the same enzyme isolated from normal rat liver.

Whereas the assumptions made by Baker might not be valid in all cases, the approach would seem to be ideally suited to the design of antiviral agents, where, as we have seen, there are large differences between a cellular enzyme and the corresponding phage-induced enzyme. Indeed, Baker (1967b) has shown up to one hundredfold differences in inhibitory effectiveness of a series of dihydrofolate reductase inhibitors when the T2 phage-induced enzyme was compared with the *E. coli* enzyme. Presumably, this difference could be magnified, if desired, by synthesis and testing of more inhibitors. In designing an antiviral agent which acted by inhibiting a virus-specific enzyme, one would, of course, have to select an enzyme whose action is essential for virus replication. The vaccinia-induced thymidine kinase would probably be a poor choice, since the presence of the cellular kinase renders it nonessential for virus replication. However, the vaccinia DNA-dependent RNA polymerase would seem to be an excellent choice, inasmuch as it is probably involved in a critical way in the transcription of uncoated viral genes. If this enzyme is really shown to be a viral gene product, the approach would be to design an active site-directed irreversible inhibitor which blocks the virus-associated enzyme but has relatively little effect on the RNA polymerase of the host cell.

Small DNA Viruses and Carcinogenesis

Some of the most exciting contemporary work in animal virology is focused upon the mechanism of malignant transformation by viruses. A primary object of this work, of course, is to learn the extent to which viruses function as a cause of cancer. Although there is still no direct evidence, to my knowledge, which demonstrates that a virus is the sole cause of any human cancer, there is much circumstantial evidence. There are, of course, numerous oncogenic viruses, known to cause tumors or leukemias in various animals. The transformation of normal cells to neoplastic cells, which accompanies infection by such viruses, can be observed in cell culture. The process bears some striking resemblances to the lysogenization of a bacterium by a temperate bacteriophage, and much of the attention toward lysogeny derives from its interest as a model system for understanding malignant transformation.

When a cell becomes transformed to the neoplastic state, the most drastic metabolic consequence is a loss of contact inhibition. Normal cells, when grown in culture, tend to cease growth when a confluent monolayer has been formed in a plate—in other words, when the cells are all in contact with their nearest neighbors. Some signal, of unknown nature, tells the cells to stop dividing once they have reached a critical concentration. This inhibitory mechanism is absent in transformed cells, which continue to grow over and around one another. This allows one to enumerate transformed cells against a background of normal cells. The transformed cells form what is called a focus, or a region of high cell density which can be easily visualized against the surrounding monolayer.

Viruses which can cause malignant transformations include both DNA and RNA viruses. Among the oncogenic DNA viruses are the small SV40 and polyoma viruses and some of the adenoviruses. The process of transformation by SV40 and polyoma has recently been reviewed by Dulbecco (1969) and by Eckhart (1969). Interestingly, some cells can be transformed by SV40 DNA (Aaronson and Todaro, 1969). Parallel to the situation we reviewed for lysogeny, infection by these viruses can lead either to a productive cycle of virus multiplication, or to transformation of the infected cell. The transformed cell retains its new properties, including loss of contact inhibition, indefinitely through subsequent generations of growth. Moreover, injection of transformed cells into a suitable laboratory animal causes the animal to develop cancer. Although for many years no trace of infective virus could be found in transformed cells, virus-specific substances were seen. These were two virus-specific proteins, the so-called T-antigen and the transplantation antigen, which could always be detected in transformed cells by immunofluorescence. The biochemical functions of these agents are still not known.

The existence of virus-specific proteins in transformed cells strongly suggested that viral genes were being expressed. Direct evidence for this was first obtained in cells transformed by adenovirus 12 (Fujinaga and Green, 1966). A tiny fraction of RNA pulse-labeled in these cells was shown to hybridize specifically with viral DNA, whereas no such virus-specific RNA could be detected in nontransformed cells. Similar viral gene expression has been detected in cells transformed by other DNA viruses, including polyoma (Martin and Axelrod, 1969). In this latter case, the fraction of the total polyoma genome which is transcribed in transformed cells is 10 to 20 percent, whereas the fraction transcribed during lytic growth is about 50 percent. Thus, just as in lysogeny, it would appear that most of the virus genome is inactive in transformed cells.

A complementary technique was used to demonstrate the presence of viral DNA in transformed cells (Westphal and Dulbecco, 1969). Labeled viral RNA was synthesized *in vitro,* by the action of RNA polymerase on viral DNA. This material was shown to hybridize with DNA from transformed, but not from normal, cells. Furthermore, the extent of hybridization provided a measure of the amount of DNA present per transformed cell. Unlikely lysogeny, where we have seen that there is usually one phage per chromosome, polyoma-transformed cells contain 5 to 10 virus-equivalents of DNA per transformed cell, while SV40-transformed cells contains as much as 20 to 60 virus-equivalents of DNA per cell. Even this relatively large amount of DNA, however, amounts to only about 5×10^{-7} of the total DNA in a cell, rendering its biochemical characterization extremely difficult.

We mentioned that for some time nobody could detect infectious virus particles in transformed cells. Does this mean that the viral DNA in such cells is present as incomplete genomes? No. Intact virus chromosomes can be rescued following interspecific hybridization, or fusion of an SV40-transformed cell with a cell of a different species (although similar results have not been observed with polyoma). One is, of course, tempted to conclude that fusion leads to the destruction or dilution of a repressor, which results in the onset of viral multiplication. However, as Eckhart (1969) has pointed out, polyoma-transformed cells can be productively superinfected by polyoma at a multiplicity as low as one; this would tend to rule against the existence of a cytoplasmic repressor of polyoma replication.

Harking back to our concept of lysogeny, we would like to know whether there exist specific chromosomal sites for the attachment of viral genomes in transformed cells. Weiss, Ephrussi, and Scaletta (1968) asked this question by taking advantage of the fact that fusion between a human cell and a mouse cell leads ultimately to selective loss of the human chromosomes of the resulting heterokaryon (the cell which is formed by fusion). Using the presence of SV40 T antigen as an index to the presence or absence of viral genetic material, Weiss et al. found that the viral genomes persisted essentially until all of the human chromosomes had been lost. This indicates either that the viral genomes are distributed among all of the human chromosomes, or, less probably, that they are associated with one or a few human chromosomes which are selectively retained following fusion.

What is the mechanism of attachment of viral DNA to the chromosomes of transformed cells? Using lysogeny as a model, one might expect that it is linearly inserted into chromosomal DNA. Whereas this has not yet been demonstrated directly, Doerfler (1968) has provided, for adenovirus 12-transformed cells, the most elegant evidence that viral DNA is covalently attached to cellular DNA. Cells labeled with BUDR were infected with H^3-labeled virus. DNA was isolated and banded in both neutral and alkaline gradients. In both cases, radioactivity was found in the position corresponding to BUDR-labeled DNA, indicating covalent attachment of viral to cellular DNA. As expected, radioactivity was not found in the dense position after the DNA had been fragmented by sonication.

What types of gene functions are involved in the establishment of the transformed state and in rescue of infective virus? Dubbs and Kit (1968) found that a certain proportion of cells transformed by uv-irradiated SV40 could not yield virus following fusion with a sensitive cell. Presence of viral genetic material in these so-called "defective lysogens" was confirmed by the presence of T-antigen. Dubbs and Kit believe that the viruses in these cells, while able to establish the transformed state, have undergone mutation in some gene whose action is essential for the initiation of virus replication.

Eckhart has provided direct evidence for the involvement of at least one polyoma gene product in the establishment of transformation. He has isolated several temperature-sensitive mutants of polyoma, which fall into three complementation groups. Mutants in one of the groups are defective in their ability to establish transformation; however, since these mutants are also defective in DNA synthesis and the formation of T-antigen, one doesn't know the primary target of mutation in this case.

RNA Viruses and Carcinogenesis

Although the oncogenic viruses present excellent systems, in many respects, for the study of viral carcinogenesis, including a rather satisfying similarity to lysogeny, there are two drawbacks: (1) there is relatively little evidence that these viruses actually do cause cancer in animals infected under natural conditions; and (2) the proportion of cells in a culture which undergo transformation is quite low, which makes it difficult to study biochemical events during this process. Much higher frequencies of transformation are encountered among the oncogenic RNA viruses. Moreover, there is abundant evidence for the involvement of these viruses in natural carcinogenesis—at least in nonhuman populations. These viruses include the Rous sarcoma virus, discovered in 1911, as well as numerous similar viruses capable of causing tumors or leukemias in a variety of birds and mammals (for review see Bader, 1969).

RNA viruses can be detected in many cells and tissues, including human, in which there are no clinical manifestations. Huebner and Todaro (1969) have compiled data from many sources which indicate that the genome of a "C-type RNA virus" (one which is similar to the Rous sarcoma virus) may be a ubiquitous component of cells, but that the genome is normally "switched off." They believe that a cell becomes neoplastic when some environmental stress, such as radiation, chemical carcinogens, or aging, activates the virus genome, or at least those genes whose action is necessary to cause a neoplastic transformation. Whereas the idea that cancer results from activation of repressed viral genes is not particularly new, the theory of Huebner and Todaro does present a unified view of carcinogenesis. If all, or nearly all, cancers result from activation of RNA virus genomes, then the oncogenic DNA viruses would be viewed largely as laboratory curiosities, at least insofar as their ability to cause cancer is concerned.

Whereas we do have some idea, from the model of lysogeny, of how a DNA virus genome might be maintained in a "switched-off" state, we have no such model for RNA viruses. Huebner (1970) proposed that the maintenance of a C-type virus in an inactive state is accomplished by a mechanism essentially the same as that proposed by Jacob and Monod. There are some difficulties with this proposal. First, as we have already stated, there is no evidence for the existence of repressor-operator mechanisms in the cells of higher organisms. Secondly, the theory of Jacob and Monod said nothing about the regulation of RNA genomes. Strict application of the Jacob-Monod model would demand that the viral genome be integrated into the chromosomes of the host cell. How is this to be accomplished, with an RNA virus genome and a chromosome containing DNA? It has been suggested that oncogenic RNA viruses direct the synthesis of a complementary DNA molecule, which can then become inserted into chromosomal DNA, by a process akin to lysogeny. Whereas there is some indirect evidence that this might occur (see Bader, 1969), the idea is far from being proved. In principle, one should be able to use DNA-RNA hybridization techniques to detect in transformed cells DNA which is complementary to viral RNA. One would merely isolate DNA from transformed cells and measure its binding to viral RNA of high specific activity. It will be recalled that this experiment was performed in an RNA phage system, by Doi and Spiegelman (1962), with negative results.

If viral RNA genomes are incorporated as complementary DNA, one should be able to detect, in infected cells, an RNA-primed DNA polymerase. No such enzyme has been detected. From the ubiquity of RNA viruses and the many known oncogenic RNA viruses, the idea of a general involvement of such viruses in carcinogenesis seems possible. However, the physical state of the virus genome in the cell is very much an open question. (Note added in proof.) Two important papers have recently described the existence of an RNA-dependent DNA polymerase, as discussed above (Baltimore, 1970; Temin and Mizutani, 1970). The enzyme is associated with virions of RNA tumor viruses, which may explain why it had not been detected earlier. Whether this enzyme actually makes a virus-specific DNA which then becomes inserted into the chromosome remains to be determined.

SOME COMMENTS ON GENETIC ENGINEERING

Whenever an important contribution to molecular genetics is announced through the mass media, it seems to provoke the appearance of a series of apprehensive articles and pronouncements about the possible misuse of knowledge in this area. "Genetic tinkering" and "breeding a race of monsters" are some of the kinder phrases which are applied. What are the possibilities of applying molecular genetics to substantially alter human heredity? Very slim, in my opinion.

Consider the means available for alteration of the hereditary properties of a cell, such as DNA-mediated transformation (as distinct from the virus-induced malignant transformation which we were just discussing). Even in the well-studied bacterial systems, the frequency of transformation is very low. In mammalian cells there have been very few reports of DNA-mediated transformation. Even if one could transform mammalian cells in a defined system, however, and even if one had a homogeneous DNA containing only the gene which one wanted to insert, I believe that the transformation frequencies would be too low to be useful. After all, one would want to alter essentially all of the cells of an animal. How would one even introduce the DNA into all of the cells, much less have it become stably integrated? And, if one proposed to transform the fertilized ovum, how would one select the 1 to 1000 or so transformants for further cultivation (assuming, of course, that one could obtain 1000 fertilized ova).

We must also consider the immense technical difficulties inherent in the preparation of homogeneous segments of DNA containing only the genetic material which we wished to insert. In Chapter 9 we discussed the heroic measures which were undergone in the isolation of part of the lactose operon in *E. coli*. This task involved the exploitation of some almost magical techniques in bacterial genetics, including the forced introduction of a prophage at any desired site on the bacterial chromosome and the isolation of deletion mutations spanning well characterized regions of the chromosome. Our knowledge of mammalian cell genetics does not begin to approach the level of sophistication needed for application of these techniques to genetic engineering, and I doubt whether it will within the foreseeable future. The synthetic approach would appear to be more feasible, although still fraught with immense technical difficulties. One would need to know the amino acid sequence of all proteins which one wished to have synthesized in the transformed cell, and one would need far more information than is presently available about the actual trinucleotide coding units which are used in the translation of messages in mammalian cells. Then there is the problem of synthesizing the required DNA, a truly Herculean task. Khorana and a large research group at the University of Wisconsin have been working for several years on the total synthesis of the structural gene for alanine tRNA of yeast. This gene is fewer than 80 nucleotides in length. Consider the extension of the techniques being used to the synthesis of the structural gene for a typical protein containing 300 amino acid residues. The gene would have to contain 900 nucleotide pairs! Even if the speed with which a DNA molecule can by synthesized is increased by one or two orders of magnitude, it strikes me that it will be quite some time before synthetic DNA molecules are available as prescription items.

Suppose that one elected to use a virus as the vehicle for transporting new genetic material into a cell. In principle this could give greater specifity and higher frequencies of hereditary alteration than a DNA-mediated transformation. Assuming that one could increase the frequency of transformation to essentially 100 percent, and assuming that oncogenic viruses could be shown to attach to specific sites on human chromosomes, one would have, in theory, an efficient means of delivering new information to cells. One proposal which is currently receiving a good deal of attention involves a search for a virus which could be used to treat xeroderma pigmentosum. This is a hereditary condition in which there is an excessive sensitivity of the skin to ultraviolet light, and a high incidence of skin cancers. Patients with this condition rarely survive to adulthood. Cleaver (1969) has recently shown that patients with this condition lack an enzyme essential to the repair of DNA, presumably an endonuclease responsible for excision of damaged regions. The proposal is to search for a naturally occurring virus which bears a gene for the missing repair enzyme and then to modify it in such a way that the structural gene for the enzyme could be introduced into the genome of the patient by viral infection under conditions promoting the integration of viral DNA. This approach is fraught with difficulties. First, the known oncogenic DNA viruses are very small, being able to code for only 5 to 10 polypeptides. As we have seen, they stimulate DNA synthesis in infected cells by activating the synthesis of cellular enzymes for DNA synthesis. It strikes me as unlikely that such small viruses would bear information for their own uv repair systems. More feasible, perhaps, is the idea of using virus capsids to enclose synthetic or cellular DNA (cf. Trilling and Axelrod, 1970). Secondly, one is faced with the problem, even if such a virus can be found, of removing unwanted genes. After all, the viruses of which we have been speaking transform infected cells to malignancy, and the induction of cancer in a treated patient would certainly be an undesirable side effect. Before we could use such a virus for genetic alteration, we would have to know the function of every gene, and we would have to be thoroughly familiar with the mechanism of viral carcinogenesis, so that we could be sure that it would not occur under our conditions of infection. There is also the problem of how we would trim away the unwanted DNA without altering the ability of the viral genome to invade the cell and insert the desired gene. Thirdly, there is the problem of directing the viral genome to integrate at a specific site. After all, the majority of the genome in any differentiated cell is in a "turned-off" condition. How will we prevent the infecting genomes from inserting at sites where they would also be "turned off," and therefore unable to synthesize the desired gene product? Finally, there is the question of the mode of treatment. Presumably one would have to alter the heredity of virtually every cell, or, in the case of xeroderma pigmentosum, at least every epidermal cell.

If a virus is being used to bear the desired information, this means that essentially every cell would have to be infected. How is this to be accomplished? Presumably infection would have to be carried out on the fertilized ovum, or else early in embryonic development. Of course, all of the above ignores the practical question of how one would find a subject willing to undergo this type of treatment. Even if xeroderma pigmentosum could be diagnosed *in utero*—and this may be feasible—it would be far safer for the mother, in my opinion, to have an abortion and become pregnant again. After all, xeroderma pigmentosum is an autosomal recessive condition, so there is only one chance in four that the offspring of two heterozygous parents will bear the disease.

However, the above discussion is directed toward but one possible application of genetic engineering. I believe that similar roadblocks stand in the way of any other means which might be proposed for experimental alteration of human heredity.

I do not make these rather skeptical remarks to discourage research on the application of microbial genetics and biochemistry to eradication of human genetic diseases. Rather, my purpose is to illustrate some of the immense technical problems which lie in the way of successful application. It is possible that these difficulties can be overcome, and I am just not sufficiently imaginative to visualize their solution. Indeed, I may be embarrassed in 25 years at having written these words—but I doubt it.

THE FUTURE

Currently we are witnessing a small but significant exodus from molecular genetics, in which some of the most productive workers are deserting the field for the new frontier of neurobiology. There is a tendency among such workers, and others, to feel that molecular genetics is "played out"—that all of the really important discoveries have been made and all that remains is the "filling in" of details. I disagree. Some of the most important discoveries, it seems to me, are going to lie in the realm of application—in treatment of viral diseases, prevention and cure of cancer, and, possibly, alteration of heredity, as discussed above. I don't see how the work yet to be done in these areas could fail to elicit important new discoveries—particularly in the area of mammalian cell genetics, a realm which remains almost completely uncharted.

Even among the fundamental questions whose basic outlines seem clear, there are still basic questions to be answered. Consider DNA replication. Even today, 18 years after Watson and Crick proposed a mechanism for the self-duplication of the DNA molecule, we do not know how this occurs, even in microbial systems. The recent isolation of viable *E. coli* mutants which lack DNA polymerase (DeLucia and Cairns, 1969) raises the question of which enzymes actually participate in this process. If cells can grow in the absence of DNA polymerase, what other enzymes could be catalyzing DNA replication *in vivo?* Our only certain knowledge of DNA replication seems to be that it is semiconservative. As illustrated by the work of Speyer (1965), we do not even know the extent to which factors other than DNA base sequence might provide information in the replication process.

Problems of gene expression and its control in cells of higher organisms also remain quite obscure. How are genes turned on and off during embryonic development? What factors determine the initiation and termination of transcription and translation? How is DNA replication initiated, and how is it coordinated with cell division in the presence of a nuclear membrane which prevents direct contact between the chromosomes and the cell surface? What is the physical state of DNA in the chromosome? It may be true that we can provide partial—or nearly complete—answers to some of these questions for microbial systems. It is certainly true, as we have seen earlier in this chapter, that these systems provide suitable models for asking the same questions in eucaryotic cells, if applied judiciously. However, I believe that it is fatuous to conclude, from the present state of our knowlelge, that all we need to do now is to fill in the details around the framework so brilliantly provided by our scientific predecessors and contemporaries. Rather, I would agree with my mentor, Seymour Cohen (1968), that "the progress of science not only reveals new specific and general truths but also shows how little we do know." Although we have learned an enormous amount in the recent explosion of molecular biology, we still know very little, and there is much work yet to be done.

References

Aaronson, S. A., and Todaro, G. J. (1969). *Science* **166,** 390.

Abelson, J., and Thomas, C. A., Jr. (1966). *J. Mol. Biol.* **18,** 262.

Adams, J. M., Jeppeson, P. G. N., Sanger, F., and Barrell, B. G. (1969). *Nature* **223,** 1009.

Adams, M. H. (1959). "Bacteriophages," New York, Interscience.

Adams, M. H., and Wade, E. (1954). *J. Bact.* **68,** 320.

Adelberg, E. A., Mandel, M., and Chen, G. C. (1965). *Biochem. Biophys. Res. Comm.* **18,** 788.

Alberts, B. M., Amodio, F. J., Jenkins, M., Gutman, E. D., and Ferris, F. L. (1968). *Cold Spr. Hbr. Symp. Quant. Biol.* **33,** 289.

Alegria A. H., and Kahan, F. M. (1968). *Biochemistry* **7,** 1132.

Alegria, A. H., Kahan, F. M., and Marmur, J. (1968). *Biochemistry* **7,** 3179.

Ames, B. N., Dubin, D. T., and Rosenthal, S. M. (1958). *Science* **127,** 814.

Ammann, J., Delius H., and Hofschneider, P. H. (1964). *J. Mol. Biol.* **10,** 557.

Anderson, T. F. (1945). *J. Cellular Comp. Physiol.* **25,** 17.

Anderson, T. F. (1948). *J. Bact.* **55,** 637.

Anderson, T. F. (1950). *J. Appl. Phys.* **21,** 70.

Anderson, T. F. (1953). *Cold Spring Harbor Symp. Quant. Biol.* **18,** 197.

Anderson, T. F. (1960). *Proc. European Regional Conf. on Electron Microscopy, Delft,* **II,** 1008.

Ando, T., Takagi, J., Kosawa, T., and Ikeda, Y. (1969). *J. Biochem.* **66,** 1.

Anfinsen, C. B. (1961). *Federation Proc.* **20,** 634.

Note: Additional references added in proofs appear on pages 354–357.

Anraku, N., and Tomizawa, J. (1965). *J. Mol. Biol.* **11,** 501.
Aposhian, H. V., and Kornberg, A. (1962). *J. Biol. Chem.* **237,** 519.
Aposhian, H. V., and Tremblay, G. Y. (1966). *J. Biol. Chem.* **241,** 5095.
Arber, W. (1965). *J. Mol. Biol.* **11,** 247.
Astrachan, L., and Volkin, E. (1958). *Biochim. Biophys. Acta* **29,** 536.
August, J. T. (1969). *Nature* **222,** 121.
August, J. T., Cooper, S., Shapiro, L., and Zinder, N. D. (1963). *Cold Spr. Hbr. Symp. Quant. Biol.* **28,** 95.
August, J. T., Banerjee, A. K., Eoyang, L., de Fernandez, M. T. F., Hori, J., Kuo, C. H., Rensing, U., and Shapiro, L. (1968). *Cold Spr. Hbr. Symp. Quant. Biol.* **33,** 73.
August, J. T., Eoyang, L., de Fernandez, M. T. F., Hasegawa, Kuo, C. H., Rensing, U., and Shapiro, S. (1969). *J. Cell. Physiol.* **74** Supp. 1, 187.
Avery, O. T., MacLeod, C. M., and McCarty, M. (1944). *J. Exp. Med.* **79,** 137.
Bachrach, U., and Friedmann, A. (1967). *Biochem. Biophys. Res. Comm.* **26,** 596.
Bader, J. P. (1969). In H. B. Levy, ed., "The Biochemistry of Viruses," New York, Marcel Dekker, p. 293.
Baguley, B. C., Bergquist, P. L., and Ralph, R. K. (1967). *Biochem. Biophys. Acta* **138,** 51.
Baker, B. R. (1967a). "Design of Active-Site-Directed Irreversible Enzyme Inhibitors," New York, John Wiley and Sons, Inc.
Baker, B. R. (1967b). *J. Med. Chem.* **10,** 912.
Baker, R., and Tessman, I. (1967). *Proc. Natl. Acad. Sci. U.S.* **58,** 1438.
Baker, R., and Tessman, I. (1968). *J. Mol. Biol.* **35,** 439.
Baldwin, R. L., Barrand, P., Fritsch, A., Goldthwait, D. A., and Jacob, F. (1966). *J. Mol. Biol.* **17,** 343.
Barner, H. D., and Cohen, S. S. (1954). *J. Bact.* **68,** 80.
Barner, H. D., and Cohen, S. S. (1959). *J. Biol. Chem.* **234,** 2987.
Barrington, L. F., and Kozloff, L. M. (1956). *J. Biol. Chem.* **223,** 615.
Basilico, C., Matsuya, Y., and Green, H. (1969). *J. Virol.* **3,** 140.
Bautz, E. K. F., Bautz, F. A., and Dunn, J. J. (1969). *Nature* **223,** 1022.
Bautz, E. K. F., and Dunn, J. J. (1969). *Biochem. Biophys. Res. Comm.* **34,** 230.
Bautz, E. K. F., Kasai, T., Reilly, E., and Bautz, F. A. (1966). *Proc. Natl. Acad. Sci. U.S.* **55,** 1081.
Bautz, E. K. F., and Reilly, E. (1966). *Science* **151,** 328.
Bautz-Freese, E. (1961). *Proc. Natl. Acad. Sci. U.S.* **47,** 540.
Bayer, M. E. (1968). *J. Virology* **2,** 346.
Bear, P. D., and Skalka, A. (1969). *Proc. Natl. Acad. Sci. U.S.* **62,** 385.
Becker, A., and Hurwitz, J. (1967). *J. Biol. Chem.* **242,** 936.
Becker, A., Lyn, G., Gefter, M., and Hurwitz, J. (1967). *Proc. Natl. Acad. Sci. U.S.* **58,** 1996.
Beckwith, J. R. (1967). *Science* **156,** 597.
Bello, L. J., and Bessman, M. J. (1963a). *J. Biol. Chem.* **238,** 1777.
Bello, L. J., and Bessman, M. J. (1963b). *Biochim. Biophys. Acta* **72,** 647.
Benzer, S. (1952). *J. Bact.* **63,** 59.

Benzer, S. (1957). In "The Chemical Basis of Heredity," W. D. McElroy and B. Glass, eds., Baltimore, Johns Hopkins Press, p. 70.

Benzer, S. (1961). *Proc. Natl. Acad. Sci. U.S.* **47,** 403.

Benzer, S., and Champe, S. P. (1961). *Proc. Natl. Acad. Sci. U.S.* **47,** 1025.

Benzer, S., and Champe, S. P. (1962). *Proc. Natl. Acad. Sci. U.S.* **48,** 1114.

Berger, H. D., and Kozinski, A. W. (1969). *Proc. Natl. Acad. Sci. U.S.* **64,** 897.

Berger, H., Warren, A. J., and Fry, K. E. (1969). *J. Virol.* **3,** 171.

Berglund, O. (1969). *J. Biol. Chem.* **244,** 6306.

Berglund, O., Karlström, O., and Reichard, P. (1969). *Proc. Natl. Acad. Sci. U.S.* **62,** 829.

Bernstein, H. (1968). *Cold Spring Harbor Symp. Quant. Biol.* **33,** 325.

Bernstein, H., Edgar, R. S., and Denhardt, G. H. (1965). *Genetics* **51,** 987.

Bertino, J. R., Hillcoat, B. L., and Johns, D. G. (1967). *Federation Proc.* **26,** 893.

Bessman, M. J. (1959). *J. Biol. Chem.* **234,** 2735.

Bessman, M. J., and Bello, L. J. (1961). *J. Biol. Chem.* **236,** PC72.

Bessman, M. J., Herriott, S. T., and Orr, M. J. V. B. (1965). *J. Biol. Chem.* **240,** 439.

Billeter, M. A., Dahlberg, J. E., Goodman, H. M., Hindley, J., and Weissmann, C. (1969). *Nature* **224,** 1083.

Bishop, D. H. L., Pace, N. R., and Spiegelman, S. (1967). *Proc. Natl. Acad. Sci. U.S.* **58,** 1790.

Bishop, J. O., and Robertson, F. W. (1969). *Biochem. J.* **115,** 353.

Biswas, C., Hardy, J., and Beck, W. S. (1965). *J. Biol. Chem.* **240,** 3631.

Bleichrodt, J. F., and van Abkoude, E. R. (1967). *Virology* **32,** 93.

Bode, V. C., and Kaiser, A. D. (1965a). *J. Mol. Biol.* **14,** 399.

Bode, V. C., and Kaiser, A. D. (1965b). *Virology* **25,** 111.

Bode, V. C., and MacHattie, L. A. (1968). *J. Mol. Biol.* **32,** 673.

Bode, W. (1967). *J. Virol.* **1,** 948.

Boedtker, H. (1967). *Biochemistry* **6,** 2718.

Bolle, A., Epstein, R. H., Salser, W., and Geiduschek, E. P. (1968a). *J. Mol. Biol.* **33,** 339.

Bolle, A., Epstein, R. H., Salser, W., and Geiduschek, E. P. (1968b). *J. Mol. Biol.* **31,** 325.

Bonhoeffer, F., and Gierer, A. (1963). *J. Mol. Biol.* **7,** 534.

Borek, E., and Ryan, A. (1958). *Proc. Natl. Acad. Sci. U.S.* **44,** 374.

Borsa, J., and Graham, A. F. (1968). *Biochem. Biophys. Res. Comm.* **33,** 895.

Borst, P., Lees, J., Christ, E., and Slater, E. (1962). *Biochim. Biophys. Acta* **62,** 509.

Bose, S. K., and Nossal, N. G. (1964). *Federation Proc.* **23,** 272.

Bose, S. K., and Warren, R. J. (1969). *J. Virol.* **3,** 549.

Botstein, D. (1968). *J. Mol. Biol.* **34,** 621.

Boyce, R. P., and Howard-Flanders, P. (1964). *Proc. Natl. Acad. Sci. U.S.* **51,** 293.

Boyce, R. P., Kraiselburd, E., Ryan, S., and Chessin, H. (1969). *Virology* **37,** 679.

Boyle, J. M., Ritchie, D. A., and Symonds, N. (1965). *J. Mol. Biol.* **14,** 48.
Bradley, D. E. (1963). *J. Gen. Microbiol.* **31,** 435.
Bradley, D. E. (1967). *Bacteriol. Revs.* **31,** 231.
Bremer, H., Konrad, M., and Bruner, R. (1966). *J. Mol. Biol.* **16,** 104.
Bremer, H., Konrad, M., Gaines K., and Stent, G. S. (1965). *J. Mol. Biol.* **13,** 540.
Bremer, H., and Yuan, D. (1968a). *J. Mol. Biol.* **38,** 163.
Bremer, H., and Yuan, D. (1968b). *J. Mol. Biol.* **34,** 527.
Brenner, S., and Barnett, L. (1959). *Brookhaven Symposia in Biology* **12,** 86.
Brenner, S., Jacob, F., and Meselson, M. (1961). *Nature* **190,** 576.
Brenner, S., Streisinger, G., Horne, R. W., Champe, S. P., Barnett, L., Benzer, S., and Rees M. W. (1959). *J. Mol. Biol.* **1,** 281.
Brenner, S., Stretton, A. O. W., and Kaplan, S. (1965). *Nature* **206,** 994.
Britten, R. J., and Kohne, D. E. (1968). *Science* **161,** 529.
Brock, M. L. (1965). *Virology* **26,** 221.
Brodetsky, A. M., and Romig, W. R. (1965). *J. Bacteriol.* **90,** 1655.
Broker, T. R., Anraku, N., and Lehman, I. R. (1969). *Federation Proc.* **28,** 348.
Brooks, K. (1965). *Virology* **26,** 489.
Brooks, K., and Clark, A. J. (1967). *J. Virol.* **1,** 283.
Brown, D. T., and Anderson, T. F. (1969). *J. Virol.* **4,** 94.
Brown, L. R., and Dowell, C. E. (1968a). *J. Virol.* **2,** 1290.
Brown, L. R., and Dowell, C. E. (1968b). *J. Virol.* **2,** 1296.
Brownlee, G. G., and Sanger, F. (1967). *J. Mol. Biol.* **23,** 337.
Buchwald, M., and Siminovitch, L. (1969). *Virology* **38,** 1.
Bujard, H. (1969). *Proc. Natl. Acad. Sci. U.S.* **62,** 1167.
Buller, C. S., and Astrachan, L. (1968). *J. Virol.* **2,** 298.
Burgess, R. R., Travers, A. A., Dunn, J. J., and Bautz, E. K. F. (1969). *Nature* **221,** 43.
Burgi, E., and Hershey, A. D. (1963). *Biophys. J.* **3,** 309.
Burgi, E. Hershey, A. D., and Ingraham, L. (1966). *Virology* **28,** 11.
Burnet, F. M., and McKie, M. (1929). *Austral. J. Exp. Biol. Med. Sci.* **6,** 277.
Burton, A., and Sinsheimer, R. L. (1963). *Science* **142,** 961.
Burton, A., and Sinsheimer, R. L. (1965). *J. Mol. Biol.* **14,** 327.
Cairns, J. (1963). *J. Mol. Biol.* **6,** 208.
Cairns, J., Stent, G. S., and Watson, J. D., eds., (1966). "Phage and the Origins of Molecular Biology," Cold Spring Harbor, N.Y., Cold Spring Harbor Laboratory of Quantitative Biology.
Calef, E., Marchelli, C., and Guerrini, F. (1965). *Virology* **27,** 1.
Campbell, A. (1959). *Virology* **9,** 293.
Campbell, A. (1961). *Virology* **14,** 22.
Campbell, A. (1962). *Adv. in Genetics* **11,** 101.
Campbell, A., and Balbinder, E. (1958). *Carnegie Institution of Washington Year Book* **57,** 386.
Capecchi, M. R. (1966). *Proc. Natl. Acad. Sci. U.S.* **55,** 1517.
Capecchi, M. R. and Gussin, G. N. (1965). *Science* **149,** 417.
Carlson, K. (1968). *J. Virol.* **2,** 1230.

Caro, L. G., and Schnös, M. (1966). *Proc. Natl. Acad. Sci. U.S.* **56,** 126.
Carter, B. J., Shaw, B. D., and Smith, M. G. (1969). *Biochim. Biophys. Acta* **195,** 494.
Celis, J. E., and Conway, T. W. (1968). *Proc. Natl. Acad. Sci. U.S.* **59,** 923.
Cerdá-Olmedo, E., Hanawalt, P. C., and Guerola, N. (1968). *J. Mol. Biol.* **33,** 705.
Chamberlin, M., and Berg, P. (1962). *Proc. Natl. Acad. Sci. U.S.* **48,** 81.
Champe, S. P. (1963). *Ann. Rev. Microbiol.* **17,** 87.
Champoux, J. J., Egan, J. B., and Hogness, D. S. (1968). *Federation Proc.* **27,** 646.
Chandler, B., Hayashi, M., Hayashi, M.N., and Spiegelman, S. (1964). *Science* **143,** 47.
Changeux, J. P., and Thiéry, J. (1967). *J. Theoret. Biol.* **17,** 315.
Chaproniere-Rickenberg, D. M., Mahler, H. R., and Fraser, D. (1964). *Virology* **23,** 96.
Chesterton, C. J., and Green, M. H. (1968). *Biochem. Biophys. Res. Comm.* **31,** 919.
Chiu, C-S., and Greenberg, G. R. (1968). *Cold Spring Harbor Symp. Quant. Biol.* **33,** 351.
Chrispeels, M. J., Boyd, R. F., Williams, L. S., and Neidhardt, F. C. (1968). *J. Mol. Biol.* **31,** 463.
Cleaver, J. E. (1969). *Proc. Natl. Acad. Sci. U.S.* **63,** 428.
Cohen, L. W., and Levine, M. (1966). *Virology* **28,** 208.
Cohen, P. S. (1968). *J. Virology* **2,** 192.
Cohen, P. S., and Ennis, H. L. (1966). *J. Bact.* **92,** 1345.
Cohen, S. N., and Hurwitz, J. (1968). *J. Mol. Biol.* **37,** 387.
Cohen, S. S. (1947). *Cold Spring Harbor Symp. Quant. Biol.* **12,** 35.
Cohen, S. S. (1948). *J. Biol. Chem.* **174,** 295.
Cohen, S. S. (1951). *Nature* **168,** 746.
Cohen, S. S. (1961). *Federation Proc.* **20,** 641.
Cohen, S. S. (1966). In J. S. Colter and W. Paranchych, eds. "The Molecular Biology of Viruses," New York, Academic Press, p. 3.
Cohen, S. S. (1968). "Virus-Induced Enzymes," New York, Columbia Univ. Press.
Cohen, S. S., and Anderson, T. F. (1946). *J. Exptl. Med.* **84,** 511.
Cohen, S. S., and Barner, H.D. (1962). *J. Biol. Chem.* **237,** PC1376.
Cohen, S. S., Barner, H. D., and Lichtenstein, J. (1961). *J. Biol. Chem.* **236,** 1448.
Cohen, S. S., Flaks, J. G., Barner, H. D., Loeb, M. R., and Lichtenstein, J. (1958). *Proc. Natl. Acad. Sci. U.S.* **44,** 1004.
Cohen, S. S., and Fowler, C. B., *J. Exptl. Med.* **85,** 771 (1947).
Cohen, S. S., and Weed, L. L. (1954). *J. Biol. Chem.* **209,** 789.
Coolsma, J., and Haselkorn, R. (1969). *Biochem. Biophys. Res. Comm.* **34,** 253.
Cooper, S., and Helmstetter, C. E. (1967). *J. Mol. Biol.* **31,** 519.
Cooper, S., and Zinder, N. D. (1962). *Virology* **18,** 405.
Cota-Robles, E. H., and Coffman, M. D. (1963). *J. Bact.* **86,** 266.
Couse, N. L. (1968). *J. Virol.* **2,** 198.
Crawford, E. M., and Gesteland, R. F. (1964). *Virology* **22,** 165.

Crick, F. H. C., Barnett, L., Brenner, S., and Watts-Tobin, R. J. (1961). *Nature* **192,** 1227.
Cronan, J. E., Jr., and Wulff, D. L. (1969). *Virology* **38,** 241.
Cross, R. A., and Lieb, M. (1967). *Genetics* **57,** 549.
Crouch, R. J., Hall, B. D., and Hager, G. (1969). *Nature* **223,** 476.
Cummings, D. J. (1963a). *Biochim. Biophys. Acta* **68,** 472.
Cummings, D. J. (1963b). *Virology* **19,** 536.
Cummings, D. J. (1964). *Virology* **23,** 408.
Cummings, D. J., Chapman, V. A., and DeLong, S. S. (1968). *J. Virol.* **2,** 610.
Cummings, D. J., Chapman, V. A., and DeLong, S. S. (1969). *Virology* **37,** 94.
Cummings, D. J., and Kozloff, L. M. (1959). *Biochim. Biophys. Acta* **44,** 445.
Cunningham, C. C., and Hager, L. P. (1968). *Bacteriol. Proc.*, p. 133.
Dalgarno, L., and Sinsheimer, R. L. (1968). *J. Virol.* **2,** 822.
Dambly, C., Couturier, M., and Thomas, R. (1968) *J. Mol. Biol.* **32,** 67.
Daniel, V., Sarid, S., and Littauer, U. Z. (1968). *Proc. Natl. Acad. Sci. U.S.* **60,** 1403.
Davis, R. W., and Davidson, N. (1968). *Proc. Natl. Acad. Sci. U.S.* **60,** 243.
Davis, R. W., and Parkinson, J. S. (1969). *Federation Proc.* **28,** 531.
Davison, P. F., Friefelder, D., Hede, R., and Levinthal, C. (1961). *Proc. Natl. Acad. Sci. U.S.* **47,** 1123.
Delbrück, M. (1940). *J. Gen. Physiol.* **23,** 643.
Delbrück, M., and Bailey, W. T. (1946). *Cold Spring Harbor Symp. Quant. Biol.* **11,** 33.
del Campillo-Campbell, A., and Campbell, A. (1965). *Biochem. Zeitschrift* **342,** 485.
Delihas, N. (1961). *Virology* **13,** 242.
DeLucia, P., and Cairns, J. (1969). *Nature* **224,** 1164.
DeMars, R. I. (1953). *Nature* **172,** 964.
Denhardt, D. T. (1968). *Nature* **220,** 131.
Denhardt, D. T., and Burgess, A. B. (1968). *Cold Spr. Hbr. Symp. Quant. Biol.* **33,** 449.
Denhardt, D. T., Dressler, D. H., and Hathaway, A. (1967). *Proc. Natl. Acad. Sci. U.S.* **57,** 813.
Denhardt, D. T., and Sinsheimer, R. L. (1965a). *J. Mol. Biol.* **12,** 647.
Denhardt, D. T., and Sinsheimer, R. L. (1965b). *J. Mol. Biol.* **12,** 663.
Denhardt, D. T., and Sinsheimer, R. L. (1965c). *J. Mol. Biol.* **12,** 674.
de Waard, A., Paul, A. V., and Lehman, I. R. (1965). *Proc. Natl. Acad. Sci. U.S.* **54,** 1241.
de Waard, A., Ubbink, T. E. C. M., and Beukman, W. (1967). *European J. Biochem.* **2,** 303.
d'Herelle, F. (1926). "The Bacteriophage and its Behavior," Baltimore, Williams and Wilkins Co.
di Mauro, E., Snyder, L., Marino, P., Lamberti, A., Coppo, A., and Tocchini-Valentini, G. P. (1969). *Nature* **222,** 533.
Dirksen, M. L., Hutson, J. C., and Buchanan, J. M. (1963). *Proc. Natl. Acad. Sci. U.S.* **50,** 507.
Dirksen, M. L., Wiberg, J. S., Koerner, J. F., and Buchanan, J. M. (1960). *Proc. Natl. Acad. Sci. U.S.* **46,** 1425.

Doerfler, W. (1968). *Proc. Natl. Acad. Sci. U.S.* **60,** 636.
Doerfler, W., and Hogness, D. S. (1968a). *J. Mol. Biol.* **33,** 635.
Doerfler, W., and Hogness, D. S. (1968b). *J. Mol. Biol.* **33,** 661.
Doerfler, W., Zillig, W., Fuchs, E., and Albers, M. (1962). *Z. physiol. Chem.* **330,** 96.
Doermann, A. H. (1948a). *Carnegie Inst. Wash. Yearbook* **47,** 176.
Doermann, A. H. (1948b). *J. Bact.* **55,** 257.
Doermann. A. H. (1952). *J. Gen. Physiol.* **35,** 645.
Doermann, A. H., and Boehner, L. (1963). *Virology* **21,** 551.
Doermann, A. H., and Parma, D. H. (1967). *J. Cell. Physiol. supp. 1* **70,** 147.
Doi, R. H., and Spiegelman, S. (1962). *Science* **138,** 1270.
Dove, W. F. (1966). *J. Mol. Biol.* **19,** 187.
Dove, W. F. (1968). *Ann. Rev. Genet.* **2,** 305.
Dove, W. F. (1969a). *Virology* **38,** 349.
Dove, W. F., and Weigle, J. J. (1965). *J. Mol. Biol.* **12,** 620.
Drake, J. W. (1963). *J. Mol. Biol.* **6,** 268.
Drake, J. W. (1966). *Proc. Natl. Acad. Sci. U.S.* **55,** 738.
Drake, J. W. (1967). *Proc. Natl. Acad. Sci. U.S.* **58,** 962.
Drake, J. W., Allen, E. F., Forsberg, S. A., Preparata, R. M., and Greening, E. O. (1969). *Nature* **221,** 1128.
Drake, J. W., and McGuire, J. (1967). *Genetics* **55,** 387.
Dressler, D. H., and Denhardt, D. T. (1968). *Nature* **219,** 346.
Dubbs, D. R., and Kit, S. (1968). *J. Virol.* **2,** 1272.
Duckworth, D. H. (1969). *J. Virol.* **3,** 85.
Duckworth, D. H., and Bessman, M. J. (1965). *J. Bact.* **90,** 724.
Duckworth, D. H., and Bessman, M. J. (1967). *J. Biol. Chem.* **242,** 2877.
Dukes, P. P., and Kozloff, L. M. (1959). *J. Biol. Chem.* **234,** 534.
Dulbecco, R. (1952). *J. Bact.* **63,** 209.
Dulbecco, R. (1969). *Science* **166,** 962.
Dulbecco, R., Hartwell, L. H., and Vogt, M. (1965). *Proc. Natl. Acad. Sci. U.S.* **53,** 403.
Dunn, J. J. and Bautz, E. K. F. (1969). *Biochem. Biophys. Res. Comm.* **36,** 925.
Dyson, R. D., and Van Holde, K. E. (1967). *Virology* **33,** 559.
Earhart, C. F., Daniels, M., Tremblay, G. Y., and Schaechter, M. (1968). *Cold Spring Harbor Symp. Quant. Biol.* **33,** 707.
Earhart, C. F., and Neidhardt, F. C. (1967). *Virology* **33,** 694.
Ebisuzaki, K. (1965). *Virology* **26,** 390.
Ebisuzaki, K. (1969). *J. Mol. Biol.* **42,** 375.
Echols, H., and Gingery, R. (1968). *J. Mol. Biol.* **34,** 239.
Echols, H., Gingery, R., and Moore, L. (1968). *J. Mol. Biol.* **34,** 251.
Echols, H., and Joyner, A. (1968). In Fraenkel-Conrat, H., ed., "Molecular Basis of Virology," New York, Reinhold Book Corporation, p. 526.
Echols, H., Pilarski, L., and Cheng, P. Y. (1968). *Proc. Natl. Acad. Sci. U.S.* **59,** 1016.
Eckhart, W. (1966). *J. Mol. Biol.* **18,** 292.
Eckhart, W. (1969). *Nature* **224,** 1069.
Eddleman, H. L., and Champe, S. P. (1966). *Virology* **30,** 471.

Edgar, R. S. (1966). In "Phage and the Origins of Molecular Biology," J. Cairns, G. S. Stent, and J. D. Watson, eds., Cold Spring Harbor, N.Y., Cold Spring Harbor Laboratory of Quantitative Biology, p. 166.

Edgar, R. S. (1968). In Sober, H. A., ed. "Handbook of Biochemistry," Cleveland, The Chemical Rubber Co., p. I-27.

Edgar, R. S., Denhardt, G. H., and Epstein, R. H. (1964). *Genetics* **49,** 635.

Edgar, R. S., and Lielausis, I. (1964). *Genetics* **49,** 649.

Edgar, R. S., and Lielausis, I. (1965). *Genetics* **52,** 1187.

Edgar, R. S., and Lielausis, I. (1968). *J. Mol. Biol.* **32,** 263.

Edgar R. S., and Wood, W. B. (1966). *Proc. Natl. Acad. Sci. U.S.* **55,** 498.

Edgell, M. H., Hutchison, C. A. III, and Sinsheimer, R. L. (1969). *J. Mol. Biol.* **42,** 547.

Edlin, G. (1965a). *J. Mol. Biol.* **12,** 363.

Edlin, G. (1965b). *J. Mol. Biol.* **12,** 356.

Edlin, G., and Broda J. (1968). *Bacteriol. Revs.* **32,** 206.

Eggen, K., Oeschger, M. P., and Nathans, D. (1967). *Biochem. Biophys. Res. Comm.* **28,** 587.

Eigner, J., and Block, S. (1968). *J. Virol.* **2,** 320.

Eikhom, T. S., and Spiegelman, S. (1967). *Proc. Natl. Acad. Sci. U.S.* **57,** 1833.

Eikhom, T. S., Stockley, D. J., and Spiegelman, S. (1968). *Proc. Natl. Acad. Sci. U.S.* **59,** 506.

Eisen, H. A., Fuerst, C. R., Siminovitch, L., Thomas, R., Lambert, L., Pereira da Silva, L., and Jacob, F. (1966). *Virology* **30,** 224.

Eisen, H., Pereira da Silva, L. H., and Jacob, F. (1968). *Cold Spr. Hbr. Symp. Quant. Biol.* **33,** 755.

Eisen, H. A., Siminovitch, L., and Mohide, P. T. (1968). *Virology* **34,** 97.

Eisen, H. A., Tallan, I., and Siminovitch, L. (1968). *Virology* **34,** 104.

Eisenstadt, J. M., and Lengyel, P. (1966). *Science* **154,** 524.

Eisenstark, A., Eisenstark, R., and Van Sickle, R. (1965). *Mutation Res.* **2,** 1.

Eiserling, F. A., Bolle, A., and Epstein, R. H. (1967). *Virology* **33,** 405.

Elford, W. J., and Andrewes, C. H. (1932). *Brit. J. Exp. Pathol.* **13,** 446.

Ellis, E. L., and Delbrück, M. (1939). *J. Gen. Physiol.* **22,** 365.

Emrich, J. (1968). *Virology* **35,** 158.

Emrich, J., and Streisinger, G. (1968). *Virology* **36,** 387.

Engelhardt, D. L., Robertson, H. D., and Zinder, N. D. (1968). *Proc. Natl. Acad. Sci. U.S.* **59,** 972.

Engelhardt, D. L., Webster, R. E., Wilhelm, R. C., and Zinder, N.D. (1965). *Proc. Natl. Acad. Sci. U.S.* **54,** 1791.

Engelhardt, D. L., Webster, R. E., and Zinder, N. D. (1967). *J. Mol. Biol.* **29,** 45.

Eoyang, L., and August, J. T. (1968). In L. Grossman and K. Moldave, eds., "Methods in Enzymology, vol. XII. B." New York, Academic Press, p. 530.

Epstein, H. T. (1967). *Biochem. Biophys. Res. Comm.* **27,** 258.

Epstein, R. H., Bolle, A., Steinberg, C. M., Kellenberger, E., Boy de la Tour, E., Chevalley, R., Edgar, R. S., Susman, M., Denhardt, G. H., and Lielausis, A. (1963). *Cold Spring Harbor Symp. Quant. Biol.* **28,** 375.

Erikson, R. L. (1968). *Ann. Rev. Microbiol.* **22,** 305.

Erikson, R. L., and Szybalski, W. (1964). *Virology* **111,** 124.
Eskridge, R. W., Weinfeld, H., and Paigen, K. (1967). *J. Bact.* **93,** 835.
Falkow, S., and Cowie, D. B. (1968). *J. Bact.* **96,** 777.
Fangman, W. L., and Feiss, M. (1969). *J. Mol. Biol.* **44,** 103.
Fareed, G. C., and Richardson, C. C. (1968). *Proc. Natl. Acad. Sci. U.S.* **58,** 665.
Farid, S. A. A., and Kozloff, L. M. (1968). *J. Virol.* **2,** 308.
Fattig, W. D., and Lanni, F. (1965). *Genetics* **51,** 157.
Feiner, L. Sadowski, P., Gold M., and Hurwitz, J. (1968). *Federation Proc.* **27,** 395.
Feix, G., Pollet, R., and Weissmann, C. (1968). *Proc. Natl. Acad. Sci. U.S.* **59,** 145.
Fenwick, M. L., Erikson, R. L., and Franklin, R. M. (1964). *Science* **146,** 527.
FerroLuzzi-Ames, G., and Ames, B. N. (1965). *Biochem. Biophys. Res. Comm.* **18,** 639.
Fielding, P. E., and Lunt, M. R. (1969). *FEBS Letters* **5,** 214.
Fiers, W., and Sinsheimer, R. L. (1962). *J. Mol. Biol.* **5,** 408.
Fischer-Fantuzzi, L. (1967). *Virology* **32,** 18.
Flaks, J. G., and Cohen, S. S. (1957). *Biochim. Biophys. Acta* **25,** 667.
Flaks, J. G., Lichtenstein, J., and Cohen, S. S. (1959). *J. Biol. Chem.* **234,** 1507.
Fleming, W. H., and Bessman, M. J. (1965). *J. Biol. Chem.* **240,** PC4109.
Fleming, W. H., and Bessman, M. J. (1967). *J. Biol. Chem.* **242,** 363.
Földes, J., and Trautner, T. A. (1964). *Z. Vererbungslehre* **95,** 57.
Fraenkel-Conrat, H., ed. (1968). "Molecular Basis of Virology," New York, Reinhold Book Corporation.
Fraenkel-Conrat, H. (1968). In "Molecular Basis of Virology." H. Fraenkel-Conrat, ed., New York, Reinhold Book Corporation, p. 134.
Fraenkel-Conrat, H., and Weissmann, C. (1968). In H. Fraenkel-Conrat, ed., "Molecular Basis of Virology," New York, Reinhold Book Corporation, p. 209.
Frankel, F. R. (1966a). *J. Mol. Biol.* **18,** 127.
Frankel, F. R. (1966b). *J. Mol. Biol.* **18,** 109.
Frankel, F. R. (1966c). *J. Mol. Biol.* **18,** 144.
Frankel, F. R. (1968a). *Cold Spring Harbor Symp. Quant. Biol.* **33,** 485.
Frankel, F. R. (1968b). *Proc Natl. Acad. Sci. U.S.* **59,** 131.
Frankel, F. R., Majumdar, C., Weintraub, S., and Frankel, D. M. (1968). *Cold Spring Harbor Symp. Quant. Biol.* **33,** 495.
Franklin, N. C. (1967). *Genetics* **57,** 301.
Franklin, N. C., Dove, W. F., and Yanofsky, C. (1965). *Biochem. Biophys. Res. Comm.* **18,** 910.
Fraser, D., and Williams, R. C. (1953). *J. Bact.* **65,** 458.
Freese, E. (1959). *J. Mol. Biol.* **1,** 87.
Freese, E., Bautz, E., and Bautz-Freese, E. (1961). *Proc. Natl. Acad. Sci. U.S.* **47,** 845.
Freese, E., Bautz-Freese, E., and Bautz, E. (1961). *J. Mol. Biol.* **3,** 133.
Freifelder, D. R., and Freifelder, D. (1968). *J. Mol. Biol.* **32,** 25.
French, R. C., and Siminovitch, L. (1955). *Can. J. Microbiol.* **1,** 757.

Friedberg, E. C., and King, J. J. (1969). *Biochem. Biophys. Res. Comm.* **37,** 646.

Friedkin, M., and Kornberg, A. (1957). In W. D. McElroy and B. Glass, eds., "The Chemical Basis of Heredity," Baltimore, Johns Hopkins Press, p. 609.

Friesen, J. D. (1969). *J. Bact.* **97,** 313.

Fuerst, C. R. (1966). *Virology* **30,** 581.

Fujinaga, K., and Green, M. (1966). *Proc. Natl. Acad. Sci. U.S.* **55,** 1567.

Furrow, M. H., and Pizer, L. I. (1968). *J. Virol.* **2,** 594.

Furth, J. J., Hurwitz, J., and Anders, M. (1962). *J. Biol. Chem.* **237,** 2611.

Furth, J. J., and Pizer, L. I. (1966). *J. Mol. Biol.* **15,** 124.

Gage, L. P., and Geiduschek, E. P. (1967). *J. Mol. Biol.* **30,** 435.

Gallucci, E., and Garen, A. (1966). *J. Mol. Biol.* **15,** 193.

Gamow, R. I. (1969). *J. Virol.* **4,** 113.

Gamow, R. I., and Kozloff, L. M. (1968). *J. Virol.* **2,** 480.

Garen, A. (1961). *Virology* **14,** 151.

Garen, A. (1968). *Science* **160,** 149.

Garen, A., Garen, S., and Wilhelm, R. C. (1965). *J. Mol. Biol.* **14,** 167.

Gariglio, P., Roodman, S. T., and Green, M. H. (1969). *Biochem. Biophys. Res. Comm.* **37,** 945.

Gefter, M. L., Hausmann, R., Gold, M., and Hurwitz, J. (1966). *J. Biol. Chem.* **241,** 1995.

Geiduschek, E. P., and Sklar, J. (1969). *Nature* **221,** 833.

Geiduschek, E. P., Snyder, L., Colvill, A. J. E., and Sarnat, M. (1966). *J. Mol. Biol.* **19,** 541.

Gelfand, D. H., and Hayashi, M. (1969). *J. Mol. Biol.* **44,** 501.

Gellert, M. (1967). *Proc. Natl. Acad. Sci. U.S.* **57,** 148.

Georgopoulos, C. P. (1967). *Biochem. Biophys. Res. Comm.* **28,** 179.

Georgopoulos, C. P. (1968). *Virology* **34,** 364.

Geraci, G., Rossi, M., and Scarano, E. (1967). *Biochemistry* **6,** 183.

Gesteland, R. F., Salser, W., and Bolle, A. (1967). *Proc. Natl. Acad. Sci. U.S.* **58,** 2036.

Gilbert, W., and Dressler, D. (1968). *Cold Spring Harbor Symp. Quant. Biol.* **33,** 473.

Gingery, R., and Echols, H. (1967). *Proc. Natl. Acad. Sci. U.S.* **58,** 1507.

Glover, S. W., Schell, J. Symonds, N., and Stacey, K. A. (1963). *Genet. Res.* **4,** 480.

Godson, G. N., and Sinsheimer, R. L. (1967), *J. Mol. Biol.* **23,** 495.

Gold L. M., Schweiger, M. (1969a). *Proc. Natl. Acad. Sci. U.S.* **62,** 892.

Gold, L. M., and Schweiger, M. (1969b). *J. Biol. Chem.* **244,** 5100

Gold, M., Hausmann, R. L., Maitra, U., and Hurwitz, J. (1964). *Proc. Natl. Sci. U.S.* **52,** 292.

Gold, M., and Hurwitz, J., *Cold Spr. Hbr. Symp. Quant. Biol.* **28,** 149 (1963).

Goldberg, A. R., and Howe, M. (1969). *Virology* **38,** 200.

Goldberg, E. B. (1966). *Proc. Natl. Acad. Sci. U.S.* **56,** 1457.

Goldthwait, D. A., and Jacob, F. (1964). *Compt. Rend.* **258,** 661.

Gottesman, M. E., and Yarmolinsky, M. B. (1968a). *J. Mol. Biol.* **31,** 487.

Gottesman, M., and Yarmolinsky, M. (1968b). *Cold Spr. Hbs. Symp. Quant. Biol.* **33,** 735.

Gottschalk, A. (1960). "The Chemistry and Biology of Sialic Acids and Related Substances," Cambridge, England, Cambridge University Press.
Gottschling, H., and Heidelberger, C. (1963). *J. Mol. Biol.* **7,** 541.
Gough, M., and Levine, M. (1968). *Genetics* **58,** 161.
Goulian, M., Kornberg, A., and Sinsheimer, R. L. (1967). *Proc. Natl. Acad. Sci. U.S.* **58,** 2321.
Goulian, M., Lucas, Z. J., and Kornberg, A. (1968). *J. Biol. Chem.* **243,** 627.
Granboulan, N., and Franklin, R. M. (1966). *J. Mol. Biol.* **22,** 173.
Grasso, R. J., and Buchanan, J. M. (1969). *Nature* **224,** 882.
Grasso, R. J., and Paigen, K. (1969). *Virology* **38,** 191.
Gratia, A. (1936). *Comptes Rendu Soc. Biol.* **122,** 812.
Grau, O., Guha, A., Geiduschek, E. P., and Szybalski, W. (1969). *Nature* **224,** 1105.
Green, D. M. (1964). *J. Mol. Biol.* **10,** 438.
Green, D. M. (1966). *J. Mol. Biol.* **22,** 1.
Green, M., and Cohen, S. S. (1957). *J. Biol. Chem.* **228,** 601.
Green, M. H. (1964). *Proc. Natl. Acad. Sci. U.S.* **52,** 1388.
Green, M. H., Gotchel, B., Hendershott, J., and Kennel, S. (1967). *Proc. Natl. Acad. Sci. U.S.* **58,** 2343.
Greenberg, G. R. (1966). *Proc. Natl. Acad. Sci. U.S.* **56,** 1226.
Greenberg, G. R., and Somerville, R. L. (1962). *Proc. Natl. Acad. Sci. U.S.* **48,** 247.
Greenberg, G. R., Somerville, R. L., and De Wolf, S. (1962). *Proc. Natl. Acad. Sci. U.S.* **48,** 262.
Greene, R., and Korn, D. (1967). *J. Mol. Biol.* **28,** 435.
Grunau, J. A., and Daniel, V. (1968). *Biochim. Biophys. Acta* **166,** 728.
Guarneros, G., and Echols, H. (1970). *J. Mol. Biol.* **47,** 565.
Guerrini, F. (1969). *J. Mol. Biol.* **46,** 523.
Guha, A., and Szybalski, W. (1968). *Virology* **34,** 608.
Guild, W. R. (1968). *Cold Spring Harbor Symp. Quan. Biol.* **33,** 142.
Gussin, G. N. (1966). *J. Mol. Biol.* **21,** 435.
Guttman, B. S., and Begley, L. (1968). *Virology* **36,** 687.
Haas, M., and Yoshikawa, H. (1969a). *J. Virol.* **3,** 233.
Haas, M., and Yoshikawa, H. (1969b). *J. Virol.* **3,** 248.
Haas, M., and Yoshikawa, H. (1969c). *J. Virol.* **4,** 844.
Haber, K. (1966). *Biochem. Biophys. Res. Commun.* **23,** 502.
Hall, B. D., and Crouch, R. (1967). In S. G. Laland and L. O. Fröholm, eds., "The Biochemistry of Virus Replication," New York, Academic Press, p. 49.
Hall, B. D., Nygaard, A. P., and Green, M. H. (1964). *J. Mol. Biol.* **9,** 143.
Hall, B. D., and Spiegelman, S. (1961). *Proc. Natl. Acad. Sci. U.S.* **47,** 137.
Hall, D. H. (1967). *Proc. Natl. Acad. Sci. U.S.* **58,** 584.
Hall, D. H., and Tessman, I. (1966). *Virology* **29,** 339.
Hall, D. H., Tessman, I., and Karlström O. (1967). *Virology* **31,** 442.
Harm, W. (1963). *Virology* **19,** 66.
Harm, W., and Rupert, C. S. (1963). *Z. Vererbungslehre* **94,** 336.
Harris, A. W., Mount, D. W. A., Fuerst, C. R., and Siminovitch, L. (1967). *Virology* **32,** 553.

Harris, R. J., Panter, R. A., and Symons, R. H. (1968). *Biochim. Biophys. Acta* **161,** 291.
Haruna, I., Nozu, K., Ohtaka, Y., and Spiegelman, S. (1963). *Proc. Natl. Acad. Sci. U.S.* **50,** 905.
Haruna, I., and Spiegelman, S. (1965a). *Proc. Natl. Acad. Sci. U.S.* **54,** 579.
Haruna, I., and Spiegelman, S. (1965b). *Science* **150,** 884.
Haselkorn, R., Baldi, M. I., and Doskoçil, J. (1967). In Laland, S. G., and Fröholm, L. O., eds., "The Biochemistry of Virus Replication," New York, Academic Press, p. 79.
Haselkorn, R., Vogel, M., and Brown, R. D. (1969). *Nature* **221,** 836.
Haslam, E. A., Roscoe, D. H., and Tucker, R. G. (1967). *Biochim. Biophys. Acta* **134,** 312.
Hatanaka, M., Twiddy, E., and Gilden, R. (1969). *J. Virol.* **4,** 801.
Hattman, S., and Fukusawa, T. (1963). *Proc. Natl. Acad. Sci. U.S.* **50,** 297.
Hattman, S., and Hofschneider, P. H. (1967). *J. Mol. Biol.* **29,** 173.
Hausmann, R. (1967). *J. Virol.* **1,** 57.
Hausmann, R. (1968). *Biochem. Biophys. Res. Comm.* **31,** 609.
Hausmann, R., and Gold, M. (1966). *J. Biol. Chem.* **241,** 1985.
Hausmann, R., and Gomez, B. (1967). *J. Virol.* **1,** 779.
Hausmann, R., and Gomez, B. (1968). *J. Virol.* **2,** 265.
Hausmann, R., Gomez, B., and Moody, B. (1968). *J. Virol.* **2,** 335.
Hausmann, R., and LaRue, K. (1969). *J. Virol.* **3,** 278.
Hayashi, M. N., and Hayashi, M. (1968). *Proc. Natl. Acad. Sci. U.S.* **61,** 1107.
Hayashi, M., Hayashi, M. N., and Spiegelman, S. (1963). *Proc. Natl. Acad. Sci. U.S.* **50,** 664.
Hayashi, M., Hayashi, M. N., and Spiegelman, S. (1964). *Proc. Natl. Acad. Sci. U.S.* **51,** 351.
Hayes, W. (1968). "The Genetics of Bacteria and their Viruses," 2nd edition, New York, John Wiley and Sons, Inc.
Hayward, W. S., and Green, M. H. (1965). *Proc. Natl. Acad. Sci. U.S.* **54,** 1675.
Hayward, W. S., and Green, M. H. (1969). *Proc. Natl. Acad. Sci. U.S.* **64,** 962.
Haywood, A. M., and Sinsheimer, R. L. (1963). *J. Mol. Biol.* **6,** 247.
Hendrickson, H. E., and McCorquodale, D. J. (1970). *Federation Proc.* **29,** 466A.
Henry, T. J., and Pratt, D. (1969). *Proc. Natl. Acad. Sci. U.S.* **62,** 800.
Hershey, A. D. (1953). *J. Gen. Physiol.* **37,** 1.
Hershey, A. D. (1955). *Virology* **1,** 108.
Hershey, A. D. (1957). *Virology* **4,** 237.
Hershey, A. D. (1964). *Carnegie Inst. Wash. Yearbook* **63,** 577.
Hershey, A. D., Burgi, E., Davern, C. I. (1965). *Biochem. Biophys. Res. Comm.* **18,** 675.
Hershey, A. D., Burgi, E., and Ingraham, L. L. (1963). *Proc. Natl. Acad. Sci. U.S.* **49,** 748.
Hershey, A. D., and Chase, M. (1951). *Cold Spring Harbor Symp. Quant. Biol.* **16,** 471.

Hershey, A. D., and Chase, M. (1952). *J. Gen. Physiol.* **36,** 39.
Hershey, A. D., Dixon, J., and Chase, M. (1953). *J. Gen. Physiol.* **36,** 777.
Hershey, A. D., Garen, A., Fraser, D. K., and Hudis, J.D. (1954). *Carnegie Inst. Wash. Yearbook* **53,** 210.
Hershey, A. D., Kalmanson, G., and Bronfenbrenner, J. (1944). *J. Immunol.* **48,** 221.
Hershey, A. D., Kamen, M. D., Kennedy, J. W., and Gest, H. (1951). *J. Gen. Physiol.* **34,** 305.
Hessler, A. Y., Barnes, M. B., and Baird, J. P. (1967). *J. Virol.* **1,** 543.
Hilleman, M. R. (1969). *Science* **164,** 506.
Hindley, J., and Staples, D. H. (1969). *Nature* **224,** 964.
Hinton, D. M., and Bode, V. C. (1969). *Federation Proc.* **28,** 532.
Hiraga, S., Igarashi, K., and Yura, T. (1967). *Biochim. Biophys. Acta* **145,** 41.
Hirsch-Kauffmann, M., and Sauerbier, W. (1968). *Molec. Gen. Genetics* **102,** 89.
Hodgetts, R. B., and Rubenstein, I. (1968). *J. Mol. Biol.* **35,** 413.
Hoffman, D. B., Jr., and Rubenstein, I. (1968a). *J. Mol. Biol.* **35,** 375.
Hoffman, D. B., Jr., and Rubenstein, I. (1968b). *J. Mol. Biol.* **35,** 401.
Hoffmann-Berling, H., and Mazé, R. (1964). *Virology* **22,** 305.
Höfschneider, P. H., and Häusen, P. (1968). In H. Fraenkel-Conrat, ed., "Molecular Basis of Virology," New York, Reinhold Book Corporation, p. 169.
Höfschneider, P. H., and Preuss, A. (1963). *J. Mol. Biol.* **7,** 450.
Hogness, D. S., Doerfler, W., Egan, J. B., and Black, L. W. (1967). In J. S. Colter and W. Paranchych, eds., "The Molecular Biology of Viruses," New York, Academic Press, p. 91.
Hogness, D. S., and Simmons, J. R. (1964). *J. Mol. Biol.* **9,** 411.
Horiuchi, K., and Adelberg, E. A. (1965). *J. Bacteriol.* **89,** 1231.
Horiuchi, K., Lodish, H. F., and Zinder, N. D. (1966). *Virology* **28,** 438.
Horiuchi, K., and Zinder, N. D. (1967). *Science* **156,** 1618.
Horowitz, N. H., and Leopold, U. (1951). *Cold Spring Harbor Symp. Quant. Biol.* **16,** 65.
Hosoda, J. (1967). *Biochem. Biophys. Res. Comm.* **27,** 294.
Hosoda, J., and Levinthal, C. (1968). *Virology* **34,** 709.
Hosoda, J., and Mathews, E. (1968). *Proc. Natl. Acad. Sci. U.S.* **61,** 997.
Howard, B. D. (1967). *Science* **158,** 1588.
Howard, B. D., and Tessman, I. (1964). *J. Mol. Biol.* **9,** 372.
Hradecna, Z., and Szybalski, W. (1967). *Virology* **32,** 633.
Hradecna, Z., and Szybalski, W. (1969). *Virology* **38,** 473.
Hsu, W-T., Foft, J. W., and Weiss, S. B. (1967). *Proc. Natl. Acad. Sci. U.S.* **58,** 2028.
Hsu, W-T., and Weiss, S. B. (1969). *Proc. Natl. Acad. Sci. U.S.* **64,** 345.
Huberman, J. (1968). *Cold Spring Harbor Symp. Quant. Biol.* **33,** 509.
Huebner, R. J. (1970). Lecture delivered at the University of Arizona, Jan. 22, 1970.
Huebner, R. J., and Todaro, G. J. (1969). *Proc. Natl. Acad. Sci. U.S.* **64,** 1087.
Huh, T. Y., and Helleiner, C. W. (1968). *Biochim. Biophys. Acta* **155,** 107.
Hung, P. P., and Overby, L. R. (1968). *J. Biol. Chem.* **243,** 5525.

Hutchison, C. A. III, and Sinsheimer, R. L. (1966). *J. Mol. Biol.* **18,** 429.
Iglewski, W. T., and Franklin, R. M. (1967). *Proc. Natl. Acad. Sci. U.S.* **58,** 1018.
Ikeda, H., and Tomizawa, J. (1965). *J. Mol. Biol.* **14,** 85.
Ikeda, H., and Tomizawa, J. (1968). *Cold Spr. Hbr. Symp. Quant. Biol.* **33,** 791.
Inouye, M., Akaboshi, E., Tsugita, A., Streisinger, G., and Okada, Y. (1967). *J. Mol. Biol.* **30,** 39.
Inouye, M., and Tsugita, A. (1968). *J. Mol. Biol.* **37,** 213.
Ippen, K. A., and Valentine, R. C. (1966). *Biochem. Biophys. Res. Comm.* **24,** 880.
Ishiwa, H., and Tessman, I. (1968). *J. Mol. Biol.* **37,** 467.
Jacob, F., Brenner, S., and Cuzin, F. (1963). *Cold Spring Harbor Symp. Quant. Biol.* **28,** 329.
Jacob, F., Fuerst, C. R., and Wollman, E. L. (1957). *Ann. Inst. Pasteur* **93,** 724.
Jacob, F., and Monod, J. (1961). *J. Mol. Biol.* **3,** 318.
Jacob, F., Sussman, R., and Monod, J. (1962). *C.R. Acad. Sci. Paris* **254,** 4214.
Jacob, F., and Wollman, E. L. (1954). *Ann. Inst. Pasteur* **87,** 653.
Jacob, F., and Wollman, E. L. (1957). In "The Chemical Basis of Heredity," W. D. McElroy and B. Glass, eds., Baltimore, Johns Hopkins Press, p. 468.
Jacob, F., and Wollman, E. L. (1961). "Sexuality and the Genetics of Bacteria," New York, Academic Press.
Jacquemin-Sablon, A., and Richardson, C. C. (1968). *Federation Proc.* **27,** 396.
Jesaitis, M. A. (1955). *Nature* **178,** 637.
Jones, O. W., and Berg, P. (1966). *J. Mol. Biol.* **22,** 199.
Jorgensen, S. E., and Koerner, J. F. (1965). *Federation Proc.* **24,** 349.
Jorgensen, S. E., Koerner, J. F., Snustad, D. P., and Warner, H. R. (1970). *Federation Proc.,* **29,** 465; Warner, H. R. *et al., J. Virol.* **5,** 738.
Josse, J., and Kornberg, A. (1962). *J. Biol. Chem.* **237,** 1968.
Joyner, A., Isaacs, L. N., Echols, H., and Sly, W. S. (1966). *J. Mol. Biol.* **19,** 174.
Kaempfer, R. O. R., and Magasanik, B. (1967). *J. Mol. Biol.* **27,** 453.
Kaempfer, R. O. R., and Sarkar, S. (1967). *J. Mol. Biol.* **27,** 469.
Kahan, F. M. (1963). *Federation Proc.* **22,** 406.
Kahan, F., Kahan, E., and Riddle, B. (1964). *Federation Proc.* **23,** 318.
Kaiser, A. D. (1957). *Virology* **3,** 42.
Kaiser, A. D. (1962). *J. Mol. Biol.* **4,** 275.
Kaiser, A. D., and Hogness, D. S. (1960). *J. Mol. Biol.* **2,** 392.
Kaiser, A. D., and Jacob, F. (1957). *Virology* **4,** 509.
Kaiser, A. D., and Wu, R. (1968). *Cold Spr. Hbr. Symp. Quant. Biol.* **33,** 729.
Kallen, R. G., Simon, M., and Marmur, J. (1962). *J. Mol. Biol.* **5,** 248.
Kamen, R. (1969). *Nature* **221,** 321.
Kammen, H. O., and Strand, M. (1967). *J. Biol. Chem.* **242,** 1854.
Kan, J., Kano-Sueoka, T., and Sueoka, N. (1968). *J. Biol. Chem.* **243,** 5584.
Kanner, L., and Kozloff, L. M. (1964). *Biochemistry* **3,** 215.

Kano-Sueoka, T., Nirenberg, M., and Sueoka, N. (1968). *J. Mol. Biol.* **35,** 1.
Kano-Sueoka, T., and Spiegelman, S. (1962). *Proc. Natl. Acad. Sci. U.S.* **48,** 1942.
Kano-Sueoka, T., and Sueoka, N. (1966). *J. Mol. Biol.* **20,** 183.
Kano-Sueoka, T., and Sueoka, N. (1968). *J. Mol. Biol.* **37,** 475.
Kano-Sueoka, T., and Sueoka, N. (1969). *Proc. Natl. Acad. Sci. U.S.* **62,** 1229.
Kara, J., and Weil, R. (1967). *Proc. Natl. Acad. Sci. U.S.* **57,** 63.
Karam, J. D. (1969). *Biochem. Biophys. Res. Comm.* **37,** 416.
Kasai, T., and Bautz, E. K. F. (1967). In H. J. Vogel, J. O. Lampen, and V. Bryson, eds., "Organizational Biosynthesis," New York, Academic Press, p. 111.
Kasai, T., and Bautz, E. K. F. (1969). *J. Mol. Biol.* **41,** 401.
Kasai, T., Bautz, E. K. F., Guha, A., and Szybalski, W. (1968). *J. Mol. Biol.* **34,** 709.
Kates, M. (1964). *Adv. Lipid Research* **2,** 17.
Kayajanian, G. (1968). *Virology* **36,** 30.
Keck, K., Mahler, H. R., and Fraser, D. (1959). *Arch. Biochem. Biophys.* **86,** 85.
Kellenberger, E. (1962). *Adv. Virus Research* **8,** 1.
Kellenberger, G., and Kellenberger, E. (1957). *Virology* **3,** 275.
Kellenberger, E., Lark, K. G., and Bolle, A. (1962). *Proc. Natl. Acad. Sci. U.S.* **48,** 1860.
Kellenberger, E., and Séchaud, J. (1957). *Virology* **3,** 256.
Kellenberger, E., Séchaud, J., and Ryter, A. (1959). *Virology* **8,** 478.
Kellenberger, G., Zichichi, M. L., and Weigle, J. (1961). *J. Mol. Biol.* **3,** 399.
Kelly, R. B., and Sinsheimer, R. L. (1964). *J. Mol. Biol.* **8,** 602.
Kelly, R. B., and Sinsheimer, R. L. (1967a). *J. Mol. Biol.* **29,** 237.
Kelly, R. B., and Sinsheimer, R. L. (1967b). *J. Mol. Biol.* **26,** 169.
Kelly, T. J., Jr. (1968). *Federation Proc.* **27,** 591.
Kelly, T. J., Jr., and Thomas, C. A., Jr. (1969). *J. Mol. Biol.* **44,** 459.
Kemp, C. L., Howatson, A. F., and Siminovitch, L. (1968). *Virology* **36,** 490.
Kennell, D. (1968). *J. Virol.* **2,** 1262.
Khesin, R. B., Shemyakin, M. F., Gorlenko, Z. M., Bogdanova, S. L., and Afanaseva, T. P. (1962). *Biokhimiya* **27,** 1092.
Khorana, H. G., Buchi, H., Ghosh, H., Gupta, N., Jacob, T. M., Kossel, H., Morgan, R., Narang, S. A., Ohtsuka, E., and Wells, R. D. (1966). *Cold Spring Harbor Symp. Quant. Biol.* **31,** 39.
Kiger, J. A., Jr., and Sinsheimer, R. L. (1969). *J. Mol. Biol.* **40,** 467.
Kiger, J. A., Jr., Young, E. T., and Sinsheimer, R. L. (1968). *J. Mol. Biol.* **33,** 395.
King, J. (1968). *J. Mol. Biol.* **32,** 231.
King, J., and Wood, W. B. (1969). *J. Mol. Biol.* **39,** 583.
Kirby, E. P., Jacob, F., and Goldthwait, D. A. (1967). *Proc. Natl. Acad. Sci. U.S.* **58,** 1903.
Kit, S. (1967). In "Subviral Carcinogenesis," Monograph of the 1st International Symposium on Tumor Viruses, p. 116.
Klein, A., Mathews, C. K., Eisenstadt, J. M., and Lengyel, P. (1967). *Biochim. Biophys. Acta* **149,** 622.

Klein, A., Mathews, C. K., Eisenstadt, J. M., and Lengyel, P. (1968). *Biochim. Biophys. Acta* **169,** 543.
Kleinschmidt, A. K., Burton, A., and Sinsheimer, R. L. (1963). *Science* **142,** 961.
Kleinschmidt, A. K., Lang, D., Jacherts, D., and Zahn, R. K. (1962). *Biochim. Biophys. Acta* **61,** 857.
Kleinschmidt, A. K., and Zahn, R. K. (1959). *Z Naturforschung* **14,** 730.
Klug, A., and Berger, J. E. (1964). *J. Mol. Biol.* **10,** 565.
Knippers, R., Komano, T., and Sinsheimer, R. L. (1968). *Proc. Natl. Acad. Sci. U.S.* **59,** 577.
Knippers, R., Razin, A., Davis, R., and Sinsheimer, R. L. (1969). *J. Mol. Biol.* **45,** 237.
Knippers, R., Salivar, W. O., Newbold, J. E., and Sinsheimer, R. L. (1969). *J. Mol. Biol.* **39,** 641.
Knippers, R., and Sinsheimer, R. L. (1968a). *J. Mol. Biol.* **35,** 591.
Knippers, R., and Sinsheimer, R. L. (1968b). *J. Mol. Biol.* **34,** 17.
Knippers, R., Whalley, J. M., and Sinsheimer, R. L. (1969). *Proc. Natl. Acad. Sci. U.S.* **64,** 275.
Knolle, P., and Kaudewitz, F. (1963). *Biochem. Biophys. Res. Comm.* **11,** 383.
Koch, G., and Weidel, W. (1956). *Z. Naturforschung* **116,** 345.
Koerner, J. F., Smith, M. S., and Buchanan, J. M. (1959). *J. Am. Chem. Soc.* **81,** 2594.
Komano, T., Knippers, R., and Sinsheimer, R. L. (1968). *Proc. Natl. Acad. Sci. U.S.* **59,** 911.
Konisky, J., and Nomura, M. (1967). *J. Mol. Biol.* **26,** 181.
Konrad, M. W. (1968). *Proc. Natl. Acad. Sci. U.S.* **59,** 171.
Korn, D., and Weissbach, A. (1963). *J. Biol. Chem.* **238,** 3390.
Kornberg, A. (1961). "Enzymatic Synthesis of DNA," New York, John Wiley and Sons.
Kornberg, A. (1969). *Science* **163,** 1410.
Kornberg, A., Zimmerman, S. B., Kornberg, S. R., and Josse, J. (1959). *Proc. Natl. Acad. Sci. U.S.* **45,** 772.
Kornberg, S. R., Zimmerman, S. B., and Kornberg, A. (1961). *J. Biol. Chem.* **236,** 1487.
Kourilsky, P., Marcaud, L., Sheldrick, P., Luzzati, D., and Gros, F. (1968). *Proc. Natl. Acad. Sci. U.S.* **61,** 1013.
Kozinski, A. W. (1961a). *Virology* **13,** 124.
Kozinski, A. W. (1961b). *Virology* **13,** 377.
Kozinski, A. W. (1969). *Biochem. Biophys. Res. Comm.* **35,** 294.
Kozinski, A. W., and Bessman, M. J. (1961). *J. Mol. Biol.* **3,** 746.
Kozinski, A. W., and Felgenhauer, Z. Z. (1967). *J. Virol.* **1,** 1193.
Kozinski, A. W., and Kozinski, P. B. (1967). *Proc. Natl. Acad. Sci. U.S.* **57,** 1705.
Kozinski, A. W., and Kozinski, P. B. (1968). *Biochem. Biophys. Res. Comm.* **33,** 670.
Kozinski, A. W., Kozinski, P. B., and James, R. (1967). *J. Virol.* **1,** 758.
Kozinski, A. W., and Lin, T. H. (1965). *Proc. Natl. Acad. Sci. U.S.* **54,** 273.

Kozinski, A. W., and Lorkiewicz, Z. K. (1967). *Proc. Natl. Acad. Sci. U.S.* **58,** 2109.
Kozinski, A. W., and Mitchell, M. (1969). *J. Virology* **4,** 823.
Kozloff, L. M. (1953). *Cold Spring Harbor Symp. Quant. Biol.* **18,** 209.
Kozloff, L. M. (1960). *Ann. Reviews of Biochem.* **29,** 475.
Kozloff, L. M. (1968). In "Molecular Basis of Virology," H. Fraenkel-Conrat, ed., New York, Reinhold Book Corporation, p. 435.
Kozloff, L. M., and Lute, M. (1959). *J. Biol. Chem.* **234,** 539.
Kozloff, L. M., and Lute, M. (1960). *Biochim. Biophys. Acta* **37,** 420.
Kozloff, L. M., and Lute, M. (1965). *J. Mol. Biol.* **12,** 780.
Kozloff, L. M., Lute, M., Crosby, L. K., Wong, R., and Stern, B. (1969). *J. Virol.* **3,** 217.
Krakoff, I. H., Brown, N. C., and Reichard, P. (1968). *Cancer Res.* **28,** 1559.
Krieg, D. R. (1959). *Virology* **8,** 80.
Krimm, S., and Anderson, T. F. (1967). *J. Mol. Biol.* **27,** 197.
Krueger, A. P. (1931). *J. Gen. Physiol.* **14,** 493.
Kumar, S., Bøvre, K., Guha, A., Hradecna, Z., Maher, Sr. V. M., and Szybalski, W. (1969). *Nature* **221,** 823.
Kunkee, R. E., and Pardee, A. B. (1956). *Biochim. Biophys. Acta* **19,** 236.
Kuno, S., and Lehman, I. R. (1962). *J. Biol. Chem.* **237,** 1266.
Kutter, E. M., and Wiberg, J. S. (1968). *J. Mol. Biol.* **38,** 395.
Kutter, E. M., and Wiberg, J. S. (1969). *J. Virology* **4,** 439.
Labaw, L. W. (1953). *J. Bact.* **66,** 429.
Landy, A., and Spiegelman, S. (1968). *Biochemistry* **7,** 585.
Lanni, Y. T. (1960). *Virology* **10,** 514.
Lanni, Y. T. (1965). *Proc. Natl. Acad. Sci. U.S.* **53,** 969.
Lanni, Y. T. (1968). *Bact. Reviews* **32,** 227.
Lanni, Y. T. (1969). *J. Mol. Biol.* **44,** 173.
Lanni, F., and Lanni, Y. T. (1966). *J. Bact.* **92,** 521.
Lanni, Y. T., Lanni, F., and Tevethia, M. J. (1966). *Science* **152,** 208.
Lanni, Y. T., and McCorquodale, D. J. (1963). *Virology* **19,** 72.
Lanni, Y. T., McCorquodale, D. J., and Wilson, C. M. (1964). *J. Mol. Biol.* **10,** 19.
Lanni, Y. T., and Szybalski, W. (1969). *Bacteriol. Proceedings,* p. 192.
Lark, C. (1968). *J. Mol. Biol.* **31,** 401.
Lark, K. G. (1966). *Bacteriol. Rev.* **30,** 1.
Lederberg, E. M. (1951). *Genetics* **36,** 560.
Lehman, I. R., and Pratt, E. A. (1960). *J. Biol. Chem.* **235,** 3254.
Lehman, I. R., Roussos, G. G., and Pratt, E. A. (1962). *J. Biol. Chem.* **237,** 829.
Lehninger, A. L. (1965). "The Mitochondrion," New York, W. A. Benjamin, Inc.
Leive, L. (1965). *Proc. Natl. Acad. Sci. U.S.* **53,** 745.
Lembach, K. J., Kuninaka, A., and Buchanan, J. M. (1969). *Proc. Natl. Acad. Sci. U.S.* **62,** 446.
Lengyel, P., and Söll, D. (1969). *Bacteriol. Rev.* **33,** 264.
Lennarz, W. (1967). *Advan. Lipid Res.* **4,** 175.
Lennox, E. S. (1955). *Virology* **1,** 190.

Le Pecq, J.-B., and Baldwin, R. L. (1968). *Cold Spr. Hbr. Symp. Quant. Biol.* **33,** 609.
Levine, A. J., and Sinsheimer, R. L. (1968). *J. Mol. Biol.* **32,** 567.
Levine, A. J., and Sinsheimer, R. L. (1969a). *J. Mol. Biol.* **39,** 619.
Levine, A. J., and Sinsheimer, R. L. (1969b). *J. Mol. Biol.* **39,** 655.
Levine, A. J., and Sinsheimer, R. L. (1969c). *Proc. Natl. Acad. Sci. U.S.* **62,** 1226.
Levine, L., Barlow, J. L., and Van Vunakis, H. (1958). *Virology* **6,** 702.
Levine, M., and Smith, H. O. (1964). *Science* **146,** 1581.
Levinthal, C., Hosoda, J., and Shub, D. (1967). In J. S. Colter and W. Paranchych, eds., "The Molecular Biology of Viruses," New York, Acad. Press, p. 71.
Levisohn, R., and Spiegelman, S. (1968). *Proc. Natl. Acad. Sci. U.S.* **60,** 866.
Levisohn, R., and Spiegelman, S. (1969). *Proc. Natl. Acad. Sci. U.S.* **63,** 805.
Lichtenstein, J., and Cohen, S. S. (1960). *J. Biol. Chem.* **235,** 1134.
Lindqvist, B. H., and Sinsheimer, R. L. (1967a), *J. Mol. Bioi.* **28,** 87.
Lindqvist, B. H., and Sinsheimer, R. L. (1967b), *J. Mol. Biol.* **30,** 69.
Lindqvist, B. H., and Sinsheimer, R. L. (1968). *J. Mol. Biol.* **32,** 285.
Ling, C. M., Hung, P. P., and Overby, L. R. (1969). *Biochemistry* **8,** 4464.
Lipton, A., and Weissbach, A. (1966). *J. Mol. Biol.* **21,** 517.
Little, J. (1967). *J. Biol. Chem.* **242,** 679.
Lodish, H. F., and Zinder, N. D. (1966a). *J. Mol. Biol.* **19,** 333.
Lodish, H. F., and Zinder, N. D. (1966b). *Science* **152,** 372.
Loeb, T. (1960). *Science* **131,** 932.
Loeb, T., and Zinder, N. D. (1961). *Proc. Natl. Acad. Sci. U.S.* **47,** 282.
Lomax, M. I. S., and Greenberg, G. R. (1967). *J. Biol. Chem.* **242,** 109.
Loveless, A., and Howarth, S. (1959). *Nature* **184,** 1780.
Lozeron, H. A., and Szybalski, W. (1969). *Virology* **39,** 373.
Luftig, R., and Haselkorn, R. (1968). *Virology* **34,** 675.
Luftig, R., and Wood, W. (1969). Unpublished data cited by R. L. Sinsheimer, *J. Cell. Physiol.* **74** Supp. 1, p. 26.
Luria, S. E. (1947). *Proc. Natl. Acad. Sci. U.S.* **33,** 253.
Luria, S. E. (1951). *Cold Spring Harbor Symp. Quant. Biol.* **16,** 463.
Luria, S. E. (1965). *Biochem. Biophys. Res. Comm.* **18,** 735.
Luria, S. E., and Anderson, T. F. (1942). *Proc. Natl. Acad. Sci. U.S.* **28,** 127.
Luria, S. E., and Darnell, J. E. (1967). "General Virology," 2nd ed. New York, John Wiley and Sons, Inc.
Luria, S. E., and Human, M. L. (1950). *J. Bact.* **59,** 551.
Luria, S. E., and Latarjet, R. (1947). *J. Bact.* **53,** 149.
Lwoff, A. (1966). *Science* **152,** 1216.
Lwoff, A., and Gutmann, A. (1950). *Ann. Inst. Pasteur* **78,** 711.
Lwoff, A., Siminovitch, L., and Kjeldgaard, N. (1951). *C. R. Acad. Sci. Paris* **231,** 190.
MacHattie, L. A., Ritchie, D. A., Thomas, C. A., Jr., and Richardson, C. C. (1967). *J. Mol. Biol.* **23,** 355.
MacHattie, L. A., and Thomas, C. A., Jr. (1964). *Science* **144,** 1142.
Mackal, R. P., and Kozloff, L. M. (1954). *J. Biol. Chem.* **209,** 83.
Mackinlay, A. G., and Kaiser, A. D. (1969). *J. Mol. Biol.* **39,** 679.

Neidhardt, F. C., and Earhart, C. F. (1966). *Cold Spr. Hbr. Symp. Quant. Biol.* **31,** 557.
Neidhardt, F. C., Marchin, G. L., McClain, W. H., Boyd, R. F., and Earhart, C. F. (1969). *J. Cell Physiol.* **74** Supp. 1, p. 87.
Neu, H. C., and Heppel, L. A. (1964). *Biochem. Biophys. Res. Comm.* **14,** 109.
Newman, J., and Hanawalt, P. (1968). *J. Mol. Biol.* **35,** 639.
Nichols, J. L. (1970). *Nature* **225,** 147.
Nirenberg, M., Leder, P., Bernfield, M., Brimacombe, R., Trupin, J., Rottman, F., and O'Neal, C. (1965). *Proc. Natl. Acad. Sci. U.S.* **53,** 1161.
Nishihara, M., Chrambach, A., and Aposhian, H. V. (1967). *Biochemistry* **6,** 1877.
Nishihara, M., Friedman, N., and Aposhian, H. V. (1969). *J. Virol.* **3,** 164.
Nomura, M. (1963). *Cold Spring Harbor Symp. Quant. Biol.* **28,** 315.
Nomura, M. (1967). *Ann. Rev. Microbiol.* **21,** 257.
Nomura, M., Hall, B. D., and Spiegelman, S. (1960). *J. Mol. Biol.* **2,** 306.
Nomura, M., Matsubara, K., Okamoto, K., and Fujimura, R. (1962). *J. Mol. Biol.* **5,** 535.
Nomura, M., Okamoto, K., and Asano, K. (1962). *J. Mol. Biol.* **4,** 376.
Nomura, M., Witten, C., Mantei, N., and Echols, H. (1966). *J. Mol. Biol.* **17,** 273.
Nossal, N. G. (1969). *J. Biol. Chem.* **244,** 218.
Novick, A., and Szilard, L. (1951). *Science* **113,** 34.
Novogrodsky, A., and Hurwitz, J. (1966). *J. Biol. Chem.* **241,** 2923.
Novogrodsky, A., Tal, M., Traub, A., and Hurwitz, J. (1966). *J. Biol. Chem.* **241,** 2933.
Nygaard, A. P., and Hall, B. D. (1964). *J. Mol. Biol.* **9,** 125.
Ogawa, H., and Tomizawa, J. (1967a). *J. Mol. Biol.* **23,** 265.
Ogawa, T., and Tomizawa, J. (1967b). *J. Mol. Biol.* **23,** 225.
Ohtaka, Y., and Spiegelman, S. (1963). *Science* **142,** 493.
Oishi, M. (1968). *Proc. Natl. Acad. Sci. U.S.* **60,** 1000.
Okada, Y., Streisinger, G., Emrich, J., Newton, J., Tsugita, A., and Inouye, M. (1968). *Science* **162,** 807.
Okada, Y., Strensinger, G., Emrich, J., Tsugita, A., and Inouye, M. (1969). *J. Mol. Biol.* **40,** 299.
Okada, Y., Terzaghi, E., Streisinger, G., Emrich, J., Inouye, M., and Tsugita, A. (1966). *Proc. Natl. Acad. Sci. U.S.* **56,** 1692.
Okamoto K., Mudd, J. A., Mangan, J., Huang, W. M., Subbaiah, T. V., and Marmur, J. (1968). *J. Mol. Biol.* **34,** 413.
Okamoto, K., Mudd, J. A., and Marmur, J. (1968). *J. Mol. Biol.* **34,** 429.
Okamoto, K., Sugino, Y., and Nomura, M. (1962). *J. Mol. Biol.* **5,** 527.
Okazaki, R., and Kornberg, A. (1964). *J. Biol. Chem.* **239,** 269.
Okazaki, R., Okazaki, T., Sakabe, K., Sugimoto, K., and Sugino, A. (1968). *Proc. Natl. Acad. Sci. U.S.* **59,** 598.
Okubo, S., Strauss, B., and Stodolsky, M. (1964). *Virology* **24,** 552.
Oleson, A. E., and Koerner, J. F. (1964). *J. Biol. Chem.* **239,** 2935.
Oleson, A. E., Pispa, J. P., and Buchanan, J. M. (1969). *Proc. Natl. Acad. Sci. U.S.* **63,** 473.

Olivera, B. M., and Lehman, I. R. (1967). *Proc. Natl. Acad. Sci. U.S.* **57,** 1700.

Olsnes, S., and Hauge, J. G. (1968). *European J. Biochem.* **7,** 128.

Orr, C. W. M., Herriott, S. T., and Bessman, M. J. (1965). *J. Biol. Chem.* **240,** 4652.

Pace, N. R., and Spiegelman, S. (1966). *Science* **153,** 64.

Packman, S., and Sly, W. S. (1968). *Virology* **34,** 778.

Palchoudhury, S. R., and Poddar, R. K. (1968). *J. Mol. Biol.* **32,** 505.

Pardee, A. B., and Williams, I. (1952). *Arch. Biochem. Biophys.* **40,** 222.

Parkinson, J. S. (1968). *Genetics* **59,** 311.

Paul, A. V., and Lehman, I. R. (1966). *J. Biol. Chem.* **241,** 3441.

Pène, J. J., and Marmur, J. (1967). *J. Virol* **1,** 86.

Pereira da Silva, L. H., and Jacob, F. (1967). *Virology* **33,** 618.

Peterson, R. H. F., and Buller, C. S. (1969). *J. Virol.* **3,** 463.

Pirrotta, V., and Ptashne, M. (1969). *Nature* **222,** 541.

Pizer, L. I., and Cohen, S. S. (1962). *J. Biol. Chem.* **237,** 1251.

Pollet, R., Knolle, P., and Weissmann, C. (1967). *Proc. Natl. Acad. Sci. U.S.* **58,** 766.

Polsinelli, M., Milanesi, G., and Ganesan, A. T. (1969). *Science* **166,** 243.

Pratt, D., and Erdahl, W. S. (1968). *J. Mol. Biol.* **37,** 181.

Pratt, D., Tzagoloff, H., and Beaudoin, J. (1969). *Virology* **39,** 42.

Price, A. R., and Warner, H. R. (1968). *Virology* **36,** 523.

Pricer, W. E., and Weissbach, A. (1967). *J. Biol. Chem.* **242,** 1701.

Protass, J. J., and Korn, D. (1966). *J. Biol. Chem.* **241,** 4175.

Ptashne, M. (1965). *J. Mol. Biol.* **11,** 90.

Ptashne, M. (1967a). *Proc. Natl. Acad. Sci. U.S.* **57,** 306.

Ptashne, M. (1967b). *Nature* **214,** 232.

Ptashne, M., and Hopkins, N. (1968). *Proc. Natl. Acad. Sci. U.S.* **60,** 1282.

Puck, T. T., and Lee, H. H. (1954). *J. Exptl. Med.* **99,** 481.

Puck, T. T., and Lee, H. H. (1955). *J. Exptl. Med.* **101,** 151.

Quigley, J. W., and Cohen, S. S. (1969). *J. Biol. Chem.* **244,** 2450.

Radding, C. M. (1964). *Proc. Natl. Acad. Sci. U.S.* **52,** 965.

Radding, C. M. (1966). *J. Mol. Biol.* **18,** 235.

Radding, C. M. (1970). *J. Mol. Biol.*, in press.

Radding, C. M., and Echols, H. (1968). *Proc. Natl. Acad. Sci. U.S.* **60,** 707.

Radding, C. M., and Shreffler, D. C. (1966). *J. Mol. Biol.* **18,** 251.

Radding, C. M., Szpirer, J., and Thomas, R. (1967). *Proc. Natl. Acad. Sci. U.S.* **57,** 277.

Rao, R. N., and Smith, H. O. (1968). *Virology* **36,** 328.

Ravin, V. K., and Poluchina, G. N. (1968). *Virology* **35,** 616.

Ray, D. S. (1968). In "Molecular Basis of Virology," H. Fraenkel-Conrat, ed., New York, Reinhold Book Corporation, p. 222.

Ray, D. S. (1969). *J. Mol. Biol.* **43,** 631.

Ray, D. S., Bscheider, H-P., and Hofschneider, P. H. (1966). *J. Mol. Biol.* **21,** 473.

Ray, D. S., and Schekman, R. W. (1969). *J. Mol. Biol.* **43,** 645.

Reichard, P. (1968). *European J. Biochem.* **3,** 259.

Reichard, P., Baldesten, A., and Rutberg, L. (1961). *J. Biol. Chem.* **236,** 1150.

Reilly, B. E., and Spizizen, J. (1965). *J. Bact.* **89,** 782.

Revel, H. R. (1967). *Virology* **31,** 1967.

Revel, H. R., Hattman, S., and Luria, S. E. (1965). *Biochem. Biophys. Res. Comm.* **18,** 545.

Reznikoff, W. S., and Thomas, C. A., Jr. (1969). *Virology* **37,** 309.

Rhoades, M., MaçHattie, L. A., and Thomas, C. A., Jr. (1968). *J. Mol. Biol.* **37,** 21.

Rhoades, M., and Thomas, C. A., Jr. (1968). *J. Mol. Biol.* **37,** 41.

Richardson, C. C. (1965). *Proc. Natl. Acad. Sci. U.S.* **54,** 158.

Richardson, C. C. (1966). *J. Biol. Chem.* **241,** 2084.

Richardson, J. P. (1966). *J. Mol. Biol.* **21,** 83.

Riva, S. C. (1969). *Biochem. Biophys. Res. Comm.* **34,** 824.

Riva, S., and Geiduschek, E. P. (1969). *Federation Proc.* **28,** 660.

Riva, S., Polsinelli, M., and Falaschi, A. (1968). *J. Mol. Biol.* **35,** 347.

Robbins, P. W., Keller, I. M., Wright, A. W., and Bernstein, R. (1965). *J. Biol. Chem.* **240,** 384.

Roberts, J. W. (1969a). *Nature* **223,** 480.

Roberts, J. W. (1969b). *Nature* **224,** 1168.

Roberts, J. W., and Steitz, J. A. (1967). *Proc. Natl. Acad. Sci. U.S.* **58,** 1416.

Roizman, B. P. (1969). In H. B. Levy, ed., "The Biochemistry of Viruses," New York, Marcel Dekker, p. 415.

Roller, A. (1964). *J. Mol. Biol.* **9,** 260.

Roscoe, D. H., and Tucker, R. G. (1964). *Biochem. Biophys, Res. Comm.* **16,** 106.

Roscoe, D. H., and Tucker, R. G. (1966). *Virology* **29,** 157.

Rosenberg, E. (1965). *Proc. Natl. Acad. Sci. U.S.* **53,** 836.

Roth, T. F., and Hayashi, M. (1966). *Science* **154,** 658.

Roth, T. F., and Helinski, D. R. (1967). *Proc. Natl. Acad. Sci. U.S.* **58,** 650.

Rothman, J. L. (1965). *J. Mol. Biol.* **12,** 892.

Rouvière, J., Wyngaarden, J., Cantoni, J., Gros, F., and Kepes, A. (1968). *Biochim. Biophys. Acta* **166,** 94.

Rubenstein, I. (1968). *Virology* **36,** 356.

Rubenstein, I., Thomas, C. A., Jr., and Hershey, A. D. (1961). *Proc. Natl. Acad. Sci. U.S.* **47,** 1113.

Rueckert, R. R., and Zillig, W. (1962). *J. Mol. Biol.* **5,** 1.

Rush, M. G., Kleinschmidt, A. K., Hellmann, W., and Warner, R. C. (1967). *Proc. Natl. Acad. Sci. U.S.* **58,** 1676.

Ruska, H. (1940). *Naturwiss.* **28,** 45.

Rüst, P., and Sinsheimer, R. L. (1967). *J. Mol. Biol.* **23,** 545.

Rutberg, B., and Rutberg, L. (1965). *J. Bact.* **90,** 891.

Rutberg, B., and Rutberg, L. (1966). *J. Bact.* **91,** 76.

Sadowski, P., Ginsberg, B., Yudelevich, A., Feiner, L., and Hurwitz, J. (1968). *Cold Spr. Hbr. Symp. Quant. Biol.* **33,** 165.

Sadowski, P. D., and Hurwitz, J. (1969a). *J. Biol. Chem.* **244,** 6182.

Sadowski, P. D., and Hurwitz, J. (1969b). *J. Biol. Chem.* **244,** 6192.

Sadowski, P. D., and Kerr, C. (1970). *J. Virol.,* **6,** 149.

Sagik, B. P., Green, M. H., Hayashi, M., and Spiegelman, S. (1962). *Biophys. J.* **2,** 409.

Salivar, W. O., Henry, T. J., and Pratt, D. (1967). *Virology* **32,** 41.
Salivar, W. O., and Sinsheimer, R. L. (1969). *J. Mol. Biol.* **41,** 39.
Salser, W., Gesteland, R. F., and Bolle, A. (1967). *Nature* **215,** 588.
Salzman, L. A., and Weissbach, A. (1967). *J. Mol. Biol.* **28,** 53.
Sambrook, J.,and Shatkin, A. (1969). *J. Virol.* **4,** 719.
Sarabhai, A., and Brenner, S. (1967). *J. Mol. Biol.* **27,** 145.
Sarabhai, A. S., Stretton, A. O. W., Brenner, S., and Bolle, A. (1964). *Nature* 201, 13.
Sarkar, N., Sarkar, S., and Kozloff, L. M. (1964). *Biochemistry* **3,** 511.
Sarkar, S., Sarkar, N., and Kozloff, L. M. (1964). *Biochemistry* **3,** 517.
Sauerbier, W., Puck, S. M., Braütigam, A. R., and Hirsch-Kauffmann, M. (1969). *J. Virol.* **4,** 742.
Schachtele, C. F., Anderson, D. L., and Rogers, P. (1968). *J. Mol. Biol.* **33,** 861.
Schildkraut, C. L., Wierzkowski, K. L., Marmur, J., Green, D. M., and Doty, P. (1962). *Virology* **18,** 43.
Schlesinger, M. (1932). *Z. Hyg. Infektionskrank.* **114,** 149. (English translation in "Papers on Bacterial Viruses," G. S. Stent, ed., Boston, Little, Brown, and Company, 1960).
Schlesinger, M. (1934). *Biochem. Z.* **273,** 306.
Schlesinger, M. (1936). *Nature* **138,** 508.
Schuster, R. C., Breitman, T. R., and Weissbach, A. (1967). *J. Biol. Chem.* **242,** 3723.
Schuster, R. C., and Weissbach, A. (1968a). *J. Virology* **2,** 1096.
Schuster, R. C., and Weissbach, A. (1968b). *Biochem. Biophys. Res. Comm.* **33,** 514.
Schwartz, F. M., and Zinder, N. D. (1963). *Virology* **21,** 276.
Schweiger, M., and Gold, L. M. (1969). *Proc. Natl. Acad. Sci. U.S.* **63,** 1351.
Scott, J. R. (1968). *Virology* **36,** 564.
Scotti, P. D. (1969). *Proc. Natl. Acad. Sci. U.S.* **62,** 1093.
Séchaud, J., and Kellenberger, E. (1956). *Ann. Inst. Pasteur* **90,** 102.
Séchaud, J., Kellenberger, E., and Streisinger, G. (1967). *Virology* **33,** 402.
Séchaud, J., Streisinger, G., Emrich, J., Newton, J., Lanford, H., Reinhold, H., and Stahl, M. M. (1965). *Proc. Natl. Acad. Sci. U.S.* **54,** 1333.
Sedat, J. W., Kelly, R. B., and Sinsheimer, R. L. (1967). *J. Mol. Biol.* **27,** 537.
Seifert, W., Qasba, P., Walter, G., Palm, P., Schachner, M., and Zillig, W. (1969). *European J. Biochem.* **9,** 319.
Sekiguchi, M. (1966). *J. Mol. Biol.* **16,** 503.
Sekiguchi, M., and Cohen, S. S. (1963). *J. Biol. Chem.* **238,** 349.
Sekiguchi, M., and Cohen, S. S. (1964). *J. Mol. Biol.* **8,** 638.
Sellin, H. G., Srinivasan, P. R., and Borek, E. (1966). *J. Mol. Biol.* **19,** 219.
Sercarz, E. (1966). *Virology* **28,** 339.
Sertic, V. (1929). *Comp. Rend. Soc. Biol.* **100,** 477.
Setlow, R. B. (1966). *Science* **153,** 379.
Setlow, R. B., and Carrier, W. L. (1964). *Proc. Natl. Acad. Sci. U.S.* **51,** 226.
Setlow, R. B., and Carrier, W. L. (1966). *Biophys. Soc. Abstr.* **6,** 68.
Shahn, E., and Kozinski, A. W. (1966). *Virology* **30,** 455.
Shalitin, C. (1967). *J. Virology* **1,** 569.

Shalitin C., and Ben-Ishai, R. (1966). *Proc. Israel J. Chem.* **4,** 61p.
Shapiro, D., and Kozloff, L. M. (1969). *Bacteriological Proc.*, p. 162.
Shapiro, D. M., Eigner, J., and Greenberg, G. R. (1965). *Proc. Natl. Acad. Sci. U.S.* **53,** 874.
Shapiro, J., MacHattie, L., Eron, L., Ihler, G., Ippen, K., and Beckwith, J. (1969). **224,** 768.
Sharp, K., and Green, M. H. (1968). *J. Bact.* **96,** 111.
Shatkin, A. J. (1968). In H. Fraenkel-Conrat, ed., "Molecular Basis of Virology," New York, Reinhold Book Corporation, p. 351.
Shedlovsky, A., and Brenner, S. (1963). *Proc. Natl. Acad. Sci. U.S.* **50,** 300.
Shleser, R., Puga, A., and Tessman, E. S. (1969). *J. Virol.* **4,** 394.
Short, E. C., Jr., and Koerner, J. F. (1965). *Proc. Natl. Acad. Sci. U.S.* **54,** 595.
Short, E. C., Jr., and Koerner, J. F. (1969). *J. Biol. Chem.* **244,** 1487.
Shortman, K., and Lehman, I. R. (1964). *J. Biol. Chem.* **239,** 2964.
Siegel, J. E. D., and Hayashi, M. (1967). *J. Mol. Biol.* **27,** 443.
Siegel, J. E. D., and Hayashi, M. (1969). *J. Virol.* **4,** 400.
Signer, E. R. (1968). *Ann. Rev. Microbiol.* **22,** 451.
Signer, E., Echols, H., Weil, J., Radding, C., and Shulman, M., Moore, L., and Manly, K. (1968). *Cold Spr. Symp. Quant. Biol.* **33,** 711.
Signer, E. R., and Weil, J. (1968a). *J. Mol. Biol.* **34,** 261.
Signer, E., and Weil, J. (1968b). *Cold Spr. Hbr. Sypm. Quant. Biol.* **33,** 715.
Signer, E. R., Weil, J., and Kimball, P. C. (1969). *J. Mol. Biol.* **46,** 543.
Silver, S. (1965). *Proc. Natl. Acad. Sci. U.S.* **53,** 24.
Silver, S. (1967). *J. Mol. Biol.* **29,** 191.
Silver, S., Levine, E., and Spielman, P. M. (1968). *J. Virol.* **2,** 763.
Simon, E. (1966). *Science* **150,** 760.
Simon, E. H., and Tessman, I. (1963). *Proc. Natl. Acad. Sci. U.S.* **50,** 526.
Simon, L. D. (1969). *Virology* **38,** 285.
Simon, L. D., and Anderson, T. F. (1967a). *Virology* **32,** 297.
Simon, L. D., and Anderson, T. F. (1967b). *Virology* **32,** 279.
Simon, M. N., and Davidson, N. (1969). *Federation Proc.* **28,** 531.
Sinsheimer, R. L. (1954). *Science* **120,** 551.
Sinsheimer, R. L. (1959). *J. Mol. Biol.* **1,** 43.
Sinsheimer, R. L. (1961). *Proc. Welch Foundation Conf. Chem. Res.* ch. V, p. 227.
Sinsheimer, R. L., Hutchison, C. A. III, and Lindqvist, B. H. (1967). In J. S. Colter and W. Paranchych, eds., "The Molecular Biology of Viruses," New York, Academic Press, p. 175.
Sinsheimer, R. L., Knippers, R., and Komano, T. (1968). *Cold Spr. Hbr. Symp. Quant. Biol.* **33,** 443.
Sinsheimer, R. L., Starman, B., Nagler, C., and Guthrie, S. (1962). *J. Mol. Biol.* **4,** 142.
Sippel, A., and Hartmann, G. (1968). *Biochem. Biophys. Acta* **157,** 218.
Skalka, A. (1966). *Proc. Natl. Acad. Sci. U.S.* **55,** 1190.
Skalka, A. (1969). *J. Virol.* **3,** 150.
Skalka, A., Burgi, E., and Hershey, A. D. (1968). *J. Mol. Biol.* **34,** 1.

Skalka, A., Butler, B., and Echols, H. (1967). *Proc. Natl. Acad. Sci. U.S.* **58,** 576.
Sköld, O., and Buchanan, J. M. (1964). *Proc. Natl. Acad. Sci. U.S.* **51,** 553.
Sly, W. S., Eisen, H. A., and Siminovitch, L. (1968). *Virology* **34,** 112.
Sly, W. S., and Rabideau, K. (1969). *J. Mol. Biol.* **42,** 385.
Smith, D. W., and Hanawalt, P. C. (1967). *Biochim. Biophys. Acta* **149,** 519.
Smith, F. L., and Haselkorn, R. (1969). *Cold Spr. Hbr. Symp. Quant. Biol.* **34,** 91.
Smith, H. O. (1968). *Virology* **34,** 203.
Smith, H. O., and Levine, M. (1965). *Virology* **25,** 585.
Smith, H. O., and Levine, M. (1967). *Virology* **31,** 207.
Smith, J. D., Abelson, J. N., Clark, B. F., Goodman, H. M., and Brenner, S. (1966). *Cold Spring Harbor Symp. Quant. Biol.* **31,** 479.
Smith, M. G., and Skalka, A. (1966). *J. Gen. Physiol.* **49** (no. 6, part 2), 127.
Snustad, D. P. (1966a). *Genetics* **54,** 923.
Snustad, D. P. (1966b). *Genetics* **54,** 937.
Snustad, D. P. (1968). *Virology* **35,** 550.
Snustad, D. P. (1969). *Virology* **38,** 361.
Snyder, L., and Geiduschek, E. P. (1968). *Proc. Natl. Acad. Sci. U.S.* **59,** 459.
Somerville, R., Ebisuzaki, K., and Greenberg, G. R. (1959). *Proc. Natl. Acad. Sci. U.S.* **45,** 1240.
Speyer, J. F. (1965). *Biochem. Biophys. Res. Comm.* **21,** 6.
Speyer, J. F., and Rosenberg, D. (1968). *Cold Spring Harbor Symp. Quant. Biol.* **33,** 345.
Spiegelman, S. (1961). *Cold Spr. Hbr. Symp. Quant. Biol.* **26,** 75.
Spiegelman, S. and Haruna, I. (1966). *Proc. Natl. Acad. Sci. U.S.* **55,** 1539.
Spiegelman, S., Haruna, I., Holland, I. B., Beaudreau, G., and Mills, D. (1965). *Proc. Natl. Acad. Sci. U.S.* **54,** 919.
Spiegelman, S., Pace, N. R., Mills, D. R., Levisohn, R., Eikhom, T. S., *Cold Spr. Hbr. Symp. Quant. Biol.* **33,** 101.
Spizizen, J. (1958). *Proc. Natl. Acad. Sci. U.S.* **44,** 1072.
Stahl, F. W., Edgar, R. S., and Steinberg, C. M. (1964). *Genetics* **50,** 539.
Stahl, F. W., and Murray, N. E. (1966). *Genetics* **53,** 569.
Stahl, F. W., Murray, N. E., Nakata, A., and Crasemann, J. M. (1966). *Genetics* **54,** 223.
Steinberg, C. M., and Stahl, F. W. (1961). *J. Theoret. Biol.* **1,** 488.
Steinberg, R. A., and Denhardt, D. T. (1968). *J. Mol. Biol.* **37,** 525.
Steitz, J. A. (1969). *Nature* **224,** 957.
Stent, G. S. (1955). *J. Gen. Physiol.* **38,** 853.
Stent, G. S. (1963). "Molecular Biology of Bacterial Viruses," San Francisco, W. H. Freeman and Company.
Stent, G. S., and Brenner, S. (1961). *Proc. Natl. Acad. Sci. U.S.* **47,** 2005.
Stent, G. S., and Fuerst, C. R. (1960). *Adv. Biol. Med. Phys.* **7,** 1.
Stent, G. S., and Maaløe, O. (1953). *Biochim. Biophys. Acta* **10,** 55.
Stent, G. S., and Wollman, E. L. (1952). *Biochim. Biophys. Acta* **8,** 260.
Steuart, C. D., Anand, S. R., and Bessman, M. J. (1968). *J. Biol. Chem.* **243,** 5308.

Stevens, A. (1960). *Biochem. Biophys. Res. Comm.* **3,** 92.
Stone, A. B. (1967). *Biochem. Biophys. Res. Comm.* **26,** 247.
Stone, A. B., and Burton, K. (1962). *Biochem. J.* **85,** 600.
Strack, H. B., and Kaiser, A. D. (1965). *J. Mol. Biol.* **12,** 36.
Streisinger, G., and Bruce, V. (1960). *Genetics* **45,** 1289.
Streisinger, G., Edgar, R. S., and Denhardt, G. H. (1964). *Proc. Natl. Acad. Sci. U.S.* **51,** 775.
Streisinger, G., Emrich, J., and Stahl, M. M. (1967). *Proc. Natl. Acad. Sci. U.S.* **57,** 292.
Streisinger, G., and Franklin, N. C. (1956). *Cold Spring Harbor Symp. Quant. Biol.* **21,** 103.
Streisinger, G., Mukai, F., Dreyer, W. J., Miller, B., and Horiuchi, S. (1961). *Cold Spring Symp. Quant. Biol.* **26,** 25.
Studier, F. W. (1965). *J. Mol. Biol.* **11,** 373.
Studier, F. W. (1969). *Virology* **39,** 562.
Studier, F. W., and Hausmann, R. (1969). *Virology* **39,** 587.
Studier, F. W., and Maizel, J. V., Jr. (1969). *Virology* **39,** 575.
Su, S., Weinberg, F., So, A. G., and Davie, E. W. (1970). *Nature* **225,** 62.
Sueoka, N., and Kano-Sueoka, T. (1964). *Proc. Natl. Acad. Sci. U.S.* **52,** 1535.
Sugimoto, K., Okazaki, T., Imae, Y., and Okazaki, R. (1969). *Proc. Natl. Acad. Sci. U.S.* **63,** 1343.
Sugimoto, K., Okazaki, T., and Okazaki, R. (1968). *Proc. Natl. Acad. Sci. U.S.* **60,** 1356.
Sugiyama, T. (1965). *J. Mol. Biol.* **11,** 856.
Sugiyama, T., Stone, H. O., Jr., and Nakada, D. (1969). *J. Mol. Biol.* **42,** 97.
Summers, W. C. (1968). *Nature* **219,** 159.
Summers, W. C. (1969). *Virology* **39,** 175.
Summers, W. C., and Siegel, R. B. (1969). *Nature* **223,** 1111.
Summers, W. C., and Szybalski, W. (1968a). *Virology* **34,** 9.
Summers, W. C., and Szybalski, W. (1968b). *Biochim. Biophys. Acta* **166,** 371.
Symonds, N., Stacey, K. A., Glover, S. W., Schell, J., and Silver, S. (1963). *Biochem. Biophys. Res. Comm.* **12,** 220.
Szybalski, W., Bøvre, K., Fiandt, M., Guha, A., Hradecna, Z., Kumar, S., Lozeron, H. A., Maher, Sr. V. M., Nijkamp, H. J. J., Summers, W. C., and Taylor, K. (1969). *J. Cell. Physiol.* **74** *Supp.* **1,** 33.
Szybalski, W., Kubinski, H., and Sheldrick, P. (1966). *Cold Spr. Hbr. Symp. Quant. Biol.* **31,** 123.
Takahashi, I., and Marmur, J. (1963a). *Nature* **197,** 794.
Takahashi, I., and Marmur, J. (1963b). *Biochem. Biophys. Res. Comm.* **10,** 289.
Takebe, H., Ichikawa, H., Iwo, K., and Kondo, S. (1967). *Virology* **33,** 638.
Takeda, Y., Oyama, Y., Nakajima, K., and Yura, T. (1969). *Biochem. Biophys. Res. Comm.* **36,** 533.
Tanyashin, V. I. (1968). *Biokhimiya* **33,** 713.
Taylor, A. L., and Trotter, C. D. (1967). *Bacteriol. Rev.* **31,** 332.
Taylor, C. A., and Lipson, H. (1964). "Optical Transforms: their Preparation and Application to X-Ray Diffraction Problems," London, G. Bell & Sons, Ltd.

Taylor, K., Hradecna, Z. and Szybalski, W. (1967). *Proc. Natl. Acad. Sci. U.S.* **57,** 1618.

Terzaghi, E., Okada, Y., Streisinger, G., Emrich, J., Inouye, M., and Tsugita, A. (1966). *Proc. Natl. Acad. Sci. U.S.* **56,** 500.

Terzi, M. (1967). *J. Mol. Biol.* **28,** 37.

Terzi, M. (1968). *J. Mol. Biol.* **34,** 165.

Terzi, M., and Levinthal, C. (1967). *J. Mol. Biol.* **26,** 525.

Tessman, I., Kumar, S., and Tessman, E. S. (1967). *Science* **158,** 267.

Thomas, C. A., Jr. (1959). *J. Gen. Physiol.* **42,** 503.

Thomas, C. A., Jr. (1965). In "Procedures in Nucleic Acid Research," G. L. Cantoni and D. R. Davies, eds., New York, Harper and Row, p. 553.

Thomas, C. A., Jr. (1967). *J. Cell Physiol.* **70,** supp. 1, p. 13.

Thomas, C. A., Jr., and MacHattie, L. A. (1964). *Proc. Natl. Acad. Sci. U.S.* **52,** 1297.

Thomas, C. A., Jr., and MacHattie, L. A. (1967). *Ann. Reviews of Biochem.* **36,** 485.

Thomas, C. A., Jr., Ritchie, D. A., and MacHattie, L. A. (1967). In Colter, J. S., and Paranchych, W., eds., "The Molecular Biology of Viruses," New York, Academic Press, p. 9.

Thomas, C. A., Jr., and Rubenstein, I. (1964). *Biophys. J.* **4,** 93.

Thomas, R. (1966). *J. Mol. Biol.* **22,** 79.

Thomas, R., and Bertani, L. E. (1964). *Virology* **24,** 241.

Tikhonenko, A. S., and Polglazov, B. F. (1963). *Biokhimiya* **28,** 274.

Tillack, T. W., and Smith, D. W. E. (1968). *Virology* **36,** 212.

Tocchini-Valentini, G. P., Marino, P., and Colvill, A. J. (1968). *Nature* **220,** 275.

Tomita, T. (1969). *Molec. Gen. Genetics* **103,** 313.

Tomita, F., and Takahaski, I. (1969). *Biochim. Biophys. Acta* **179,** 18.

Tomizawa, J. (1967). *J. Cell. Physiol.* **70** Supp. 1, 201.

Tomizawa, J., and Anraku, N. (1964a). *J. Mol. Biol.* **8,** 508.

Tomizawa, J., and Anraku, N. (1964b). *J. Mol. Biol.* **8,** 516.

Tomizawa, J., and Anraku, N. (1965). *J. Mol. Biol.* **11,** 509.

Tomizawa, J., Anraku, N., and Iwama, Y. (1966). *J. Mol. Biol.* **21,** 247.

Tomizawa, J., and Ogawa, H. (1967). *J. Mol. Biol.* **30,** 7.

Tomizawa, J., and Ogawa, T. (1968). *Cold Spr. Hbr. Symp. Quant. Biol.* **33,** 525.

Tomizawa, J., and Sunakawa, S. (1956). *J. Gen. Physiol.* **39,** 553.

Tomkins, G. M., Gelehrter, T. D., Granner, D., Martin, D., Jr., Samuels, H. H., and Thompson, E. B. (1969). *Science* **166,** 1474.

Tonegawa, S., and Hayashi, M. (1968). *Proc. Natl. Acad. Sci. U.S.* **61,** 1320.

Tooze, J., and Weber, K. (1967). *J. Mol. Biol.* **28,** 311.

Travers, A. A. (1969). *Nature* **223,** 1107.

Tremblay, G. Y., Daniels, M. J., and Schaechter, M. (1969). *J. Mol. Biol.* **40,** 65.

Trenkner, E., Bonhoeffer, F., and Gierer, A. (1967). *Biochem. Biophys. Res. Comm.* **28,** 932.

Trilling, D. M., and Aposhian, H. V. (1965). *Proc. Natl. Acad. Sci. U.S.* **54,** 622.

Trilling, D. M., and Aposhian, H. V. (1968). *Proc. Natl. Acad. Sci. U.S.* **60,** 214.
Tsugita, A., and Inouye, M. (1968). *J. Mol. Biol.* **37,** 201.
Tsugita, A., Inouye, M., Terzaghi, E., and Streisinger, G. (1968). *J. Biol. Chem.* **243,** 391.
Tucker, R. G. (1969). *J. Gen. Virol.* **4,** 489.
Valentine, R. C., and Strand, M. (1965). *Science* **148,** 511.
Valentine, R. C., and Wedel, H. (1965). *Biochem. Biophys. Res. Comm.* **21,** 106.
van de Pol, J. H., Veldhuisen, G., and Cohen, J. A. (1961). *Biochim. Biophys. Acta* **48,** 417.
Van Vunakis, H., Baker, W. H., and Brown, R. K. (1958). *Virology* **5,** 327.
Veldhuisen, G., Poelman, M. C., and Cohen, J. A. (1968a). *Biochim. Biophys. Acta* **161,** 94.
Veldhuisen, G., Poelman, M. C., and Cohen, J. A. (1968b). *Biochim. Biophys. Acta* **161,** 109.
Verses, C., Lute, M., Crosby, L., and Kozloff, L. M. (1968). *Bacteriological Proc.*, p. 153.
Vidaver, G. A., and Kozloff, L. M. (1957). *J. Biol. Chem.* **225,** 335.
Viñuela, E., Algranati, I. D., and Ochoa, S. (1967). *European J. Biochem.* **1,** 3.
Volkin, E. L. (1954). *J. Am. Chem. Soc.* **76,** 5892.
Volkin, E., and Astrachan, L. (1956). *Virology* **2,** 149.
Volkin, E., and Riggsby, W. S. (1969). *Genetics Supplement* **61,** 1.
Volkin, E., and Ruffilli, A. (1962). *Proc. Natl. Acad. Sci. U.S.* **48,** 2193.
Wahl, R., and Kozloff, L. M. (1962). *J. Biol. Chem.* **237,** 1953.
Wainfan, E. (1968). *Virology* **35,** 282.
Wainfan, E., and Visser, D. W. (1969). *Virology* **37,** 148.
Walter, G., Seifert, W. and Zillig, W. (1968). *Biochem. Biophys. Res. Comm.* **30,** 240.
Wang, J. C. and Schwarz, H. (1967). *Biopolymers* **5,** 953.
Ward, R., Shive, K., and Valentine, R. C. (1967). *Biochem. Biophys. Res. Comm.* **29,** 8.
Warnaar, S. O., Mulder, G., van den Sigtenhorst-van der Sluis, I., van Kestern, L. W., and Cohen, J. A. (1969). *Biochim. Biophys. Acta* **174,** 239.
Warner, H. R., and Barnes, J. E. (1966a). *Proc. Natl. Acad. Sci. U.S.* **56,** 1233.
Warner, H. R., and Barnes, J. E. (1966b). *Virology* **28,** 100.
Warner, H. R., and Hobbs, M. D. (1967). *Virology* **33,** 376.
Warner, H. R., and Hobbs, M. D. (1968). *Virology* **36,** 527.
Warner, H. R., and Hobbs, M. D. (1969). *J. Virol.* **3,** 331.
Warner, H. R., and Lewis, N. (1966). *Virology* **29,** 172.
Warren, R. J., and Bose, S. K. (1968). *J. Virol.* **2,** 327.
Watanabe, I. (1957). *Biochim. Biophys. Acta* **25,** 665.
Waters, L. C., and Novelli, G. D. (1967). *Proc. Natl. Acad. Sci. U.S.* **57,** 979.
Watson, J. D. (1965). "Molecular Biology of the Gene," New York, W. A. Benjamin, Inc., p. 99.
Watson, J. D., and Crick, F. H. (1953). *Cold Spring Harbor Symp. Quant. Biol.* **18,** 123.
Watson, J. D., and Maaløe, O. (1953). *Biochim. Biophys. Acta* **10,** 432.
Weber, K. (1967). *Biochemistry* **6,** 3144.

Weber, K., and Konigsberg, W. (1967). *J. Biol. Chem.* **242,** 3563.
Wehrli, W., Nüesch, J., Knüsel, F., and Staehelin, M. (1968). *Biochim. Biophys. Acta* **157,** 215.
Weidel, W., Frank, H., and Martin, H. H. (1960). *J. Gen. Microbiol.* **22,** 158.
Weidel, W., and Kellenberger, E. (1955). *Biochim. Biophys. Acta* **17,** 1.
Weidel, W., and Primosigh, J. (1958). *J. Gen. Microbiol.* **18,** 513.
Weigert, M. G., and Garen, A. (1965). *Nature* **206,** 992.
Weigert, M. G., Lanka, E., and Garen, A. (1967). *J. Mol. Biol.* **23,** 391.
Weigle, J. (1966). *Proc. Natl. Acad. Sci. U.S.* **55,** 1462.
Weil, J., and Signer, E. R. (1968). *J. Mol. Biol.* **34,** 273.
Weil, R., Michel, M. R., and Ruschman, G. K. (1965). *Proc. Natl. Acad. Sci. U.S.* **53,** 1468.
Weisberg, R. A., and Gottesman, M. E. (1969). *J. Mol. Biol.* **46,** 565.
Weisblum, B., Benzer, S., and Holley, R. W. (1962). *Proc. Natl. Acad. Sci. U.S.* **48,** 1449.
Weiss, B., Jacquemin-Sablon, A., Live, T. R., Fareed, G. C., and Richardson, C. C. (1968). *J. Biol. Chem.* **243,** 4543.
Weiss, B., Live, T. R., and Richardson, C. C. (1968). *J. Biol. Chem.* **243,** 4530.
Weiss, B., and Richardson, C. C. (1967a). *Proc. Natl. Acad. Sci. U.S.* **57,** 1021.
Weiss, B., and Richardson, C. C. (1967b). *J. Biol. Chem.* **242,** 4270.
Weiss, B., and Richardson, C. C. (1967c). *J. Mol. Biol.* **23,** 405.
Weiss, B., Thompson, A., and Richardson, C. C. (1968). *J. Biol. Chem.* **243,** 4556.
Weiss, M. C., Ephrussi, B., and Scaletta, L. J. (1968). *Proc. Natl. Acad. Sci. U.S.* **59,** 1132.
Weiss, S. B., Hsu, W-T., Foft, J. W., and Scherberg, N. H. (1968). *Proc. Natl. Acad. Sci. U.S.* **61,** 114.
Weissbach, A., Bartl, P., and Salzman, L. A. (1968). *Cold Spr. Hbr. Symp. Quant. Biol.* **33,** 525.
Weissbach, A., and Salzman, L. A. (1967). *Proc. Natl. Acad. Sci. U.S.* **58,** 1096.
Weissmann, C., and Borst, P. (1963). *Science* **142,** 1188.
Weissmann, C., Feix, G., and Slor, H. (1968). *Cold Spr. Hbr. Symp. Quant. Biol.* **33,** 83.
Weissmann, C., Simon, L., Borst, P., and Ochoa, S. (1963). *Cold Spr. Hbr. Symp. Quant. Biol.* **28,** 99.
Weith, H. L., and Gilham, P. T. (1969). *Science* **166,** 1004.
Wells, M. A., and Dittmer, J. C. (1966). *Biochemistry* **5,** 3405.
Werner, R. (1968a). *J. Mol. Biol.* **33,** 679.
Werner, R. (1968b). *Cold Spring Harbor Symp. Quant. Biol.* **33,** 501.
Westmoreland, B. C., Szybalski, W., and Ris. H. (1969). *Science* **163,** 1343.
Westphal, H., and Dulbecco, R. (1968). *Proc. Natl. Acad. Sci. U.S.* **59,** 158.
Wiberg, J. S. (1966). *Proc. Natl. Acad. Sci. U.S.* **55,** 614.
Wiberg, J. S. (1967). *J. Biol. Chem.* **242,** 5824.
Wiberg, J. S., and Buchanan, J. M. (1964). *Proc. Natl. Acad. Sci. U.S.* **51,** 421.

Wiberg, J. S., Dirksen, M-L., Epstein, R. H., Luria, S. E., and Buchanan, J. M. (1962). *Proc. Natl. Acad. Sci. U.S.* **48,** 293.
Wildy, P., and Anderson, T. F. (1964). *J. Gen. Microbiol.* **34,** 273.
Williams, R. C., and Fraser, D. (1956). *Virology* **2,** 289.
Wilson, D. L., and Geiduschek, E. P. (1969). *Proc. Natl. Acad. Sci. U.S.* **62,** 514.
Winslow, R. M., and Lazzarini, R. A. (1969). *J. Biol. Chem.* **244,** 1128.
Wise, E. J., and Park, J. T. (1965). *Proc. Natl. Acad. Sci. U.S.* **54,** 75.
Wolf, B., and Meselson, M. (1963). *J. Mol. Biol.* **7,** 636.
Wollman, E. (1928). *Bull. Inst. Pasteur* **26,** 1.
Wood, W. B. (1966). *J. Mol. Biol.* **16,** 118.
Wood, W. B., Edgar, R. S., King, J. Lielausis, I., and Henninger, M. (1968). *Federation Proc.* **27,** 1160.
Wood, W. B., and Henninger, M. (1969). *J. Mol. Biol.* **39,** 603.
Wu, R., and Kaiser, A. D. (1967). *Proc. Natl. Acad. Sci. U.S.* **57,** 170.
Wu, R., and Kaiser, A. D. (1968). *J. Mol. Biol.* **35,** 523.
Wyatt, G. R., and Cohen, S. S. (1952). *Nature* **170,** 1072.
Yamagami, H., and Endo, H. (1969). *J. Virol.* **3,** 343.
Yarmolinsky, M., and Korn, D. (1968). *J. Mol. Biol.* **32,** 475.
Yarus, M. J., and Sinsheimer, R. L. (1967). *J. Virol.* **1,** 135.
Yeh, Y. C., Dubovi, E. J., and Tessman, I. (1969). *Virology* **37,** 615.
Yeh, Y. C., and Greenberg, G. R. (1967). *J. Biol. Chem.* **242,** 1307.
Young, C. W., Schoechetman, G., and Karnofsky, D. A. (1967). *Cancer Research* **27,** 526.
Young, E. T., and Sinsheimer, R. L. (1964). *J. Mol. Biol.* **10,** 562.
Young, E. T., and Sinsheimer, R. L. (1967). *J. Mol. Biol.* **30,** 147.
Young, E. T., and Sinsheimer, R. L. (1968). *J. Mol. Biol.* **33,** 49.
Zimmerman, S. B., and Kornberg, A. (1961). *J. Biol. Chem.* **236,** 1480.
Zimmerman, S. B., Kornberg, S. R., and Kornberg, A. (1962). *J. Biol. Chem.* **237,** 512.
Zimmerman, S. B., Little, J. W., Oshinsky, C. K., and Gellert, M. (1967. *Proc. Natl. Acad. Sci. U.S.* **57,** 1841.
Zinder, N. D. (1965). *Ann. Rev. Microbiol.* **19,** 455.
Zinder, N. D., and Lederberg, J. (1952). *J. Bacteriol.* **64,** 679.
Zinder, N. D., and Lyons, L. B. (1968). *Science* **159,** 84.
Zipser, D. (1967). *J. Mol. Biol.* **29,** 441.
Zipser, D. (1969). *Nature* **221,** 1.
Zissler, J. (1967). *Virology* **31,** 189.
Zissler, J., and Campbell, A. (1969). *Virology* **37,** 318.
Zograf, Y. N., Nikiforov, V. G., and Shemyakin, M. F. (1967). *Molekulyarnaya Biologiya* **1,** 94.

ADDITIONAL REFERENCES

Adesnik, M., and Levinthal, C. (1970). *J. Mol. Biol.* **48,** 187.
Alberts, B., and Frey, L. (1970). *Federation Proc.* **29,** 466A.
Altman, S., and Meselson, M. (1970). *Proc. Natl. Acad. Sci. U.S.* **66,** 716.
Ando, T., Takagi, J. Kosawa, T., and Ikeda, Y. (1970). *J. Biochemistry* **67,** 497.

Anraku, N., Anraku, Y., and Lehman, I. R. (1969). *J. Mol. Biol.* **46,** 481.
Baltimore, D. (1970). *Nature* **226,** 1209.
Bautz, E. K. F., and Bautz, F. A. (1970). *Nature* **226,** 1219.
Bayer, M. E., and Remsen, C. C. (1970). *Virology* **40,** 703.
Benz, W. C., Baumann, L., Wright, A., and Goldberg, E. B. (1970). *Bact. Proc.* p. 167.
Brody, E. N., Diggelmann, H., and Geiduschek, E. P. (1970). *Biochem.* **9,** 1289.
Brody, E. N., and Geiduschek, E. P. (1970). *Biochem.* **9,** 1300.
Burgess, A. B., and Denhardt, D. T. (1969). *J. Mol. Biol.* **45,** 377.
Butler, B., and Echols, H. (1970). *Virology* **40,** 212.
Center, M. S., Studier, F. W., and Richardson, C. C. (1970). *Proc. Natl. Acad. Sci. U.S.* **65,** 242.
Chan, V. L., Shugar, S., and Ebisuzaki, K. (1970). *Virology* **40,** 403.
Cohen, S. N., and Chang, A. C. Y. (1970). *J. Mol. Biol.* **49,** 557.
Daniel, V., Sarid, S., and Littauer, U. Z. (1970). *Science* **167,** 1682.
Diggelmann, H., Pulitzer, J. F., and Geiduschek, E. P. (1970). *J. Mol. Biol.* **49,** 509.
Dove, W. F. (1969b). XII International Congress of Genetics, *Japan J. Genet.* **44,** 11.
Dove, W. F. (1970). *J. Mol. Biol.* **47,** 585.
Drake, J. W., and Greening, E. O. (1970). *Proc. Natl. Acad. Sci. U.S.* **66,** 823.
Duckworth, D. H. (1970). *Virology* **40,** 673.
Ebisuzaki, K., and Campbell, L. (1969). *Virology* **38,** 72.
Eisen, H., Brachet, P., Pereira da Silva, L., and Jacob, F. (1970). *Proc. Natl. Acad. Sci. U.S.* **66,** 855.
Ennis, H. L. (1970). *Virology* **40,** 727.
Forrest, G. L., and Cummings, D. J. (1970). *J. Virol.* **5,** 398.
Freifelder, D., and Meselson, M. (1970). *Proc. Natl. Acad. Sci. U.S.* **65,** 200.
Gold, L. M., and Schweiger, M. (1970). *J. Biol. Chem.* **245,** 2255.
Hattman, S. (1970). *J. Mol. Biol.* **47,** 599.
Hercules, K., Munro, J. L. Mendelsohn, S., and Wiberg, J. S. (1970). *Federation Proc.* **29,** 465A.
Hendrickson, H. E., McCorquodale, D. J., and Anderson, G. C. (1970). *Bacteriol. Proc.* p. 168.
Hinkle, D., and Chamberlin, M. (1970). *Cold Spr. Hbr. Symp. Quant. Biol.* **35,** in press.
Hopkins, N. (1970). *Virology* **40,** 223.
Israeli, M., and Artman, M. (1970). *J. Gen. Virol.* **7,** 137.
Jacquemin-Sablon, A., and Richardson, C. C. (1970). *J. Mol. Biol.* **47,** 477.
Jeng, Y., Gelfand, D., Hayashi, M., Shleser, R., and Tessman, E. S. (1970). *J. Mol. Biol.* **49,** 521.
Jeppesen, P. G. N., Argetsinger-Steitz, J., Gesteland, R. F., and Spahr, P. F. (1970). *Nature* **226,** 230.
Josslin, R. (1970). *Virology* **40,** 719.
Kaiser, A. D., and Masuda, T. (1970). *J. Mol. Biol.* **47,** 557.
Klagsbrun, M., and Rich, A. (1970). *J. Mol. Biol.* **48,** 421.
Knippers, R., and Strätling, W. (1970). *Nature* **226,** 713.

Konings, R. N. H., Ward, R., Francke, B., and Hofschneider, P. H. (1970). *Nature* **226,** 604.
Kozloff, L. M., Verses, C., Lute, M., and Crosby, L. K. (1970). *J. Virol.* **5,** 740.
Kuwano, M., Kwan, C. N., Apirion, D. and Schlessinger, D. (1969). *Proc. Natl. Sci. U.S.* **64,** 693.
Laemmli, U.K., Beguin, F., and Gujer-Kellenberger, G. (1970). *J. Mol. Biol.* **47,** 69.
Lee, C. S., Davis, R. W., and Davidson, N. (1970). *J. Mol. Biol.* **48,** 1.
Lembach, K. J., and Buchanan, J. M. (1970). *J. Biol. Chem.* **245,** 1575.
Lodish, H. F. (1970). *Nature* **226,** 705.
Marchin, G. L., and Neidhardt, F. C. (1970). *Bacteriol. Proc.* p. 183.
Mathews, C. K. (1970a). *J. Virol.* **5,** 659.
Mathews, C. K. (1970b). *J. Virol.,* **6,** 163.
McClain, W. H. (1970). *FEBS Letters* **6,** 99.
McCorquodale, D. J., and Lanni, Y. T. (1970). *J. Mol. Biol.* **48,** 133.
McMacken, R., Mantei, N., Butler, B., Joyner, A., and Echols, H. (1970). *J. Mol. Biol.* **49,** 639.
Mechler, B., and Arber, W. (1969). *J. Mol. Biol.* **45,** 443.
Milanesi, G., Brody, E. N., Grau, O., and Geiduschek, E. P. (1970). *Proc. Natl. Acad. Sci. U.S.* **66,** 181.
Miller, R. C., Jr. (1970a). *J. Virol.* **5,** 536.
Miller, R. C., Jr. (1970b). *J. Virol.* **5,** 533.
Miller, R. C., Jr., and Buckley, P. (1970). *J. Virol.* **5,** 502.
Miller, R. C., Jr., and Kozinski, A. W. (1970). *J. Virol.* **5,** 490.
Miller, R. C., Jr., Kozinski, A. W., and Litwin, S. (1970). *J. Virol.* **5,** 368.
Morse, D. E., and Yanofsky, C. (1969). *Nature* **224,** 329.
Moyer, R. W., and Buchanan, J. M. (1969). *Proc. Natl. Acad. Sci. U.S.* **64,** 1249.
Newlon, M. C., and Marcus, M. (1970). *Bacteriol. Proc.* p. 175.
Oda, K., Sakakibara, Y., and Tomizawa, J. (1969). *Virology* **39,** 901.
Okazaki, T., and Okazaki, R. (1969). *Proc. Natl. Acad. Sci. U.S.* **64,** 1242.
Oppenheim, A. B. (1969). *Virology* **32,** 832.
Oppenheim, A. B., Neubauer, Z., and Calef, E. (1970). *Nature* **226,** 31.
Pero, J. (1970). *Virology* **40,** 65.
Pulitzer, J. F. (1970). *J. Mol. Biol.* **49,** 473.
Pulitzer, J. F., and Geiduschek, E. P. (1970). *J. Mol. Biol.* **49,** 489.
Riva, S., Cascino, A., and Geiduschek, E. P. (1970). *Federation Proc.* **29,** 465A.
Salser, W., Bolle, A., and Epstein, R. (1970). *J. Mol. Biol.* **49,** 271.
Salser, W. Gesteland, R. F., and Ricard, B. (1969). *Cold Spring Harbor Symp. Quant. Biol.* **34,** 771.
Sauerbier, W., and Braütigam, A. R. (1970). *J. Virol.* **5,** 179.
Schedl, P., Singer, R. E., and Conway, T. W. (1970). *Biochem. Biophys. Res. Comm.* **38,** 631.
Schmidt, D. A., Mazaitis, A. J., Kasai, T., and Bautz, E. K. F., (1970). *Nature* **225,** 1012.

Scocca, J. J., Panny, S. R., and Bessman, M. J. (1969). *J. Biol. Chem.* **244,** 3698.
Sekiguchi, M., Yasuda, S., Okubo, S., Nakayama, H., Shimada, K., and Takagi, Y. (1970). *J. Mol. Biol.* **47,** 231.
Siegel, R. B., and Summers, W. C. (1970). *J. Mol. Biol.* **49,** 115.
Signer, E. R. (1970). *Virology* **40,** 624.
Simon, L. D., Swan, J. G., and Flatgaard, J. E. (1970). *Virology* **41,** 77.
Sirbasku, D. A., and Buchanan, J. M. (1970). *J. Biol. Chem.* **245,** 2679.
Smith, D. W., Schaller, H., and Bonhoeffer, F. (1970). *Nature* **226,** 711.
Stahl, F. W., Crasemann, J. M., Yegian, C., Stahl, M. M. and Nakata, A. (1970). *Genetics* **64,** 157.
Sternberg, N., and Champe, S. P. (1969). *J. Mol. Biol.* **46,** 377.
Stone, A. B. (1970). *J. Mol. Biol.* **47,** 215.
Summers, W. C. and Siegel, R. B. (1970). *Cold Spr. Hbr. Symp. Quant. Biol.* **35,** in press.
Szybalski, W., Bøvre, K., Kumar, S., Lozeron, H. A., Nijkamp, H. J. J., and Stevens, W. F. (1970). *Cold Spr. Hbr. Symp. Quant. Biol.* **35,** in press .
Temin, H. M., and Mizutani, S. (1970). *Nature* **226,** 1211.
Thirion, J. P., and Kaesberg, P. (1970). *J. Mol. Biol.* **47,** 193.
Thomas, R., and Mousset, S. (1970). *J. Mol. Biol.* **47,** 179.
To, C. M., Kellenberger, E., and Eisenstark, A. (1969). *J. Mol. Biol.* **46,** 493.
Travers, A. A. (1970). *Nature* **225,** 1009.
Trilling, D. M., and Axelrod, D. (1970). *Science* **168,** 268.
Wilhelm, J. M., and Haselkorn, R. (1970). *Proc. Natl. Acad. Sci. U.S.* **65,** 388.
Yamazaki, Y. (1969). *Biochim. Biophys. Acta* **178,** 542.
Yasuda, S., and Sekiguchi, M. (1970). *J. Mol. Biol.* **47,** 243.
Zissler, J., and Signer, E. (1970). *Bacteriol. Proc.* p. 201.

Author Index

Subject Index

Note. References to the T-even phages are too numerous for listing in this Index.